ELEMENTARY LINEAR ALGEBRA

ELEMENTARY LINEAR ALGEBRA

John K. Luedeman
Stanley M. Lukawecki

Clemson University

West Publishing Company
St. Paul
New York
Los Angeles
San Francisco

Credits

Copyediting: Linda Thompson
Interior Design: Laura Ierardi
Artwork: Anco/Boston
Composition: Science Typographers
Cover Design: Delor Erickson

50 West Kellogg Boulevard
P.O. Box 64526
St. Paul, MN 55164-1003

Printed in the United States of America

Library of Congress Cataloging-in-Publication Data

Luedeman, John K.
Elementary linear algebra.

Includes index.
1. Algebras, Linear. I. Lukawecki, Stanley M.
II. Title.
QA184.L84 1986 512′.5 85-20332
ISBN 0-314-85259-X

TABLE OF CONTENTS

PREFACE

Students majoring in mathematics, the sciences, or engineering generally take a first-level one semester linear algebra course as an undergraduate. Commonly, these students all take the same course, but for different reasons. Some want to build a foundation in theorem proving for use in later courses in abstract algebra and discrete mathematics; some need to understand the use of Gauss elimination with backsolving in solving systems of linear equations for later courses in numerical analysis; some want to learn the LDU-decomposition of a matrix for use in data analysis courses; while for others the study of eigenvalues is paramount for later studies in applied analysis and differential equations. The list could go on and on. Many textbooks on linear algebra concentrate on developing skills in thorem proving. Several of the newer books deal with applied linear algebra which require too much mathematical maturity on the part of the student while interrupting the flow of the development of linear algebra with the applied material. We have tried to write a text which lies between these two types of books.

In Chapter One we develop the Gauss elimination method to solve systems of linear equations. The basics of matrix arithmetic are also developed and, in an optional section, the LDU-decomposition of a matrix is discussed and applied to the solution of linear systems. Chapter Two begins our study of vector spaces by developing vector operations in R^2 and R^3 as well as the geometry of R^n. If students have studied this material previously in a multi-variable calculus course or a physics course, this chapter may be omitted. In Chapter Three the theory of abstract vector spaces is developed, using examples from R^n or P_n, the space of real polynomials of degree less than or equal to n. Examples which deal with concepts from calculus are clearly marked as such. In an optional section, infinite dimensional vector spaces are considered. In Chapter Four, matrices are developed as representations of linear transformations, leading to similar matrices and change of basis matrices. The null space and range space of a linear transformation are also discussed. Chapter Five discusses determinants and their uses. With a mathematically well-prepared class, this chapter may be assigned as outside reading. In Chapter Six, real eigenvalues and eigenvectors are considered and applied to the diagonalization of matrices and quadratic forms. We have chosen not to include vectors with complex numbers as entries. Admittedly, such vectors are mathematically useful and interesting. However, one of their main uses in a linear algebra text is in developing the Jordan Canonical Form and other canonical forms of a matrix. We feel that such topics are best left to a more advanced linear algebra course. Moreover, their inclusion would require the deletion of some other important topics if the material in this text is to be covered in one semester. The core of this course are Chapters One, Three, Four, and Six, in that order.

Throughout the text we have developed many applications of linear algebra to statistics, biology, and to other areas of mathematics. These applications are considered immediately after we develop the mathematical concepts required by these applications. If so desired, the instructor may omit these sections (marked optional) with no loss of continuity in the course.

We have developed the theory of linear algebra throughout the text. Most theorems and techniques are motivated and illustrated by several examples. In Chapter One, the theory problems in the exercise sets are relatively straightforward, but by the final chapter, the student will have increased in mathematical maturity to the extent that more challenging exercises can be mastered. To facilitate the students' mastery of linear algebra, each section ends with a summary of the main ideas and theorems developed in that section. The exercise sets contain both computational drill exercises as well as more challenging exercises and theorems to prove. The Instructor's Manual and Student's Manual available for this book contain complete solutions to most exercises as well as sample tests for each chapter. Answers to the odd-numbered numerical exercises may be found at the end of this text. The answers have been calculated by both authors and independently verified.

We would like to acknowledge and thank Ms. Dianne Haselton, for the excellent typing of the various drafts of this manuscript; our families, who gave us their time and support; Gary Woodruff, and our developmental editor Phyllis Mueller, for her help and confidence in the writing of this text; as well as Kim Bornhoft, production editor; and Charlie Caswell, our problem checker.

No text is successful without the constructive criticism of the colleagues who review the manuscript. We benefited from having an excellent group to help us: Daniel Anderson, University of Iowa; William Anderson, East Tennessee State University; Richard Ball, Auburn University; Gerald Bradley, Claremont-McKenna College; Tom Brown, Simon Fraser University; Daniel Curtin, Northern Kentucky University; Garret Etgen, University of Houston; Michael Evans, North Carolina State University; George Graham, Indiana State University; John Gregory, Southern Illinois University; Howard Holcomb, Monroe Community College; Terry Lawson, Tulane University; Anne Leggett McDonald, Loyola University of Chicago; Michael Penna, Indiana University—Purdue University—Indianapolis; David Schedler, Virginia Commonwealth University; and Mark Watkins, Syracuse University.

Finally, we thank you, the instructor and student, for using our text. We welcome any comments on how we can improve and extend the coverage of the material in this text.

ELEMENTARY LINEAR ALGEBRA

CHAPTER ONE

SYSTEMS OF LINEAR EQUATIONS

Arthur Cayley (1821–1895) and James Joseph Sylvester (1814–1897), two English mathematicians, invented the matrix (plural, matrices) in the 1850s. An early use and probably the most popular use of a matrix today is as a convenient representation of a rectangular array of numbers. The operations of addition and multiplication of matrices were later defined and the algebra of matrices was then developed. In 1925, Werner Heisenberg, a German physicist, used matrices in developing his theory of quantum mechanics, extending the role of matrices from algebra to the area of applied mathematics. Matrices are useful in many situations where there are a large number of interrelationships, such as economic theory. Wassily Leontief won the 1973 Nobel Prize in economics for the development of an input-output model of the United States economy using matrix theory. At the request of the United Nations, Leontief later developed the first input-output model of the world economy in order to suggest how a system of international economic relations featuring a partial disarmament could narrow the gap between the rich and the poor. Matrices are useful in the biological and social sciences, business, forestry, textiles and engineering, and mathematical and computer sciences. In this text, many applications of matrix theory are developed. We begin by introducing a matrix notation helpful in reducing the time it requires to determine the solution of simultaneous linear equations (i.e., a system of linear equations).

SECTION 1.1 INTRODUCTION

There are many situations in management and economics in which one sets up an array of numbers to tabulate and analyze numerical data. A **matrix** A is defined to be a rectangular array of real numbers enclosed by a bracket. (Matrices whose entries are complex numbers are considered in more advanced courses.) If the matrix has m rows and n columns, the matrix is said to be an $m \times n$ (read "m by n") matrix. If $m = n$, the matrix is said to be a *square matrix* of order n.

Example 1.1.1 a. $A = \begin{bmatrix} 1 & 2 \\ 3 & 7 \end{bmatrix}$ is a 2×2 matrix.

b. $A = [1 \quad 2 \quad \sqrt{5} \quad 1.45]$ is a 1×4 matrix.

c. $A = \begin{bmatrix} 8 & 9 \\ 2 & 7 \\ 1 & 4 \end{bmatrix}$ is a 3×2 matrix.

d. $A = \begin{bmatrix} 1 \\ 0 \\ 0 \\ 0 \end{bmatrix}$ is a 4×1 matrix.

e. $A = \begin{bmatrix} 0 & 0 & 0 \\ 0 & 0 & 0 \\ 0 & 0 & 0 \end{bmatrix}$ is a 3×3 matrix.

f. $A = \begin{bmatrix} 1 & 0 & 0 & 0 \\ 0 & 1 & 0 & 0 \\ 0 & 0 & 1 & 0 \\ 0 & 0 & 0 & 1 \end{bmatrix}$ is a 4×4 matrix. ■

The central topic in this chapter is the solution of systems of linear equations. Matrices and matrix theory are used not only in solving a given system of linear equations but also in determining whether a given system of this type has a solution.

In general, a **linear equation** is an equation that is of degree 1 or 0 in every unknown—that is, a linear equation has the form

$$a_1x_1 + a_2x_2 + \cdots + a_nx_n = b.$$

A **system of linear equations** is one or more linear equations.

Example 1.1.2 a.

$$\begin{aligned} 3x_1 + 4x_2 - 7x_3 &= 7 \\ 5x_1 + 4x_2 - 2x_4 &= 6 \end{aligned}$$

is a system of two equations in four unknowns.

b.

$$\begin{aligned} \tfrac{1}{3}x_1 + 2x_2 \phantom{- \tfrac{1}{2}x_3} &= 8 \\ x_1 + x_2 - \tfrac{1}{2}x_3 &= 2 \\ x_2 - x_3 &= 5 \\ x_1 + x_2 + x_3 &= 0 \end{aligned}$$

is a system of four equations in three unknowns.

c. A manufacturing firm produces three products, each of which requires a fixed number of hours in each of five processing operations, as given in the table.

		Process				
		1	2	3	4	5
Product	1	1	2	1	3	5
	2	3	1	2	2	4
	3	4	2	1	4	3

Assume that during the production run, the manufacturing firm will have 46 hours available in process 1, 73 hours in process 2, 40 hours in process 3, 24 hours in process 4, and 88 hours in process 5. Let x_i, $i = 1, 2, 3$, denote the number of units produced of product i. Then

$$\begin{aligned} x_1 + 3x_2 + 4x_3 &= 46 \\ 2x_1 + x_2 + 2x_3 &= 73 \\ x_1 + 2x_2 + x_3 &= 40 \\ 3x_1 + 2x_2 + 4x_3 &= 24 \\ 5x_1 + 4x_2 + 3x_3 &= 88 \end{aligned}$$

is a system of five equations in three unknowns. ■

One of the major uses of the matrix is as a notational aid in describing a system of linear equations. To the system of linear equations

$$\begin{aligned} 3x + 4y &= 7 \\ 2x + 7y &= 0 \end{aligned}$$

we associate two matrices. The matrix

$$\begin{bmatrix} 3 & 4 \\ 2 & 7 \end{bmatrix}$$

is the *coefficient matrix* of the system, whereas the matrix

$$\left[\begin{array}{cc|c} 3 & 4 & 7 \\ 2 & 7 & 0 \end{array}\right]$$

is the *augmented matrix* of the system. The **coefficient matrix** of a system of linear equations consists of the coefficients of the variables in the same order as they appear in the equations. The **augmented matrix** of a system of linear equations is obtained by adjoining the column of constants to the right of the coefficient matrix.

Example 1.1.3 The matrices

$$\begin{bmatrix} 3 & 2 & 1 & 4 \\ -1 & 1 & -3 & 2 \\ -2 & 1 & -5 & 1 \end{bmatrix} \quad \text{and} \quad \left[\begin{array}{cccc|c} 3 & 2 & 1 & 4 & -1 \\ -1 & 1 & -3 & 2 & 0 \\ -2 & 1 & -5 & 1 & 6 \end{array}\right]$$

are the coefficient matrix and augmented matrix, respectively, of the following system of linear equations:

$$\begin{aligned} 3x_1 + 2x_2 + x_3 + 4x_4 &= -1 \\ -x_1 + x_2 - 3x_3 + 2x_4 &= 0 \\ -2x_1 + x_2 - 5x_3 + x_4 &= 6 \end{aligned}$$

■

In Example 1.1.2(a), the system of linear equations consisted of two equations in four unknowns. In Example 1.1.2b, the system contained four equations in three unknowns, and in Example 1.1.2(c), five equations and three unknowns. In general, an $m \times n$ *system of linear equations* is written in the form

$$\begin{aligned} a_{11}x_1 + a_{12}x_2 + \cdots + a_{1n}x_n &= b_1 \\ a_{21}x_1 + a_{22}x_2 + \cdots + a_{2n}x_n &= b_2 \\ \vdots \qquad \vdots \qquad\qquad \vdots \quad &\quad \vdots \\ a_{m1}x_1 + a_{m2}x_2 + \cdots + a_{mn}x_n &= b_m \end{aligned} \tag{1.1.1}$$

The coefficient matrix for system (1.1.1) is

$$A = \begin{bmatrix} a_{11} & a_{12} & \cdots & a_{1n} \\ a_{21} & a_{22} & \cdots & a_{2n} \\ \vdots & \vdots & & \vdots \\ a_{m1} & a_{m2} & \cdots & a_{mn} \end{bmatrix} \tag{1.1.2}$$

The corresponding augmented matrix is

$$[A|b] = \left[\begin{array}{cccc|c} a_{11} & a_{12} & \cdots & a_{1n} & b_1 \\ a_{21} & a_{22} & \cdots & a_{2n} & b_2 \\ \vdots & \vdots & & \vdots & \vdots \\ a_{m1} & a_{m2} & \cdots & a_{mn} & b_m \end{array}\right] \tag{1.1.3}$$

Hence to each *system of m linear equations in n unknowns* is associated its $m \times n$ coefficient matrix and its $m \times (n + 1)$ augmented matrix. Conversely, to each $m \times (n+1)$ matrix

$$\left[\begin{array}{ccc|c} a_{11} & \cdots & a_{1n} & b_1 \\ a_{21} & \cdots & a_{2n} & b_2 \\ \vdots & & \vdots & \vdots \\ a_{m1} & \cdots & a_{mn} & b_m \end{array}\right]$$

is associated the system

$$\begin{aligned} a_{11}x_1 + \cdots + a_{1n}x_n &= b_1 \\ a_{21}x_1 + \cdots + a_{2n}x_n &= b_2 \\ \vdots \qquad\qquad \vdots \quad &\quad \vdots \\ a_{m1}x_1 + \cdots + a_{mn}x_n &= b_m \end{aligned}$$

of m linear equations in n unknowns.

Definition 1.1.4 A system of m linear equations in n unknowns is said to be **homogeneous** if $b_1 = b_2 = \cdots = b_m = 0$. Otherwise, the system is said to be **nonhomogeneous**.

Example 1.1.5 The system in (a) is nonhomogeneous, whereas the system in (b) is homogeneous.

(a)
$$\begin{aligned} x_1 + 2x_2 + 3x_3 + x_4 + x_5 &= 2 \\ x_1 + x_2 - x_3 + x_4 - x_5 &= 0 \\ 2x_1 + x_3 + x_5 &= 0 \\ 3x_1 + x_2 + x_4 - 3x_5 &= 1 \end{aligned}$$

(b)
$$\begin{aligned} x_1 + 2x_2 + 3x_3 + x_4 + x_5 &= 0 \\ x_1 - x_2 - x_3 + x_4 - x_5 &= 0 \\ 2x_1 + x_3 + x_5 &= 0 \\ 3x_1 + x_2 + x_4 - 3x_5 &= 0 \end{aligned}$$

Notice that each entry in the matrix

$$A = \begin{bmatrix} a_{11} & a_{12} & \cdots & a_{1n} \\ a_{21} & a_{22} & \cdots & a_{2n} \\ \vdots & \vdots & & \vdots \\ a_{m1} & a_{m2} & \cdots & a_{mn} \end{bmatrix}$$

has two subscripts. The first subscript of each entry in the matrix A denotes the row in which the entry lies, whereas the second subscript denotes the column containing the entry. For example, a_{23} is the entry in the second row and third column. In order to save time and space, the notation $[a_{ij}]_{m\times n}$ is also used for the matrix A, that is, $A = [a_{ij}]_{m\times n}$. The entry a_{ij} is the element in the ith row and jth column. When programming a computer, most languages write $a(i, j)$ in place of a_{ij}. When the size of the matrix is clear from the context of the discussion, we denote A by $[a_{ij}]$. ■

SUMMARY

Terms to know:

- $m \times n$ matrix
- Linear equation
- System of m linear equations in n unknowns
- Coefficient matrix
- Augmented matrix
- Homogeneous system of equations
- Nonhomogeneous system of equations

EXERCISE SET 1.1

For each of the following systems of linear equations, write its coefficient matrix and augmented matrix and determine whether or not the system is homogeneous.

1. $$\begin{aligned} 3x + 4y + 7z &= 4 \\ 2x + 6y \phantom{{}+7z} &= 3 \end{aligned}$$

2. $$\begin{aligned} \pi x + ey + \sqrt{2}a \phantom{{}+(1+\sqrt{3})z} &= 0 \\ 4x + \sqrt{2}y + (1+\sqrt{3})z &= 0 \end{aligned}$$

3. $$\begin{aligned} 2x + 4y &= 7 \\ 8x + 3y &= 4 \end{aligned}$$

4. $3x - 2y + z = 7$

5. $$\begin{aligned} 7x_1 + 9x_2 + 4x_3 &= 3 \\ 7x_1 + \phantom{9x_2 +{}} 4x_3 &= 6 \\ 2x_2 + 7x_3 &= 5 \\ 8x_1 + 3x_2 \phantom{{}+4x_3} &= 2 \end{aligned}$$

6. $$\begin{aligned} x + 2y + z &= 1 \\ y - z &= 2 \\ z &= 3 \end{aligned}$$

7. $$\begin{aligned} 6x_1 + 4x_2 &= 0 \\ 7x_1 + 2x_2 &= 6 \end{aligned}$$

8. $$\begin{aligned} x_1 + x_2 + x_3 + x_4 &= 0 \\ x_2 + x_3 + x_4 &= -1 \\ x_3 + x_4 &= 2 \end{aligned}$$

9. $$\begin{aligned} 3x + 4y &= 0 \\ 2x + 7y &= 0 \end{aligned}$$

10. $$\begin{aligned} x_1 + 2x_2 + 3x_3 + x_4 - x_5 &= 1 \\ x_1 - x_2 + x_3 - x_4 + x_5 &= 0 \\ 2x_1 + x_2 - x_3 + x_4 - x_5 &= 2 \\ 3x_1 + 21x_2 + 6x_3 + 2x_4 - 3x_5 &= 0 \end{aligned}$$

11. $$\begin{aligned} 3x + 4y &= 7 - 6z \\ y + 8 &= x - z \end{aligned}$$

12. $$\begin{aligned} 2x_1 - x_2 + x_3 - x_4 &= x_1 \\ x_1 + 2x_2 - x_3 + 2x_4 &= x_2 \\ -x_1 - x_2 - x_3 + 6x_4 &= x_3 \\ 4x_1 + 3x_2 + 2x_3 - 5x_4 &= x_4 \end{aligned}$$

13. $$\begin{aligned} 2x_1 + 3x_2 - \phantom{x_3 +{}} x_4 - 2 &= 0 \\ x_1 - \phantom{3x_2 +{}} x_3 + x_4 + 1 &= 0 \\ x_2 + 2x_3 - 3x_4 - 1 &= 0 \end{aligned}$$

14. $$\begin{aligned} (2 - t)x_1 + 3x_2 - 4x_3 &= 0 \\ 2x_1 + (1 - t)x_2 + x_3 &= 0 \\ x_1 - x_2 - (3 - t)x_3 &= 0 \end{aligned}$$

For each of the following augmented matrices, give the corresponding linear system.

15. $$\left[\begin{array}{ccc|c} 3 & 2 & 1 & 2 \\ 1 & 2 & -1 & 0 \\ 3 & 2 & 1 & 2 \end{array}\right]$$

16. $$\left[\begin{array}{ccc|c} 1 & 0 & 1 & -1 \\ 2 & 3 & -1 & 2 \end{array}\right]$$

17. $$\left[\begin{array}{ccc|c} -2 & 4 & 2 & 0 \\ -1 & 2 & 1 & 0 \end{array}\right]$$

18. $$\left[\begin{array}{ccc|c} 1 & 2 & -3 & 2 \\ 0 & 1 & 2 & 1 \\ 0 & 0 & 2 & 3 \end{array}\right]$$

19. $$\left[\begin{array}{cccccc|c} 2 & 3 & -1 & 4 & 2 & 5 & 7 \\ 6 & -5 & 3 & 0 & 2 & 0 & 6 \\ 1 & 3 & 5 & 7 & 9 & 7 & 2 \\ 4 & 2 & 6 & 8 & 2 & 4 & 1 \end{array}\right]$$

20. $$\left[\begin{array}{cccc|c} 3 & 2 & 1 & 4 & 0 \\ 0 & 2 & 1 & 2 & 0 \\ 0 & 0 & 1 & -1 & 0 \\ 0 & 0 & 0 & 3 & 0 \end{array}\right]$$

21. $$\left[\begin{array}{cccc|c} 3 & -4 & 6 & 2 & 7 \end{array}\right]$$

22. $$\left[\begin{array}{ccc|c} 1 & -1 & 0 & 2 \\ 0 & 1 & -1 & -3 \\ 2 & 0 & -3 & 0 \\ -3 & 4 & 6 & -10 \end{array}\right]$$

23. List the entries a_{11}, a_{21}, and a_{32} for each of the coefficient matrices in Exercises 1–14, provided they exist.

24. A straight line of the form $y = mx + b$ is determined by two given points (x_1, y_1) and (x_2, y_2) when $x_1 \neq x_2$. In that case,
$$\begin{aligned} y_1 &= mx_1 + b \\ y_2 &= mx_2 + b \end{aligned}$$
Write the coefficient and augmented matrix that result when this system is solved for the unknowns m and b.

25. Three noncollinear points determine the plane $z = ax + by + c$. If the three given points are (x_1, y_1, z_1), (x_2, y_2, z_2), and (x_3, y_3, z_3), then
$$\begin{aligned} z_1 &= ax_1 + by_1 + c \\ z_2 &= ax_2 + by_2 + c \\ z_3 &= ax_3 + by_3 + c \end{aligned}$$
Write the coefficient and augmented matrix for this system in unknowns a, b, and c.

26. The equation of a parabola with vertical axis is given by $y = ax^2 + bx + c$. The unknowns a, b, and c for any given parabola can be determined by giving three points, (x_1, y_1), (x_2, y_2) and (x_3, y_3). Substituting (x_1, y_1), (x_2, y_2), and (x_3, y_3) into the equation $y = ax^2 + bx + c$, gives a system of three equations in the unknowns a, b, and c. Write the coefficient and augmented matrix for this system.
27. A circle, $x^2 + y^2 + Ax + By + C = 0$, is determined by three noncollinear points (x_1, y_1), (x_2, y_2), and (x_3, y_3). Set up a system of equations that would give the coefficients in the circle determined by the points $(1, 2)$, $(3, 4)$, and $(1, 4)$.

SECTION 1.2 THE SOLUTION OF SYSTEMS OF EQUATIONS

Given a system of linear equations, we wish to find all solutions to this system or, if there are no solutions, to discover this fact. We need an organized systematic method which, when applied to any linear system, gives us all solutions to the system. The method should also be easy to program. One candidate for this method, which we use in this text, is the method of elimination. You have probably seen a variation of this method in high school algebra. We begin by defining the concept of a solution to a linear system of equations.

Definition 1.2.1 A **solution** of system

$$\begin{aligned} a_{11}x_1 + \cdots + a_{1n}x_n &= b_1 \\ a_{21}x_1 + \cdots + a_{2n}x_n &= b_2 \\ \vdots \qquad\qquad \vdots\quad &\quad \vdots \\ a_{m1}x_1 + \cdots + a_{mn}x_n &= b_m \end{aligned}$$

is an ordered set $s_1, s_2, \ldots, s_n$ of numbers such that the equations of the system become statements of numerical facts when each variable x_i is replaced by its corresponding value s_i.

Example 1.2.2 The ordered pair $(3, 0)$ is a solution to the system

$$x_1 + x_2 = 3$$

$$2x_1 - x_2 = 6$$

since when x_1 is replaced by 3 and x_2 is replaced by 0 in the system, we obtain the following true sentences:

$$3 + 0 = 3$$

$$2(3) - 0 = 6$$

■

We demonstrate the method of elimination in solving the following system of linear equations:

$$\begin{aligned} -2x + 2y - 2z &= 10 \\ 3x - y - 2z &= -1 \\ 2x - y - z &= -6 \end{aligned} \qquad (1.2.1)$$

1. First, multiply equation 1 by $-\frac{1}{2}$ to obtain a new system.

$$\begin{aligned} x - y + z &= -5 \\ 3x - y - 2z &= -1 \\ 2x - y - z &= -6 \end{aligned}$$

2. Second, multiply equation 1 by -3 and add the result to equation 2.

$$\begin{aligned} x - y + z &= -5 \\ 2y - 5z &= 14 \\ 2x - y - z &= -6 \end{aligned}$$

3. Next, multiply equation 1 by -2 and add the result to equation 3.

$$\begin{aligned} x - y + z &= -5 \\ 2y - 5z &= 14 \\ y - 3z &= 4 \end{aligned}$$

4. Next, for simplicity, interchange equations 2 and 3.

$$\begin{aligned} x - y + z &= -5 \\ y - 3z &= 4 \\ 2y - 5z &= 14 \end{aligned}$$

5. Finally, multiply equation 2 by -2 and add to equation 3.

$$\begin{aligned} x - y + z &= -5 \\ y - 3z &= 4 \\ z &= 6 \end{aligned}$$

6. From the last equation, we begin the process of **backsolving**:

$$z = 6$$

7. Substituting $z = 6$ into the second equation, $y - 3z = 4$, yields $y - 3(6) = 4$, or

$$y = 22$$

8. Finally, substituting $z = 6$ and $y = 22$ into the first equation, $x - y + z = -5$, yields $x - (22) + (6) = -5$, or

$$x = 11$$

We now claim that the triple $x = 11$, $y = 22$, and $z = 6$ is a solution to the original system of equations. Substituting $x = 11$, $y = 22$, and $z = 6$ into the original system yields the true statements

$$\begin{aligned} -2(11) + 2(22) - 2(6) &= 10 \\ 3(11) - (22) - 2(6) &= -1 \\ 2(11) - (22) - (6) &= -6 \end{aligned}$$

Notice that only three distinct operations were used in solving system (1.2.1):

1. An equation was multiplied by a nonzero number [as in (1)].
2. A multiple of one equation was added to another equation [i.e., an equation was multiplied by a number and the resulting equation added to another equation, as in (2), (3), and (5)].
3. Two equations were interchanged [as in (4)].

In Section 1.3 we verify that each of these operations transforms a linear system into an equivalent system. By an **equivalent system of linear equations** we mean a second system, each solution of which is a solution of the first system, and vice versa.

Example 1.2.3 The linear systems

$$\begin{aligned} x_1 + 2x_2 &= -3 \\ 3x_1 + 7x_2 &= -11 \\ 2x_1 + 3x_2 &= -4 \end{aligned} \qquad \text{and} \qquad \begin{aligned} x_1 - x_2 &= 3 \\ 2x_1 + x_2 &= 0 \end{aligned}$$

are equivalent since each system has $x_1 = 1$, $x_2 = -2$ as its only solution. ■

The coefficients in the equations determine the solution; the variables serve only as placeholders. Therefore, the solution of a linear system of equations can be reached by working only with the augmented matrix of the given system and row operations that correspond to the three distinct operations used to solve system (1.2.1). We will show that these operations, called **elementary row operations**, do not change the solution of a system.

1. The entries in any row may be multiplied by a nonzero number.
2. The entries in any row may be multiplied by a number and added to the corresponding entries of another row.
3. Any two rows may be interchanged.

Utilizing only the augmented matrix and the three elementary row operations, we illustrate the **Gauss reduction method** for solving system (1.2.1),

$$\left[\begin{array}{rrr|r} -2 & 2 & -2 & 10 \\ 3 & -1 & -2 & -1 \\ 2 & -1 & -1 & -6 \end{array}\right]$$

1. Multiply row 1 by $-\frac{1}{2}$ [write $(-\frac{1}{2})$R1] to obtain:

$$\left[\begin{array}{rrr|r} 1 & -1 & 1 & -5 \\ 3 & -1 & -2 & -1 \\ 2 & -1 & -1 & -6 \end{array}\right]$$

2. Next add (-3) times row 1 to row 2 [(-3)R1 to R2]:

$$\left[\begin{array}{rrr|r} 1 & -1 & 1 & -5 \\ 0 & 2 & -5 & 14 \\ 2 & -1 & -1 & -6 \end{array}\right]$$

3. Add (-2) times row 1 to row 3 [(-2)R1 to R3]:

$$\left[\begin{array}{rrr|r} 1 & -1 & 1 & -5 \\ 0 & 2 & -5 & 14 \\ 0 & 1 & -3 & 4 \end{array}\right]$$

4. Interchange row 2 and row 3 (R2 $\leftrightarrow$ R3):

$$\left[\begin{array}{rrr|r} 1 & -1 & 1 & -5 \\ 0 & 1 & -3 & 4 \\ 0 & 2 & -5 & 14 \end{array}\right]$$

5. Add (-2) times row 2 to row 3 [(-2)R2 to R3]:

$$\left[\begin{array}{rrr|r} 1 & -1 & 1 & -5 \\ 0 & 1 & -3 & 4 \\ 0 & 0 & 1 & 6 \end{array}\right]$$

The resulting augmented matrix represents the system of equations

$$\begin{aligned} x - y + z &= -5 \\ y - 3z &= 4 \\ z &= 6 \end{aligned}$$

As before, we backsolve to find the solutions. (Gauss reduction may be stopped when any equivalent system is in a form that is convenient to backsolve.)

We solve for z using the third equation. Knowing z, we find y from the second equation. Finally, we use these values of y and z in the first equation and determine x. As a check, we substitute the values of x, y, and z in all three of the equations in the original system. Since the left-hand side of each equation is equal to the right-hand side, we have a solution.

When is it convenient to stop and backsolve? It is convenient to stop and backsolve when you feel that the linear system is in a form you can solve without too much trouble by backsolving as in 5 of the preceding example.

We make the idea of solving a system precise by defining *row echelon form* and noting that the process stops when the augmented matrix of the system has been reduced to this state.

Definition 1.2.4 The matrix

$$\begin{bmatrix} c_{11} & c_{12} & \cdots & c_{1n} \\ c_{21} & c_{22} & \cdots & c_{2n} \\ \vdots & \vdots & & \vdots \\ c_{m1} & c_{m2} & \cdots & c_{mn} \end{bmatrix}$$

is in **row echelon form** when it satisfies the following conditions:

1. If row i consists of all zeros and $j > i$, then row j consists of all zeros (rows below an all-zero row are all zeros).

The first nonzero entry in each row is called the **leading entry** of that row. Let the leading entry of row i occur in column $L(i)$.

2. If $i < j$ and row j has a leading entry, then $L(i) < L(j)$ (every entry below a leading entry is zero).

The leading entry in each row is called a **pivot**. The variable corresponding to column $L(i)$ of the pivot in row i is called a **pivot variable**.

Example 1.2.5 a. The matrix

$$\begin{bmatrix} 1 & 3 & 5 & 0 & -4 & 1 \\ 0 & 1 & 3 & 0 & 0 & 2 \\ 0 & 0 & 0 & 0 & 1 & 5 \\ 0 & 0 & 0 & 0 & 0 & 0 \end{bmatrix}$$

is in row echelon form. The pivots are $a_{11} = 1$, $a_{22} = 1$, and $a_{35} = 1$.

b. The matrix

$$\begin{bmatrix} 0 & 0 & 2 & 4 & 9 \\ 0 & 0 & 0 & 1 & 3 \end{bmatrix}$$

is in row echelon form. The pivots are $a_{13} = 2$ and $a_{24} = 1$.

c. The matrix

$$\begin{bmatrix} 1 & 3 & 4 & 1 & 9 & -2 \\ 0 & 0 & 0 & 0 & 0 & 0 \\ 0 & 1 & 2 & 4 & 1 & 0 \end{bmatrix}$$

is not in row echelon form. Condition (1) is violated.

d. The matrix

$$\begin{bmatrix} 0 & 3 & 1 & 4 & 7 \\ 0 & 1 & 0 & 0 & 1 \end{bmatrix}$$

is not in row echelon form. Column 2 violates Condition (2). ■

We can carry the augmented matrix given earlier,

$$\left[\begin{array}{ccc|c} 1 & -1 & 1 & -5 \\ 0 & 1 & -3 & 4 \\ 0 & 0 & 1 & 6 \end{array}\right]$$

to an even simpler echelon form.

6. (3)R3 to R2 and (-1)R3 to R1 yields

$$\left[\begin{array}{ccc|c} 1 & -1 & 0 & -11 \\ 0 & 1 & 0 & 22 \\ 0 & 0 & 1 & 6 \end{array}\right]$$

7. (1)R2 to R1 gives

$$\left[\begin{array}{ccc|c} 1 & 0 & 0 & 11 \\ 0 & 1 & 0 & 22 \\ 0 & 0 & 1 & 6 \end{array}\right]$$

The resulting system has the solution

$$x = 11$$
$$y = 22$$
$$z = 6$$

since x, y, and z are uniquely determined. To extend the augmented matrix given in (5) to the one in (7) requires more row operations, but the solution (provided one exists) is immediate; that is, no backsolving is needed. The matrix given in (7) is called a *reduced row echelon matrix*. A **reduced row echelon matrix** is a row echelon matrix with 1s as leading entries and 0s above and below each 1. Many authors use the terminology **Gauss-Jordan method** of elimination when the process is carried on until the final augmented matrix is in reduced row echelon form.

Using the augmented matrix for a system of equations plus the three row operations can save much time in solving systems and can avoid circularity. Moreover, the notation makes clear exactly what operations have been used. (The value of such a notation may appear to be small; however, after solving some systems of equations using the Gauss reduction method, you should be convinced of the time-saving value of this notation.)

Remember, the *elementary row operations* are:

1. Multiply a row by a nonzero number;
2. Add a multiple of one row to another row;
3. Interchange two rows.

Example 1.2.6 Solve the system

$$3x + 4y = 2$$
$$7x + 5y = 1$$

by the Gauss reduction method.

Solution The augmented matrix of this system is

$$\left[\begin{array}{cc|c} 3 & 4 & 2 \\ 7 & 5 & 1 \end{array}\right]$$

$(-\frac{7}{3})$R1 to R2 yields

$$\left[\begin{array}{cc|c} 3 & 4 & 2 \\ 0 & -\frac{13}{3} & -\frac{11}{3} \end{array}\right]$$

We stop, since this matrix is in row echelon form. This echelon matrix is the augmented matrix of the system

$$3x + \quad 4y = 2$$

$$-\tfrac{13}{3}y = -\tfrac{11}{3}$$

The second equation yields

$$y = \tfrac{11}{13}$$

Rewriting the first equation as $x = \frac{2}{3} - \frac{4}{3}y$ and substituting $y = \frac{11}{13}$ yields

$$x = \frac{2}{3} - \left(\frac{4}{3}\right)\left(\frac{11}{13}\right)$$

$$= \frac{26 - 44}{39} = -\frac{6}{13}$$

Substituting $x = -\frac{18}{39}$ and $y = \frac{11}{13}$ into the original system, we obtain the true statements

$$3\left(-\tfrac{6}{13}\right) + 4\left(\tfrac{11}{13}\right) = 2$$

$$7\left(-\tfrac{6}{13}\right) + 5\left(\tfrac{11}{13}\right) = 1$$

Notice that x and y are uniquely determined. This system has exactly one solution. ■

Example 1.2.7 Solve the following system of equations:

$$3x + 4y + z = 7$$

$$y + z = 3$$

Solution The augmented matrix of this system is

$$\left[\begin{array}{ccc|c} 3 & 4 & 1 & 7 \\ 0 & 1 & 1 & 3 \end{array}\right]$$

The matrix is in row echelon form. This matrix is the augmented matrix of the system

$$3x + 4y + 1z = 7$$

$$y + \quad z = 3$$

To find the solution(s) of this system, we solve for y and x (the pivot variables), treating z as an unspecified constant:

$$y = 3 - z$$

and

$$\begin{aligned} x &= \tfrac{7}{3} - (\tfrac{4}{3})y - (\tfrac{1}{3})z \\ &= \tfrac{7}{3} - (\tfrac{4}{3})(3 - z) - (\tfrac{1}{3})z \\ &= -\tfrac{5}{3} + z \end{aligned}$$

A convenient way to record the solution is

$$\begin{aligned} x &= -\tfrac{5}{3} + z \\ y &= \quad 3 - z \\ z &= z \quad \text{where } z \text{ is arbitrary} \end{aligned}$$

This means there are infinitely many solutions to this system, all determined by the selection of a value for z.

For example, if $z = 0$, then $y = 3$ and $x = -\frac{5}{3}$, and if $z = 1$, then $y = 2$ and $x = -\frac{2}{3}$. You can easily verify that both sets of values are solutions.

$$\begin{aligned} 3(-\tfrac{5}{3}) + 4(3) + (0) &= 7 \\ (3) + (0) &= 3 \end{aligned}$$

and

$$\begin{aligned} 3(-\tfrac{2}{3}) + 4(2) + (1) &= 7 \\ (2) + (1) &= 3 \end{aligned}$$

Also, if you want a solution with $y = 17$, then $z = 3 - y = 3 - 17 = -14$, so that $x = -\frac{5}{3} + z = -\frac{5}{3} - 14 = -\frac{47}{3}$.

To check that for any arbitrarily chosen value of z there is a solution for x and y given by the equations

$$\begin{aligned} y &= \quad 3 - z \\ x &= -\tfrac{5}{3} + z \end{aligned}$$

we have, upon substituting in the original system,

$$3(-\tfrac{5}{3} + z) + 4(3 - z) + (z) = 7$$

and

$$(3 - z) + (z) = 3 \quad \text{for all } z$$ ■

Example 1.2.8 Solve the following system of equations:

$$\begin{aligned} 2x - 3y &= 1 \\ -6x + 9y &= 4 \end{aligned}$$ ■

The augmented matrix of this system is

$$\left[\begin{array}{rr|r} 2 & -3 & 1 \\ -6 & 9 & 4 \end{array}\right]$$

(3)R1 to R2 yields

$$\left[\begin{array}{rr|r} 2 & -3 & 1 \\ 0 & 0 & 7 \end{array}\right]$$

This matrix is in row echelon form and is the augmented matrix of the system

$$2x - 3y = 1$$

$$0 = 7$$

This last system clearly has no solution, and therefore the original system has none. (Notice that the last row in the augmented matrix is nonzero, whereas the last row of coefficient matrix is not.) ■

We have seen that for a given system of linear equations there are three possibilities:

1. There is a unique solution (Example 1.2.6).
2. There are infinitely many solutions (Example 1.2.7).
3. There are no solutions (Example 1.2.8).

If there is at least one solution, the system is said to be **consistent**; if no solutions exist, the system is **inconsistent**. If the system is homogeneous, there is always the trivial solution obtained by setting each variable to zero. Thus a homogeneous system has two possibilities:

1. There is a unique solution.
2. There are infinitely many solutions.

Example 1.2.9 Solve the following homogeneous system of equations:

$$x_1 - 4x_2 - 6x_3 = 0$$

$$3x_1 + 10x_2 + 4x_3 = 0$$

$$3x_1 - x_2 - 7x_3 = 0$$

Solution The augmented matrix of this system is

$$\left[\begin{array}{rrr|r} 1 & -4 & -6 & 0 \\ 3 & 10 & 4 & 0 \\ 3 & -1 & -7 & 0 \end{array}\right]$$

Note that the 1 in the (1, 1) position of this matrix makes it easier to obtain zeros in the first column.

1. (-3) to R2 yields

$$\left[\begin{array}{ccc|c} 1 & -4 & -6 & 0 \\ 0 & 22 & 22 & 0 \\ 3 & -1 & -7 & 0 \end{array}\right]$$

2. (-3) to R3 yields

$$\left[\begin{array}{ccc|c} 1 & -4 & -6 & 0 \\ 0 & 22 & 22 & 0 \\ 0 & 11 & 11 & 0 \end{array}\right]$$

3. $(\frac{1}{22})$R2 and $(\frac{1}{11})$R3 yields

$$\left[\begin{array}{ccc|c} 1 & -4 & -6 & 0 \\ 0 & 1 & 1 & 0 \\ 0 & 1 & 1 & 0 \end{array}\right]$$

4. (-1)R2 to R3 yields

$$\left[\begin{array}{ccc|c} 1 & -4 & -6 & 0 \\ 0 & 1 & 1 & 0 \\ 0 & 0 & 0 & 0 \end{array}\right]$$

which is in row echelon form.

The system associated with the last augmented matrix is

$$\begin{aligned} x_1 - 4x_2 - 6x_3 &= 0 \\ x_2 + \ \ x_3 &= 0 \end{aligned}$$

For a solution of this system, let x_3 be arbitrary and solve $x_2 + x_3 = 0$ for x_2, obtaining $x_2 = -x_3$. Then

$$\begin{aligned} x_1 - 4x_2 - 6x_3 &= x_1 - 4(-x_3) - 6x_3 \\ &= x_1 + \quad 4x_3 - 6x_3 \\ &= x_1 - \quad 2x_3 = 0 \end{aligned}$$

or $x_1 = 2x_3$.

Thus the solution to the original system is

$$\begin{aligned} x_1 &= 2x_3 \\ x_2 &= -x_3 \\ x_3 &= \ \ x_3 \quad \text{where } x_3 \text{ is arbitrary} \end{aligned}$$

since for all values of x_3,

$$\begin{aligned} 2x_3 \quad - \ \ 4(-x_3) - 6x_3 &= 0 \\ 3(2x_3) + 10(-x_3) + 4x_3 &= 0 \\ 3(2x_3) - \quad (-x_3) - 7x_3 &= 0 \end{aligned}$$

Consequently, as in Example 1.2.6, this system has infinitely many solutions. (We could also have solved $x_2 + x_3 = 0$ for x_3, obtaining $x_3 = -x_2$, and then have solved $x_1 - 4x_2 - 6x_3 = x_1 - 4x_2 + 6x_2 = x_1 + 2x_2 = 0$ for x_1, obtaining $x_1 = -2x_2$.) ■

A proof of the fact that the Gauss method of elimination yields all the possible solutions of a given system of equations is discussed in the next section.

REMARK You may have seen other ways to solve certain systems of linear equations, such as Cramer's rule. You may also be concerned about the detail in which we explain row echelon form.

In real-world applications, most solutions of linear systems take place on high-speed digital computers. The main virtue of digital computers is that, with suitable programming, they can do many tedious calculations very swiftly. However, the computer must be given very explicit instructions; these can allow no misunderstanding about when a matrix is in row echelon form, that is, when the program should halt. For this reason, we give great detail in our definitions.

We discuss Cramer's rule later in this text. However, the coefficient matrix of a linear system must be square and invertible before we can apply Cramer's rule. Moreover, in Cramer's rule, more operations of multiplication and addition are used than in the Gauss method of elimination. Each operation has inherent error due to rounding when a digital computer is used in the solution of a linear system. Since the Gauss method applies to more linear systems and has fewer operations (in which round-off error is introduced), it is superior to most other methods.

It takes approximately $n^3/3$ operations to reduce an $n \times n$ matrix to row echelon form. To use Cramer's rule, it takes approximately $n^3/3$ steps to reduce a system to triangular form, and so it would take approximately $(n + 1)n^3/3$ steps to solve an $n \times n$ system using Cramer's rule.

SUMMARY

Terms to know:

Equivalent system of linear equations
Elementary row operations
Gauss reduction method
Row echelon form of a matrix
Gauss-Jordan method
Consistent system of linear equations
Inconsistent system of linear equations

Methods to know:

Gauss reduction method

EXERCISE SET 1.2

Using the Gauss method of elimination, find a solution (provided one exists) for each of these systems of equations

1. $$\begin{aligned} 2x + 3y &= 7 \\ 3y &= 6 \end{aligned}$$

2. $$\begin{aligned} 4x + \qquad 3z &= 6 \\ 4x + 2y + z &= 7 \\ 3y + z &= 1 \end{aligned}$$

3. $$\begin{aligned} 2x + 7y + z &= 4 \\ x + y + z &= 2 \\ y + z &= 1 \end{aligned}$$

4. $$\begin{aligned} 7x + 2y &= \pi \\ x - y &= 3 \end{aligned}$$

5. $$\begin{aligned} 8x + 4y + 2z \qquad &= 7 \\ y + z + w &= 3 \\ x + y + \qquad w &= 2 \\ x + \qquad z + w &= 0 \end{aligned}$$

6. $$\begin{aligned} 2x + 3y &= 0 \\ 3y &= 0 \end{aligned}$$

7. $$\begin{aligned} 4x \qquad + 3z &= 0 \\ 4x + 2y + z &= 0 \\ 3y + z &= 0 \end{aligned}$$

8. $$\begin{aligned} 2x + 7y + z &= 0 \\ x + y + z &= 0 \\ y + z &= 0 \end{aligned}$$

9. $$\begin{aligned} 7x + 2y &= 0 \\ x - y &= 0 \end{aligned}$$

10. $$\begin{aligned} 8x + 4y + 2z \qquad &= 0 \\ y + z + w &= 0 \\ x + y + \qquad w &= 0 \\ x + \qquad z + w &= 0 \end{aligned}$$

11. $$\begin{aligned} 2x - 3y &= 8 \\ 3x \qquad &= \tfrac{1}{2}(24 + 9y) \end{aligned}$$

12. $$\begin{aligned} x_1 + 2x_2 - 3x_3 &= -5 \\ 2x_1 - 3x_2 + x_3 &= 11 \\ 3x_1 + 5x_2 + 4x_3 &= 0 \end{aligned}$$

13. $$\begin{aligned} -2x_1 + x_2 + \qquad x_4 &= -4 \\ -x_2 + x_3 + 2x_4 &= -3 \\ x_1 + 2x_2 - 3x_3 \qquad &= -14 \\ -3x_1 + 2x_2 + \qquad 3x_4 &= -13 \end{aligned}$$

14. $$\begin{aligned} 3x_1 + 2x_2 - x_3 &= \tfrac{1}{2} \\ 2x_1 + 3x_2 - 4x_3 &= 27 \\ 4x_1 - x_2 + x_3 &= 2 \end{aligned}$$

15. $$\begin{aligned} 2x_1 - x_2 \qquad &= x_1 \\ -3x_1 + 2x_2 - 2x_3 &= x_2 \\ 8x_1 + \qquad 15x_3 &= x_3 \end{aligned}$$

16. $$\begin{aligned} -x_1 + 11x_2 - 3x_3 &= 9 \\ 4x_1 \qquad + x_3 &= -14 \\ 4x_2 + x_3 &= 2 \\ x_1 - 3x_2 - x_3 &= -5 \end{aligned}$$

17. $$\begin{aligned} x_1 + 4x_2 + 5x_3 &= 0 \\ 2x_1 + x_2 - 4x_3 &= 0 \\ 3x_1 - 2x_2 - 13x_3 &= 0 \end{aligned}$$

18. $$\begin{aligned} x + 3y + z &= 0 \\ 2x - 4y - 3z &= 0 \\ 3x - y - 2z &= 0 \end{aligned}$$

19. $$\begin{aligned} x_1 - x_2 + x_3 &= 0 \\ x_1 - 5x_2 - 3x_3 &= 0 \\ 3x_1 - 4x_2 + 2x_3 &= 0 \end{aligned}$$

20. Determine any k, k a constant, for which the linear system of equations of the following form has solutions.

$$\begin{aligned} x + y \qquad &= 1 \\ y + z &= 1 \\ x \qquad + kz &= 1 \end{aligned}$$

21. Given an equation of a line, $y = mx + b$, and the two distinct points $(1, -2)$ and $(-3, 5)$, find m and b.

22. Given an equation of a plane, $z = ax + by + c$, and the three distinct points $(1, 1, 1)$, $(2, -1, 3)$, and $(3, -2, 1)$, find a, b, and c.

23. Given an equation of a parabola, $y = ax^2 + bx + c$, find a, b, and c given that the parabola passes through the points $(-2, 5)$, $(0, 0)$ and $(3, 1)$.

24. Solve

$$\begin{aligned} \frac{6}{x} - \frac{8}{y} + \frac{12}{z} &= 2 \\ \frac{2}{y} - \frac{3}{z} &= 0 \\ \frac{3}{x} + \frac{4}{y} \qquad &= 2 \end{aligned}$$

25. Solve

$$\begin{aligned} x^2 + 3y^2 - 3z^2 &= -2 \\ 2x^2 + 2y^2 - 2z^2 &= 1 \\ 3x^2 + 5y^2 - 5z^2 &= -1 \end{aligned}$$

26. Solve

$$\begin{aligned} -x^2 + y^3 - z^4 &= 6 \\ 2x^2 + y^3 + 3z^4 &= 12 \\ 4x^2 + 2y^3 + 2z^4 &= 21 \end{aligned}$$

27. Determine all λ for which the following system will have infinitely many solutions:

$$\begin{aligned} \lambda x + 2y &= 0 \\ 2x + \lambda y &= 0 \end{aligned}$$

28. Find values of λ for which the system is consistent:

$$\begin{aligned} \lambda x - 3y &= 4 \\ 2x + y &= 2 \end{aligned}$$

29. Find values of λ for which the system has an infinite number of solutions:

$$\begin{aligned} -y - 2z &= 0 \\ 3x + 7y + 4z &= 0 \\ \lambda x + y + 2z &= 0 \end{aligned}$$

These exercises are best solved using the accompanying software or a hand-held calculator.

30.
$$\begin{aligned} 1.01x_1 - 4.65x_2 - 6.30x_3 &= 0 \\ 2.99x_1 + 9.98x_2 + 4.01x_3 &= 0 \\ 3.02x_1 - 0.99x_2 - 7.31x_3 &= 0 \end{aligned}$$

31.
$$\begin{aligned} 3.10x + 1.99y - z + 3.10w &= -7.99 \\ 2.01x - 4.02y + z + w &= 7.11 \\ -1.02x - 3.02y + 2.21z + 3.10w &= 2.02 \\ -1.99x + 1.99y - 0.01z - 0.02w &= -4.99 \end{aligned}$$

32.
$$\begin{aligned} 1.03x_1 - 0.98x_2 + 1.10x_3 - 0.97x_4 + 0.99x_5 &= 1.01 \\ 2.02x_1 - 1.01x_2 + 2.98x_3 + 5.01x_5 &= 1.99 \\ 2.98x_1 - 2.02x_2 + 1.98x_3 + 1.11x_4 + 1.01x_5 &= 0.99 \\ 1.10x_1 + 1.01x_3 + 1.96x_4 + 1.04x_5 &= 0 \end{aligned}$$

33. $5.1x_1 + 29.9x_2 - 29.9x_3 + 20.1x_4 + 5.1x_5 = 0$
$20.2x_1 - 20.2x_2 + 38.8x_3 + 38.8x_4 - 38.8x_5 = -82.2$
$22.8x_1 - 18.6x_2 + 10.1x_3 + 30.2x_4 + 9.8x_5 = -7.6$

34. $2.1x_1 - 1.2x_2 + 3.4x_3 - 5.2x_4 + 2.4x_5 = 0$
$4.2x_1 - 3.6x_2 - 1.1x_3 - 4.8x_4 + 1.4x_5 = 0$

35. Find the equation of the circle $x^2 + y^2 + Ax + By + C = 0$ containing the points $(0.1, 0.1)$, $(5.4, 0.1)$ and $(0, 0.75)$.

SECTION 1.3 A LITTLE BIT OF THEORY

In this section we prove that the following statements are true:

1. The Gauss method yields all solutions to a linear system of equations.
2. A homogeneous system of equations has either the trivial solution or an infinite number of solutions.

In order to prove these statements, we must be precise in the meaning of words we use. Recall from Section 1.2 that a *solution* to the system

$$\begin{aligned} a_{11}x_1 + a_{12}x_2 + \cdots + a_{1n}x_n &= b_1 \\ a_{21}x_1 + a_{22}x_2 + \cdots + a_{2n}x_n &= b_2 \\ \vdots \quad\quad \vdots \quad\quad\quad\quad \vdots \quad &\quad \vdots \\ a_{m1}x_1 + a_{m2}x_2 + \cdots + a_{mn}x_n &= b_m \end{aligned} \tag{1.3.1}$$

is a sequence $s_1, s_2, \ldots, s_n$ of real numbers for which

$$\begin{aligned} a_{11}s_1 + a_{12}s_2 + \cdots + a_{1n}s_n &= b_1 \\ a_{21}s_1 + a_{22}s_2 + \cdots + a_{2n}s_n &= b_2 \\ \vdots \quad\quad \vdots \quad\quad\quad\quad \vdots \quad &\quad \vdots \\ a_{m1}s_1 + a_{m2}s_2 + \cdots + a_{mn}s_n &= b_m \end{aligned}$$

We first examine the effect of the elementary row operation cRi (multiply row i by the nonzero constant c) on the augmented matrix A associated with the linear system (1.3.1).

Since

$$A = \begin{bmatrix} a_{11} & \cdots & a_{1n} & b_1 \\ a_{21} & \cdots & a_{2n} & b_2 \\ \vdots & & \vdots & \vdots \\ a_{m1} & \cdots & a_{mn} & b_m \end{bmatrix}$$

c Ri applied to A yields the matrix

$$B = \begin{bmatrix} a_{11} & a_{12} & \cdots & a_{1n} & b_1 \\ \vdots & \vdots & & \vdots & \vdots \\ ca_{i1} & ca_{i2} & \cdots & ca_{in} & cb_i \\ \vdots & \vdots & & \vdots & \vdots \\ a_{m1} & a_{m2} & \cdots & a_{mn} & b_m \end{bmatrix}$$

whose associated linear system is

$$\begin{aligned} a_{11}x_1 + \cdots + a_{1n}x_n &= b_1 \\ \vdots \qquad\qquad \vdots \quad &\quad \vdots \\ ca_{i1}x_1 + \cdots + ca_{in}x_n &= cb_i \\ \vdots \qquad\qquad \vdots \quad &\quad \vdots \\ a_{m1}x_1 + \cdots + a_{mn}x_n &= b_m \end{aligned} \tag{1.3.2}$$

Let $s_1, \ldots, s_n$ be a solution of (1.3.1), then $a_{i1}s_1 + \cdots + a_{in}s_n = b_i$ is a true statement about numbers, as is $3 \cdot 2 + 4 \cdot 7 - 1 \cdot 5 = 29$. Just as $8(3 \cdot 2 + 4 \cdot 7 - 1 \cdot 5) = 8 \cdot 29 = 8 \cdot 3 \cdot 2 + 8 \cdot 4 \cdot 7 - 8 \cdot 1 \cdot 5$, so also does

$$ca_{i1}s_1 + ca_{i2}s_2 + \cdots + ca_{in}s_n = c(a_{i1}s_1 + a_{i2}s_2 + \cdots + a_{in}s_n) = cb_i$$

Since this equation is the only change between (1.3.1) and (1.3.2), with the s_i replacing the x_i, the sequence $s_1, \ldots, s_n$ is also a solution of (1.3.2).

Conversely, if the sequence $s_1, \ldots, s_n$ is a solution of (1.3.2), applying $(1/c)$Ri to (1.3.2) shows that $s_1, \ldots, s_n$ is a solution of (1.3.1).

REMARK The key idea in this proof is that when the s_i replaces the x_i, the systems (1.3.1) and (1.3.2) become statements about numbers.

These paragraphs constitute a proof of Lemma 1.3.1.

Lemma 1.3.1

Applying the row operation c Ri *to the augmented matrix A with the associated linear system* (1.3.1) *yields the augmented matrix B of a linear system having exactly the same solutions as* (1.3.1).

In the exercises for this section, we ask you to prove in special cases Lemmas 1.3.2 and 1.3.3.

Lemma 1.3.2
Applying the row operation c$\mathrm{R}i$ *to* $\mathrm{R}j$ (*multiply row i by c and add to row j*) *to the augmented matrix A associated with the linear system* (1.3.1) *yields the augmented matrix B of a linear system having exactly the same solutions as* (1.3.1).

Lemma 1.3.3
Applying the row operation $\mathrm{R}i \leftrightarrow \mathrm{R}j$ (*interchange row i and row j*) *to the augmented matrix A associated with the linear system* (1.3.1) *yields the augmented matrix B of a linear system having exactly the same solutions as* (1.3.1).

These lemmas enable us to prove the next theorem.

Theorem 1.3.4 *If B is a matrix in row echelon form obtained from the augmented matrix A of the linear system* (1.3.1), *then* (1.3.1) *and the linear system of B have the same solutions.*

Proof. Let B be obtained from $A = A_0$ by applying, in order, elementary row operations $E_1, E_2, \ldots, E_n$. Let A_i be the matrix obtained from A_{i-1} by applying elementary row operation E_i. Applying the appropriate lemma (depending on which type of elementary row operation E_i is), we know that the systems of A and A_1 have the same solutions. Next, again applying the appropriate lemma, the systems of A_1 and A_2 have the same solutions. Likewise for A_2 and A_3, A_3 and $A_4, \ldots,$ and A_{n-1} and $A_n = B$ (obtained from A_{n-1} by E_n). Hence the systems of A and B have the same solutions.

Since B is obtained from A by the Gauss method, this theorem says that *all solutions to a linear system of equations can be obtained by the Gauss method.*

Of course, this presupposes that you can determine all solutions to a linear system of equations whose augmented matrix is in row echelon form. However, it is clear from the examples in Section 1.2 that this can be done.

Theorem 1.3.5 *Let*

$$\begin{aligned} a_{11}x_1 + \cdots + a_{1n}x_n &= 0 \\ a_{21}x_1 + \cdots + a_{2n}x_n &= 0 \\ \vdots \qquad\qquad \vdots \quad &\quad \vdots \\ a_{m1}x_1 + \cdots + a_{mn}x_n &= 0 \end{aligned} \tag{1.3.3}$$

be a homogeneous system of equations with solution $s_1, s_2, \ldots, s_n$, where not all s_i are zero. Then this system has an infinite number of nonzero solutions.

Proof. Suppose that $s_t \neq 0$. We will show that for each positive integer k, the sequence $ks_1, ks_2, \ldots, ks_n$ is a solution to (1.3.3).

Clearly, since $k > 0$ and $s_t \neq 0$, $ks_t \neq 0$. By Theorem 1.3.4, $s_1, \ldots, s_n$ is a solution of the system

$$\begin{array}{ccccccc} ka_{11}x_1 & + & ka_{12}x_2 & + \cdots + & ka_{1n}x_n & = & 0 \\ ka_{21}x_1 & + & ka_{22}x_2 & + \cdots + & ka_{2n}x_n & = & 0 \\ \vdots & & \vdots & & \vdots & & \vdots \\ ka_{m1}x_1 & + & ka_{m2}x_2 & + \cdots + & ka_{mn}x_n & = & 0 \end{array}$$

obtained from (1.3.3) by the sequence of elementary operations $k\mathrm{R}1, k\mathrm{R}2, \ldots, k\mathrm{R}n$. Thus

$$\begin{array}{ccccccc} ka_{11}s_1 & + & ka_{12}s_2 & + \cdots + & ka_{1n}s_n & = & 0 \\ \vdots & & \vdots & & \vdots & & \vdots \\ ka_{m1}s_1 & + & ka_{m2}s_2 & + \cdots + & ka_{mn}s_n & = & 0 \end{array}$$

but for $i = 1, 2, \ldots, m$,

$$0 = ka_{i1}s_1 + \cdots + ka_{in}s_n = a_{i1}(ks_1) + \cdots + a_{in}(ks_n)$$

which says that $ks_1, \ldots, ks_n$ is a solution to (1.3.3).

Since for $k_1 \neq k_2$, $k_1s_t \neq k_2s_t$, these solutions $ks_1, \ldots, ks_n$ are all different, and there are an infinite number of these.

Thus the theorem is proved.

Actually, $ks_1, \ldots, ks_n$ is a solution of (1.3.3) for any real number k.

While Theorem 1.3.5 is true and the proof is valid, the proof assumes that we have one nonzero solution. What is worse, the proof does not give us all the solutions. The next theorem and its proof remedy this situation. Notice that the proof gives us an algorithm for obtaining all solutions to the homogeneous system of equations.

Theorem 1.3.6 *A homogeneous system of m equations in n unknowns always has the trivial solution $s_1 = \cdots = s_n = 0$. If it has a nontrivial solution, it has infinitely many. In particular, if $m < n$ it has a nontrivial solution.*

Proof. The proof of this theorem exhibits a five step method for finding all solutions to a homogeneous system of equations.

1. Use the Gauss method to obtain an augmented matrix in row echelon form whose associated linear system has the same solution as the original system. The non-pivot variables, if any, can be assigned to be any real numbers. Then backsolve.
2. In any row echelon matrix, call variables which are not pivot variables **free variables**.
3. If $n > m$, there are more unknown $x_1, \ldots, x_n$ than equations in the original system. Since the Gauss method adds no more equations, the system associated with the row echelon matrix has more unknowns than equations.

Example 1.3.7 The free variables in

$$\left[\begin{array}{ccccc|c} 1 & 2 & 3 & 1 & 5 & 0 \\ 0 & 0 & 3 & 4 & 2 & 0 \\ 0 & 0 & 0 & 7 & 1 & 0 \end{array}\right]$$

are x_2 and x_5, since columns 2 and 5 contain no leading entries. The leading entries (pivots) are 1 in row 1 column 1, 3 in row 2 column 3, and 7 in row 3 column 4. ■

4. Now backsolve for the remaining variables in terms of the free variables, x_2 and x_5.

Example 1.3.7 (continued) We solve for x_4 in terms of x_2 and x_5 and obtain

$$x_4 = -\tfrac{1}{7}x_5$$

Likewise,

$$\begin{aligned} x_3 &= \tfrac{1}{3}(-4x_4 - 2x_5) = \tfrac{1}{3}\left[-4(-\tfrac{1}{7}x_5) - 2x_5\right] \\ &= -\tfrac{10}{21}x_5 \end{aligned}$$

whereas

$$\begin{aligned} x_1 &= -2x_2 - 3x_3 - x_4 - 5x_5 \\ &= -2x_2 - 3(-\tfrac{10}{21}x_5) - (-\tfrac{1}{7}x_5) - 5x_5 \\ &= -2x_2 - \tfrac{24}{7}x_5 \end{aligned}$$

■

5. We may choose any values s_2 and s_5 for x_2 and x_5. Each choice determines s_1, s_3, and s_4, yielding one solution s_1, s_2, s_3, s_4, s_5. Since we can choose any of the infinite collection of pairs of real

numbers for s_2 and s_5, there are infinitely many solutions to the homogeneous system.

Example 1.3.7 (continued) Choosing $x_2 = 1$ and $x_5 = 0$ gives the solution $s_1 = -2$, $s_2 = 1$, $s_3 = 0$, $s_4 = 0$, $s_5 = 0$.

Choosing $x_2 = 0$ and $x_5 = 1$ gives the solution $-\frac{24}{7}, 0, -\frac{10}{21}, -\frac{1}{7}, 1$.

Choosing $x_2 = -1$ and $x_5 = 1$ gives the solution $-\frac{10}{7}, -1, -\frac{10}{21}, -\frac{1}{7}, 1$. ■

SUMMARY

Term to know:

Free unknowns

Theorems to know:

The Gauss method yields all solutions to a linear system of equations.

If the number of variables exceeds the number of equations in a homogeneous linear system of equations, the system has infinitely many solutions.

Notation to know:

$c\mathrm{R}i$

$c\mathrm{R}i$ to $\mathrm{R}j$

$\mathrm{R}i \leftrightarrow \mathrm{R}j$

EXERCISE SET 1.3

1. Prove Lemma 1.3.2 for $n = 3$ and $m = 4$.
2. Prove Lemma 1.3.3 for $n = 3$ and $m = 4$.
3. If a homogeneous linear system of equations has exactly one solution, what is that solution?
4. Prove that a linear system of equations is inconsistent if and only if the row echelon form of the system's augmented matrix has at least one more nonzero row than the row echelon form of the coefficient matrix of the system.

As in the proof of Theorem 1.3.6, find five distinct solutions to the homogeneous system associated with each coefficient matrix.

5. $\begin{bmatrix} 1 & 0 & 0 & 1 \\ 0 & 0 & 0 & 1 \\ 0 & 0 & 0 & 0 \end{bmatrix}$

6. $\begin{bmatrix} 3 & 0 & 4 & 1 \\ 0 & 5 & 0 & 1 \\ 0 & 0 & 0 & 1 \end{bmatrix}$

7. $\begin{bmatrix} 2 & 0 & 3 & 1 \\ 0 & 0 & 0 & 0 \end{bmatrix}$

8. $\begin{bmatrix} 2 & -1 & 4 & 0 & 5 \\ 0 & -1 & 0 & -2 & 0 \\ 0 & 0 & -2 & 1 & 1 \end{bmatrix}$

SECTION 1.4 ECONOMIC MODELS (OPTIONAL)

In 1973, Wassily Leontief won the Nobel Prize for Economics. The prize was awarded for his use of linear algebra to model the world economy. Such a model, by its very nature, is complicated and well beyond the scope of this text. However, in this section we discuss several simple input-output models of the economy to give you an indication of Leontief's ideas.

Example 1.4.1 Consider a simple economy that produces only food, clothing, and shelter, the **outputs** of this economy. The food is measured in pounds, the clothing in yards, and the shelter in houses. The outputs are made from land (measured in acres), labor (measured in hours), and capital (measured in tens of dollars). The **inputs** of this economy are land, labor, and capital. The inputs must be combined in a fixed proportion to produce one unit of output. For example, it may take 3 acres of land, 8 hours of labor, and one $10 bill to produce 1 pound of food. The information for each output is given in the accompanying table.

		Outputs	
Inputs	*Food (pounds)*	*Clothing (yards)*	*Shelter (houses)*
Land (acres)	3 acres/pound	1 acre/yard	10 acre/house
Labor (hours)	8 hours/pound	4 hours/yard	1000 hours/house
Capital (tens of dollars)	1 ten/pound	2 ten/yard	5 ten/house

Note the units of measure on each number in the table. The **technology** is represented by the **input-output matrix**

$$A = \begin{bmatrix} 3 & 1 & 10 \\ 8 & 4 & 1000 \\ 1 & 2 & 5 \end{bmatrix}$$

Note that the (i, j)-entry of A indicates the amount of input i used to produce 1 unit of output j. For example, $a_{23} = 1000$ indicates that it takes 1000 worker hours to produce 1 house.

Suppose this economy has only 150 acres, 2364 worker hours, and $750.00 (75 tens of dollars) available. How many pounds of food, yards of clothing, and units of shelter can be produced using all the available inputs?

Solution Since it takes 3 acres to produce 1 pound of food, 1 acre to produce 1 yard of clothing, and 10 acres to produce 1 house, we obtain the equation

$$3x_1 + 1x_2 + 10x_3 = 150$$

where

$$x_1 = \text{number of units of food}$$

$$x_2 = \text{number of units of clothing}$$

$$x_3 = \text{number of units of shelter}$$

Note that the units agree in this equation, i.e.,

$$\left(3\,\frac{\text{acres}}{\text{pound}}\right)(x_1 \text{ pounds}) + \left(1\,\frac{\text{acre}}{\text{yard}}\right)(x_2 \text{ yards}) + \left(10\,\frac{\text{acres}}{\text{hour}}\right)(x_3 \text{ hours}) = 150 \text{ acres}$$

Likewise, for labor,

$$8x_1 + 4x_2 + 1000x_3 = 2364$$

and, for capital,

$$1x_1 + 2x_2 + 5x_3 = 75$$

The system of linear equations

$$\begin{aligned} 3x_1 + x_2 + 10x_3 &= 150 \\ 8x_1 + 4x_2 + 1000x_3 &= 2364 \\ x_1 + 2x_2 + 5x_3 &= 75 \end{aligned}$$

can be solved using the Gauss method of elimination. The associated augmented matrix is

$$\left[\begin{array}{ccc|c} 3 & 1 & 10 & 150 \\ 8 & 4 & 1000 & 2364 \\ 1 & 2 & 5 & 75 \end{array}\right]$$

Applying the Gauss method, we obtain

$$\left[\begin{array}{ccc|c} 1 & 2 & 5 & 75 \\ 0 & 1 & 1 & 15 \\ 0 & 0 & 972 & 1944 \end{array}\right]$$

whose associated linear system is

$$\begin{aligned} x_1 + 2x_2 + 5x_3 &= 75 \\ x_2 + x_3 &= 15 \\ 972x_3 &= 1944 \end{aligned}$$

Backsolving yields

$$x_3 = 2$$

$$x_2 = 13$$

$$x_1 = 39$$

Using all the available resources produces 39 units of food, 13 units of clothing, and 2 units of shelter. ■

The terms input *and* output *are only technical terms, not statements of physical fact.* This should be made clear in Example 2, where the inputs and outputs of Example 1 are interchanged.

Example 1.4.2 Consider a simple economy with outputs land (acres), labor (hours), and capital (tens of dollars) and inputs food (pounds), clothing (yards), and shelter (houses). The information about this economy is contained in the following table.

	Land (acres)	*Labor (hours)*	*Capital (tens of dollars)*
Food (pounds)	1 pound/acre	3 pounds/hour	2 pounds/ten
Clothing (yards)	2 yards/acre	7 yards/hour	5 yards/ten
Shelter (houses)	3 houses/acre	2 houses/hour	1 house/ten

The technology is represented by the input-output matrix

$$B = \begin{bmatrix} 1 & 3 & 2 \\ 2 & 7 & 5 \\ 3 & 2 & 1 \end{bmatrix}$$

Suppose this economy needs 400 pounds of food, 925 yards of clothing, and 475 houses. How many acres of land, hours of labor, and tens of dollars are required to fulfill all the needs of the economy?

Solution Let

$$y_1 = \text{acres of land}$$

$$y_2 = \text{hours of labor}$$

$$y_3 = \text{tens of dollars}$$

required to fulfill the needs of this economy. Since 1 pound of food is produced per acre, 3 pounds of food are produced per hour of labor, and 2 pounds of food are produced for each \$10 of capital, we obtain the equation

$$1y_1 + 3y_2 + 2y_3 = 400$$

Note that the units agree in this equation,

$$\left(1\frac{\text{pound}}{\text{acre}}\right)(y_1 \text{ acres}) + \left(3\frac{\text{pounds}}{\text{hour}}\right)(y_2 \text{ hours})$$
$$+\left(3\frac{\text{pounds}}{\text{ten}}\right)(y_3 \text{ tens}) = 400 \text{ pounds}$$

Likewise, for clothing,

$$2y_1 + 7y_2 + 5y_3 = 925$$

and for houses,

$$3y_1 + 2y_2 + 1y_3 = 475$$

The system of linear equations

$$\begin{aligned} 1y_1 + 3y_2 + 2y_3 &= 400 \\ 2y_1 + 7y_2 + 5y_3 &= 925 \\ 3y_1 + 2y_2 + 1y_3 &= 475 \end{aligned}$$

can be solved using the Gauss method of elimination. The associated augmented matrix is

$$\left[\begin{array}{ccc|c} 1 & 3 & 2 & 400 \\ 2 & 7 & 5 & 925 \\ 3 & 2 & 1 & 475 \end{array}\right]$$

Applying the Gauss method of elimination we obtain

$$\left[\begin{array}{ccc|c} 1 & 3 & 2 & 400 \\ 0 & 1 & 1 & 125 \\ 0 & 0 & 2 & 150 \end{array}\right]$$

whose associated linear system is

$$\begin{aligned} y_1 + 3y_2 + 2y_3 &= 400 \\ y_2 + y_3 &= 125 \\ 2y_3 &= 150 \end{aligned}$$

Backsolving yields

$$\begin{aligned} y_3 &= 75 \\ y_2 &= 50 \\ y_1 &= 100 \end{aligned}$$

Satisfying all the needs of the economy requires 100 acres of land, 50 hours of labor, and \$750.00. ■

To close out this section, we return to Example 1.4.1 and ask the same question as in Example 1.4.2.

Example 1.4.3 We revisit Example 1.4.1 and ask a different question. We require 4000 pounds of food, 5000 yards of clothing, and 6000 houses. How many acres of land (y_1), hours of labor (y_2) and tens of dollars (y_3) are needed to meet our requirements?

Solution This problem is straightforward to solve if we examine the units used in the input-output matrix of Example 1.4.1,

$$A = \begin{bmatrix} 3 & 1 & 10 \\ 8 & 4 & 1000 \\ 1 & 2 & 5 \end{bmatrix}$$

The entry a_{11} represents 3 acres to obtain 1 pound of food, so the units are acres/pound. Thus we obtain $\frac{1}{3}$ pound of food per acre. Similarly, since it takes 8 hours to produce 1 pound of food, we obtain $\frac{1}{8}$ pound of food per hour. It takes 1 \$10 bill to produce 1 pound of food, or 1 pound of food per \$10 bill. Thus we obtain the equation

$$\tfrac{1}{3}y_1 + \tfrac{1}{8}y_2 + 1y_3 = 4000$$

where

$$y_1 = \text{number of acres}$$

$$y_2 = \text{number of hours}$$

$$y_3 = \text{number of \$10 bills}$$

For yards of clothing,

$$1y_1 + \tfrac{1}{4}y_2 + \tfrac{1}{2}y_3 = 5000$$

and for houses,

$$\tfrac{1}{10}y_1 + \tfrac{1}{1000}y_2 + \tfrac{1}{5}y_3 = 6000$$

We solve this system using the Gauss method of elimination. From the augmented matrix

$$\left[\begin{array}{ccc|c} \frac{1}{3} & \frac{1}{8} & 1 & 4000 \\ 1 & \frac{1}{4} & \frac{1}{2} & 5000 \\ \frac{1}{10} & \frac{1}{1000} & \frac{1}{5} & 6000 \end{array}\right]$$

we obtain the matrix

$$\left[\begin{array}{ccc|c} 1 & 0.25 & 0.5 & 5000 \\ 0 & 0.0416 & 0.8333 & 2333.3333 \\ 0 & 0.0 & 0.6269 & 6842.9266 \end{array}\right]$$

After backsolving, we obtain the solution (by hand-held calculator)

$$y_3 = 10875.6$$

$$y_2 = -1611.94$$

$$y_1 = 39860.7$$

Since $y_2 < 0$ and it is physically impossible to use a negative amount of hours of labor, our answer says that it is *impossible* to obtain 4000 pounds of food, 5000 yards of clothing, and 6000 houses from this economy.

SUMMARY

Terms to know:

Input
Output
Technology
Input-output matrix

EXERCISE SET 1.4

1. Tom, Dick, and Harry drive trucks that haul coal. On the first two days of the job, the three trucks hauled 158 and 143 cubic yards of coal. On the third day, Tom was sick and, with extra effort, Dick and Harry were able to haul 119 cubic yards. The number of trips made each day by each truck is given in the table.

	1st Day	*2nd Day*	*3rd Day*
Tom	6	6	0
Dick	5	4	7
Harry	6	5	9

   What is the minimum capacity of each truck?

2. Rufus Roofer Company has three machines for manufacturing roofing shingles. Since all the shingles come out of a chute onto the same conveyor belt, the quality controller wanted to determine whether each machine was producing up to its capacity. For 3 days, the quality controller kept

careful records of the hours the machines were actually in operation and the number of shingles produced. The data are given in the table.

	Machine I	*Machine II*	*Machine III*	*Roofing Shingles Manufactured*
1st Day	7	8	6	12,622
2nd Day	7	6	5	10,420
3rd Day	7	4	6	9,646

What was the average hourly production of shingles for each machine?

3. Ms. Louis has three sums of money invested. The first amount, on which she can write checks, yields only 5% interest. The other two amounts yield 8% and 12%, respectively. Her total return from the three investments is \$14,000. Ms. Louis finds that her return from the third amount is \$7500 less than the combined return from the other two. If all of the money were invested at 7%, her return would increase by \$700. How much is invested at each rate?

4. An oil slick is spreading out in a circular pattern. The Slick Out Company needs to know more about the behavior of the oil slick. They need to find numbers a, b, and c such that

$$x^2 + y^2 + ax + by + c = 0$$

is the equation of a circle that goes through the points $(1,0)$, $(0,1)$, and $(1,-1)$. Determine a, b, and c.

5. Jerome Brown owns a 1000-acre farm. He wants to plant soybeans and corn. The seed and other costs for soybeans is \$40 per acre and for corn \$50 per acre. Brown has \$48,500 as capital. How many acres of each crop could he plant if he spends all of his capital?

6. Dotti Hu, a dairy farmer, is given three foods that must be used to feed her cattle. Each unit of food A contains 10 ounces of protein, 4 ounces of fat, and 6 ounces of carbohydrate. A unit of food B has 4 ounces of protein, 3 ounces of fat, and 7 ounces of carbohydrate. A unit of food C contains 6 ounces of protein, 7 ounces of fat and 12 ounces of carbohydrate. Each time the food is mixed, it must contain at least 78 ounces of protein, 58 ounces of fat, and 75 ounces of carbohydrate. How much of each food is used to provide exactly the minimum requirements?

7. If the time that a computer requires for one addition or subtraction is m microseconds and the time for one multiplication or division is n microseconds, what is the longest time it can take a computer to solve each linear system?

   a. Two equations in two unknowns

   b. Three equations in three unknowns

8. Suppose the government in charge of the economy outlined in Example 1.4.1 decides to increase productivity by reducing the number of worker hours that are needed to produce 1 unit of either food, clothing, or shelter. They still have only 150 acres, 2364 worker hours and \$750. The technology matrix is given by

$$A = \begin{bmatrix} 3 & 1 & 10 \\ 5 & 2 & 840 \\ 1 & 2 & 5 \end{bmatrix}$$

How many units of food, clothing, and shelter can be produced using all the available resources?

9. Suppose the economic situation in Exercise 8 is on the upswing so that the government is able to increase the resources to 225 acres, 3670 worker hours, and $900. Using the technology matrix from Exercise 8, how many units of food, clothing, and shelter can be produced using all the present available resources?

These exercises are best solved using the accompanying computer software or a hand-held calculator.

10. Consider a simple economy that produces only five commodities C_1, C_2, C_3, C_4, and C_5. Assume (1) C_1 requires 3.1 units of I_2, 1.99 units of I_4, and 1.10 units of I_5: (2) C_2 requires 1.2 units of I_5 and 2.1 units of I_3; (3) C_3 requires 3.01 units of I_5, 1.12 units of I_4, and 2.96 units of I_1; (4) C_4 requires 1.02 units of I_3 and 2.97 units of I_5; and C_5 requires 1.1 units of I_2 and 3.11 units of I_4. Suppose this economy requires 5.5 units of C_1, 10.2 units of C_2, 6.4 units of C_3, 8.8 units of C_4 and 12.5 units of C_5. How can these requirements be met?

11. Miss Amy has three sums of money invested. The first amount on which she can write checks yields only 5.1% interest. The other two yield 8.2% and 12.1% respectively. Her total return from the three investments is $14,500. Miss Amy finds that her return from the third amount is $7,600 less than the combined returns of the other two. If all the money were invested at 7%, her return would increase by $15,200. How much is invested at each rate?

12. Laura Luke, a dairy farmer, is given three types of feed that must be mixed to feed her cattle. Each unit of the first feed contains 10.1 ounces protein, 4.1 ounces carbohydrates, and 6.1 ounces fat. The second contains 3.9 ounces protein, 3.3 ounces carbohydrates, and 7.6 ounces fat per unit of feed. The third contains 2.1 ounces protein, 12.1 ounces carbohydrates, and 11.9 ounces fat per unit of feed. How many units of each should be used to make a feed that contains 78 ounces protein, 58 ounces carbohydrates, and 75 ounces fat?

13. Miss Amy deposits a fixed amount (measured in dollars) at bank X that pays $7\frac{1}{4}\%$ interest, a fixed amount at bank Y paying $8\frac{3}{4}\%$ interest, and a fixed amount at bank Z paying $9\frac{1}{2}\%$ interest. At the end of the year her total interest from banks X and Y is $40 more than the interest from bank Z. And the total interest from banks X and Z is $500 more than that from bank Y. How much does Miss Amy have in each bank and how much is her total interest?

SECTION 1.5 PHYSICAL MODELS (OPTIONAL)

ELECTRIC CIRCUITS

Figures 1.1 and 1.2 represent the current loops (or meshes) of two- and three-loop electrical circuits, respectively.

Before we calculate currents in an electrical circuit, we first introduce some terminology and state two laws pertaining to electrical circuits.

1. **Current**, denoted by I, is the flow of electrons and is measured in amperes (amps).

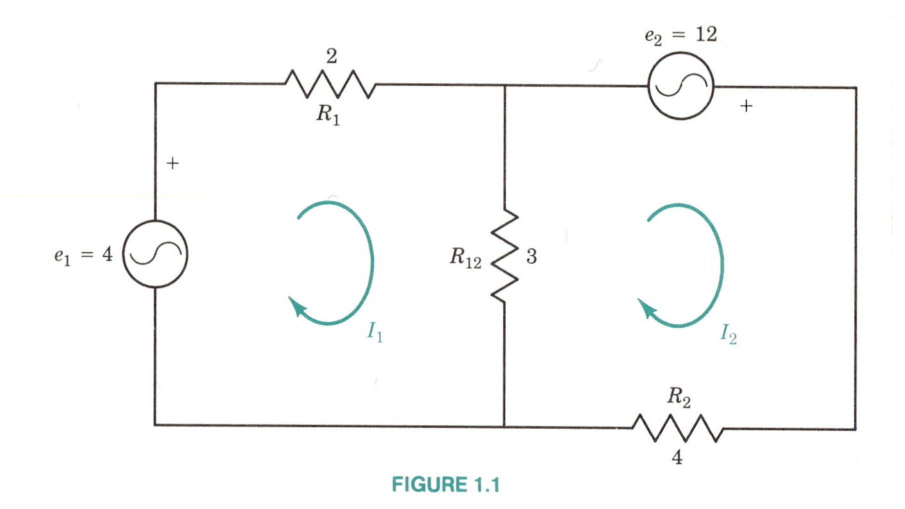

FIGURE 1.1

2. **Voltage**, denoted by e, is a source of energy (e.g., as from a battery) and is measured in volts.
3. **Resistance**, denoted by R, is a part in an electric circuit that resists the flow of a current (such as a light or heater) and is measured in ohms.

As the electric current passes through a resistor, there is a decrease, called a **voltage drop**, in the force generated by the voltage source. If E denotes the voltage drop, then by **Ohm's law**

$$E = RI$$

Another needed law is **Kirchhoff's law**: The algebraic sum of all the voltage drops around any loop is zero (i.e., the voltage supplied by the voltage source is equal to the sum of the voltage drops).

Applying Ohm's and Kirchhoff's laws to the current in the first loop of the two-current loop (Figure 1.1), where the arrow indicates the

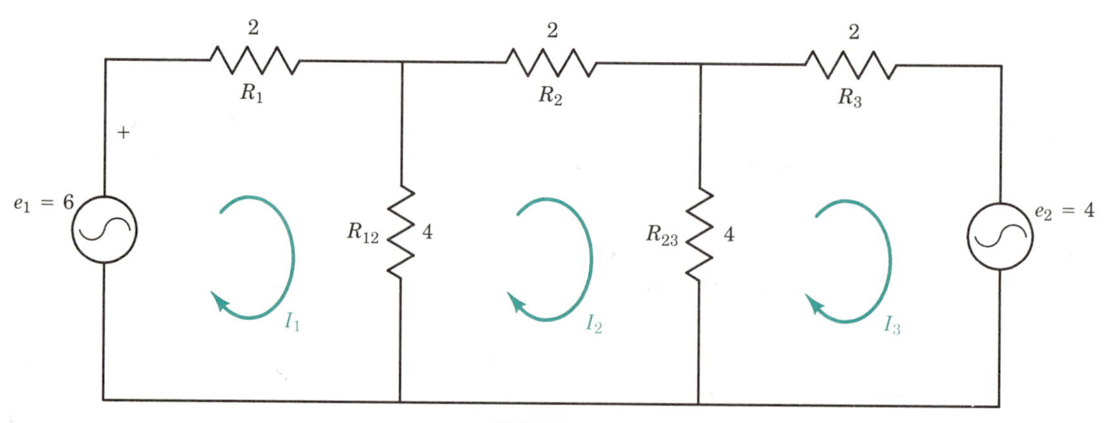

FIGURE 1.2

direction of flow, we have the following:

1. The voltage drop across resistance R_1 is

$$2I_1 \qquad \text{[Ohm's law]}$$

2. The current through the common resistance R_{12} is $I_1 - I_2$ (the currents pass through R_{12} in opposite directions), so the voltage drop across R_{12} is

$$3(I_1 - I_2) \qquad \text{[Ohm's law]}$$

3. $$3(I_1 - I_2) + 2I_1 = 4 \qquad \text{[Kirchhoff's law]}$$

   or

$$5I_1 - 3I_2 = 4$$

Applying Ohm's and Kirchhoff's laws to the second loop, we have the following:

1. The voltage drop across resistance R_2 is

$$4I_2 \qquad \text{[Ohm's law]}$$

2. The current through the common resistance R_{12} is $I_2 - I_1$, so the voltage drop across R_{12} is

$$3(I_2 - I_1) \qquad \text{[Ohm's law]}$$

3. $$3(I_2 - I_1) + 4I_2 = 12 \qquad \text{[Kirchhoff's law]}$$

   or

$$-3I_1 + 7I_2 = 12$$

Thus the currents I_1 and I_2 satisfy the linear system

$$\begin{aligned} 5I_1 - 3I_2 &= 4 \\ -3I_1 + 7I_2 &= 12 \end{aligned}$$

with augmented matrix

$$\left[\begin{array}{rr|r} 5 & -3 & 4 \\ -3 & 7 & 12 \end{array}\right]$$

Solving this system yields $I_1 = \frac{32}{13}$ and $I_2 = \frac{36}{13}$.

Example 1.5.1 Given the circuit in Figure 1.1 with $R_1 = R_2 = R_{12} = 2$ ohms and $e_1 = e_2 = 8$ volts.

a. Use Ohm's and Kirchhoff's laws to determine the linear system for finding currents I_1 and I_2.

b. Solve the system for I_1 and I_2.

Solution a. First loop:

1. The voltage drop across R_1 is $2I_1$.
2. The voltage drop across R_{12} is $2(I_1 - I_2)$.
3.
$$\begin{aligned} 2I_1 + 2(I_1 - I_2) &= 8 \\ 4I_1 - 2I_2 &= 8 \\ 2I_1 - I_2 &= 4 \end{aligned}$$

Second loop:

1. The voltage drop across R_2 is $2I_2$.
2. The voltage drop across R_{12} is $2(I_2 - I_1)$.
3.
$$\begin{aligned} 2I_2 + 2(I_2 - I_1) &= 8 \\ 4I_2 - 2I_1 &= 8 \\ I_1 - 2I_2 &= -4 \end{aligned}$$

The linear system is

$$\begin{aligned} 2I_1 - I_2 &= 4 \\ I_1 - 2I_2 &= -4 \end{aligned}$$

b. The augmented matrix for this system is

$$\left[\begin{array}{cc|c} 2 & -1 & 4 \\ 1 & -2 & -4 \end{array}\right]$$

Applying row operations yields

$$\left[\begin{array}{cc|c} 1 & -2 & -4 \\ 0 & 3 & 12 \end{array}\right]$$

The linear system associated with the last augmented matrix is

$$\begin{aligned} I_1 - 2I_2 &= -4 \\ 3I_2 &= 12 \end{aligned}$$

Backsolving yields $I_2 = 4$ and $I_1 = 4$. Thus $I_1 = I_2 = 4$ amps. ■

Example 1.5.2 Given the three-loop circuit in Figure 1.2, use Ohm's and Kirchhoff's laws to determine the linear system involving currents I_1, I_2, and I_3 and then solve the system.

Solution First loop:

$$2I_1 + 4(I_1 - I_2) = 6$$
$$6I_1 - 4I_2 = 6$$

or

$$3I_1 - 2I_2 = 3$$

Second loop:

$$2I_2 + 4(I_2 - I_3) + 4(I_2 - I_1) = 0$$
$$-4I_1 + 10I_2 - 4I_3 = 0$$

or

$$2I_1 - 5I_2 + 2I_3 = 0$$

Third loop:

$$\begin{aligned} 2I_3 + 4(I_3 - I_2) &= 4 \\ -4I_2 + 6I_3 &= 4 \\ 2I_2 - 3I_3 &= -2 \end{aligned}$$

The linear system is

$$\begin{aligned} 3I_1 - 2I_2 &= 3 \\ 2I_1 - 5I_2 + 2I_3 &= 0 \\ 2I_2 - 3I_3 &= -2 \end{aligned}$$

Applying row operations to the augmented matrix

$$\left[\begin{array}{ccc|c} 3 & -2 & 0 & 3 \\ 2 & -5 & 2 & 0 \\ 0 & 2 & -3 & -2 \end{array}\right]$$

yields

$$\left[\begin{array}{ccc|c} 3 & -2 & 0 & 3 \\ 0 & -\frac{11}{3} & 2 & -2 \\ 0 & 0 & 21 & 34 \end{array}\right]$$

The linear system associated with the last augmented matrix is

$$\begin{aligned} 3I_1 - 2I_2 &= 3 \\ -\tfrac{11}{3}I_2 + 2I_3 &= -2 \\ 21I_3 &= 34 \end{aligned}$$

Backsolving, we obtain

$$I_1 = \tfrac{451}{231}, \qquad I_2 = \tfrac{110}{77}, \quad \text{and} \quad I_3 = \tfrac{34}{21}$$ ■

HEAT CONDUCTION

We begin with the thin rectangular plate $ABCD$ shown in Figure 1.3.

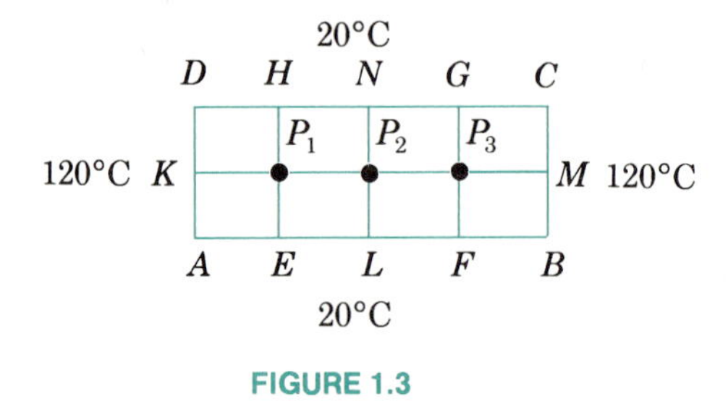

FIGURE 1.3

The top and bottom are kept at 20°C and the sides kept at 120°C. There is a practical problem in maintaining these temperatures near the corner points, but this problem is not considered in this model.

We need one law, the **heat conduction law**: At equilibrium, the temperature at any interior point P is the average of the temperatures at the four neighboring points. Thus the temperature at P_1 is the average of the temperatures at H, K, E, and P_2.

An elementary physics problem is to determine the equilibrium temperatures t_1, t_2, and t_3 at the interior points P_1, P_2, and P_3. To solve this problem, we use the law just stated to compute the average temperatures t_1, t_2, and t_3 near P_1, P_2, and P_3.

1. At P_1, the neighboring points are K, E, P_2, and H. Thus

$$t_1 = \frac{120 + 20 + t_2 + 20}{4}$$

$$4t_1 = 160 + t_2$$

$$4t_1 - t_2 = 160$$

2. At P_2, the neighboring points are P_1, L, P_3, and N. Thus

$$t_2 = \frac{t_1 + 20 + t_3 + 20}{4}$$

$$4t_2 = t_1 + t_3 + 40$$

$$t_1 - 4t_2 + t_3 = -40$$

3. At P_3, the neighboring points are P_2, F, M, and G. Thus

$$t_3 = \frac{t_2 + 20 + 120 + 20}{4}$$

$$4t_3 = t_2 + 160$$

$$t_2 - 4t_3 = -160$$

We need to solve the linear system

$$\begin{aligned} 4t_1 - t_2 \qquad &= 160 \\ t_1 - 4t_2 + t_3 &= -40 \\ t_2 - 4t_3 &= -160 \end{aligned}$$

whose augmented matrix is

$$\left[\begin{array}{ccc|c} 4 & -1 & 0 & 160 \\ 1 & -4 & 1 & -40 \\ 0 & 1 & -4 & -160 \end{array}\right]$$

Applying row operations to the augmented matrix yields

$$\left[\begin{array}{ccc|c} 1 & -4 & 1 & -40 \\ 0 & 1 & -4 & -160 \\ 0 & 0 & 56 & 2720 \end{array}\right]$$

The linear system associated with the last augmented matrix is

$$\begin{aligned} t_1 - 4t_2 + t_3 &= -40 \\ t_2 - 4t_3 &= -160 \\ 56t_3 &= 2720 \end{aligned}$$

Backsolving, we obtain

$$\begin{aligned} t_3 &= \tfrac{2720}{56} = \tfrac{340}{7} \\ t_2 - 4\left(\tfrac{340}{7}\right) &= -160 \\ t_2 &= \tfrac{240}{7} \\ t_1 - 4\left(\tfrac{240}{7}\right) + \tfrac{340}{7} &= -40 \\ t_1 &= \tfrac{340}{7} \end{aligned}$$

SUMMARY

Terms to know:

- Current
- Voltage
- Resistance
- Voltage drop
- Ohm's law
- Kirchhoff's law

Laws to know:

Ohm's law: $E = IR$

Kirchhoff's law: The algebraic sum of all the voltage drops around any loop is zero.

Heat conduction law: At equilibrium, the temperature at any interior point P is the average of the temperatures at the neighboring points.

EXERCISE SET 1.5

Given the circuit in Figure 1.1, determine I_1 and I_2 for each set of data.

1. $e_1 = e_2 = 2,\ R_1 = R_{12} = R_2 = 8$
2. $e_1 = 8,\ e_2 = 4,\ R_1 = 2,\ R_{12} = 4,\ R_2 = 2$
3. $e_1 = 6,\ e_2 = 2,\ R_1 = 1,\ R_{12} = 2,\ R_2 = 4$

Given the circuit in Figure 1.2, determine I_1, I_2, and I_3 for each set of data.

4. $e_1 = e_2 = 2,\ R_1 = R_{12} = R_2 = R_{23} = R_3 = 1$
5. $e_1 = 2,\ e_2 = 4,\ R_1 = 4,\ R_{12} = 2,\ R_2 = 4,\ R_{23} = 2,\ R_3 = 4$
6. $e_2 = 4,\ e_1 = 2,\ R_1 = 2,\ R_{12} = 4,\ R_2 = 2,\ R_{23} = 4,\ R_3 = 6$
7. For the given three-loop circuit, determine I_1, I_2, and I_3.

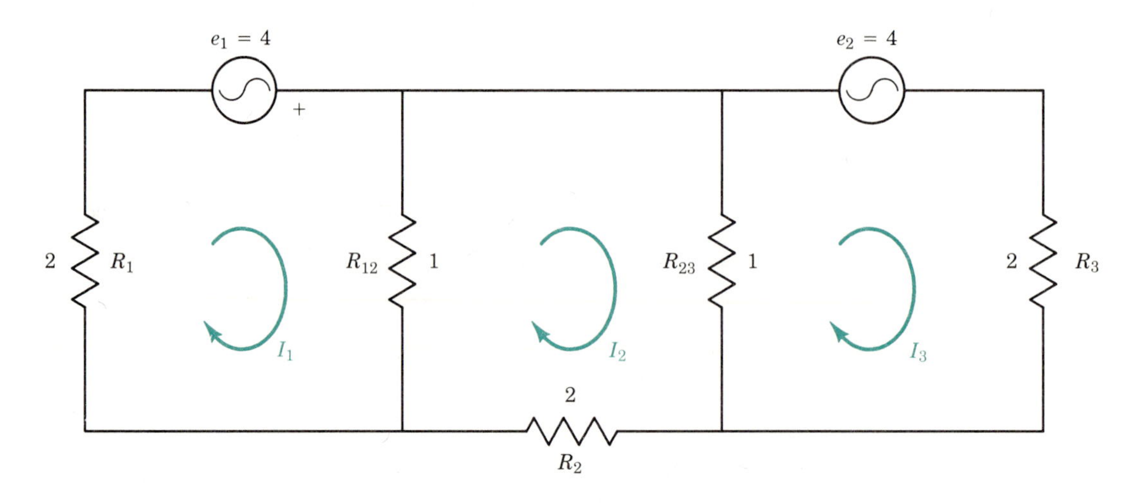

Given the circuit in Problem 7, determine I_1, I_2, and I_3 for each set of data.

8. $e_1 = 6,\ e_2 = 4,\ R_1 = R_2 = R_3 = 2,\ R_{12} = R_{23} = 1$
9. $e_1 = 4,\ e_2 = 8,\ R_1 = 2,\ R_2 = 4,\ R_3 = 8,\ R_{12} = 4,\ R_{23} = 2$
10. $e_1 = 12,\ e_2 = 6,\ R_1 = 4,\ R_2 = 2,\ R_3 = 4,\ R_{12} = 6,\ R_{23} = 3$
11. For the given three-loop circuit, determine I_1, I_2, and I_3.

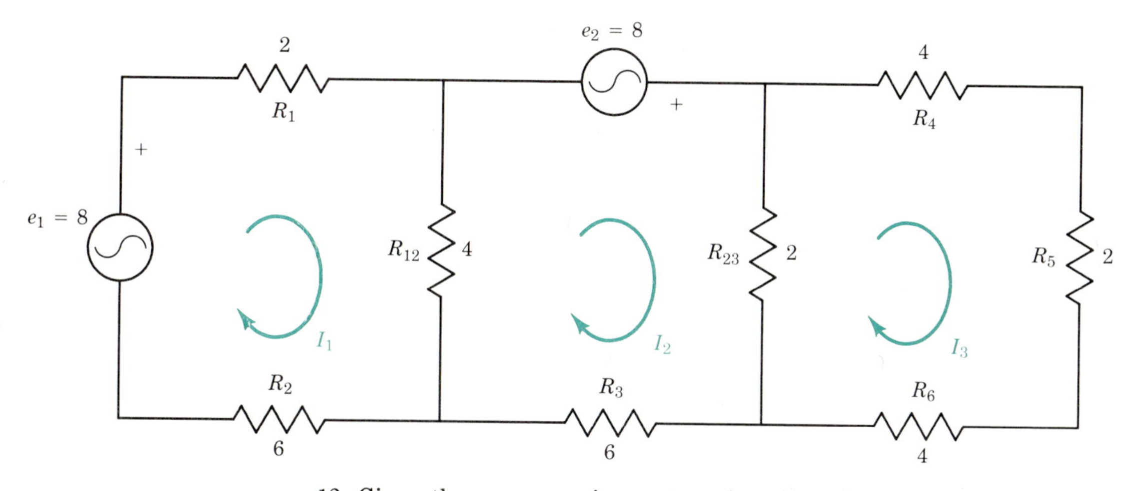

12. Given the accompanying rectangular plate, determine the equilibrium temperatures t_1 and t_2 at P_1 and P_2, respectively.

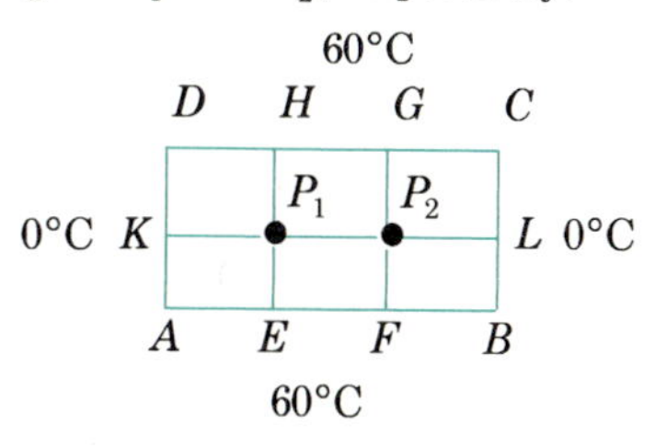

13. Given the accompanying rectangular plate, determine the equilibrium temperatures t_1, t_2, and t_3 at P_1, P_2, and P_3, respectively.

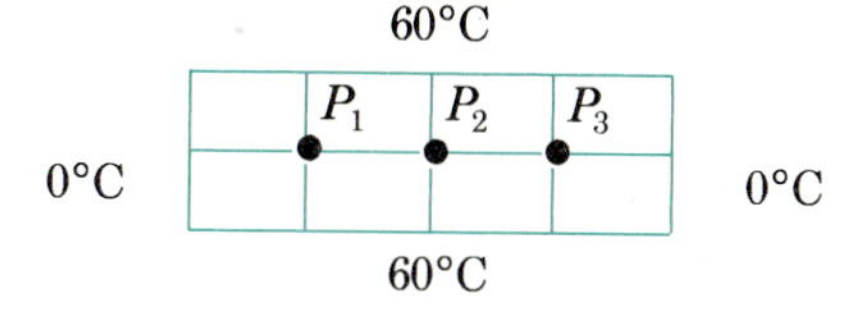

These exercises are best solved using the accompanying software or a hand-held calculator.

14. For the given rectangular plate, determine the equilibrium temperatures t_1, t_2, t_3, t_4, t_5, and t_6 at P_1, P_2, P_3, P_4, P_5, and P_6, respectively.

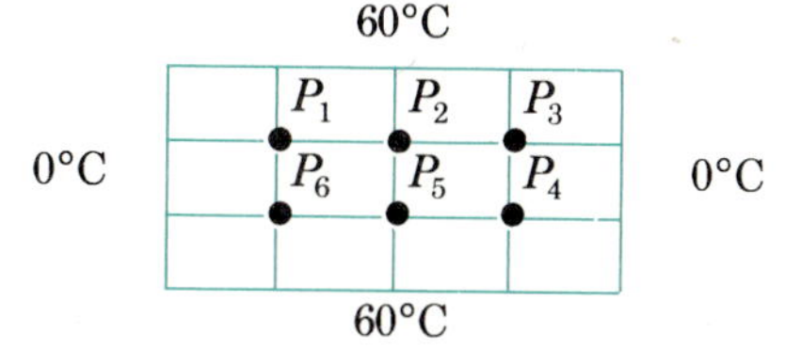

15. For the given plate, find the equilibrium temperatures t_1, t_2, and t_3 at P_1, P_2, and P_3, respectively.

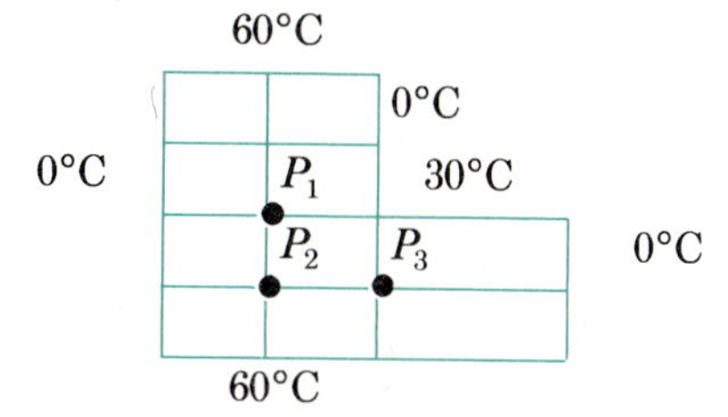

16. For the given plate, find the equilibrium temperatures t_1, t_2, t_3, t_4, t_5, t_6, t_7, and t_8 at P_1, P_2, P_3, P_4, P_5, P_6, P_7, and P_8, respectively.

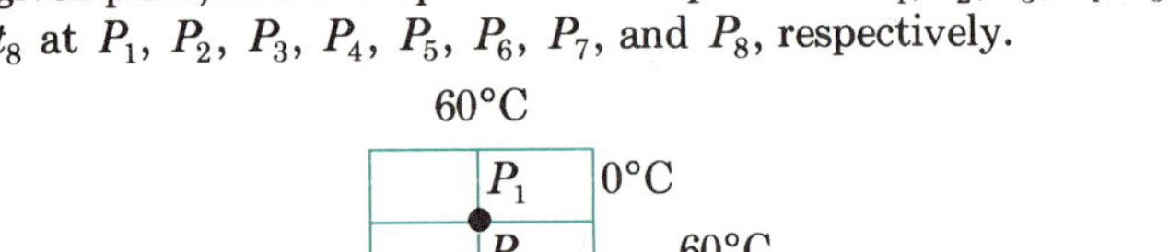
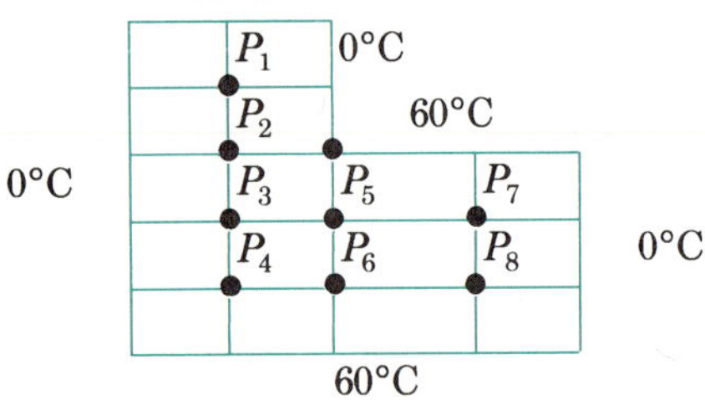

17. In the following rectangular plate

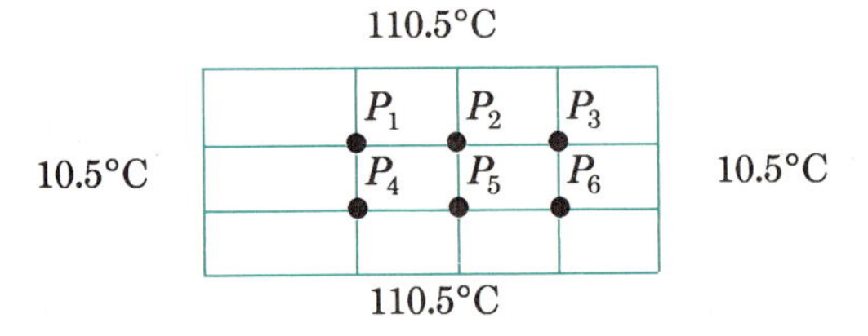

the top and bottom are kept at 110.5° and the sides at 10.5° centigrade. Determine the equilibrium temperatures at t_1, t_2, t_3, t_4, t_5, and t_6 at the interior points P_1, P_2, P_3, P_4, P_5, and P_6.

18. Given the circuit in the following figure

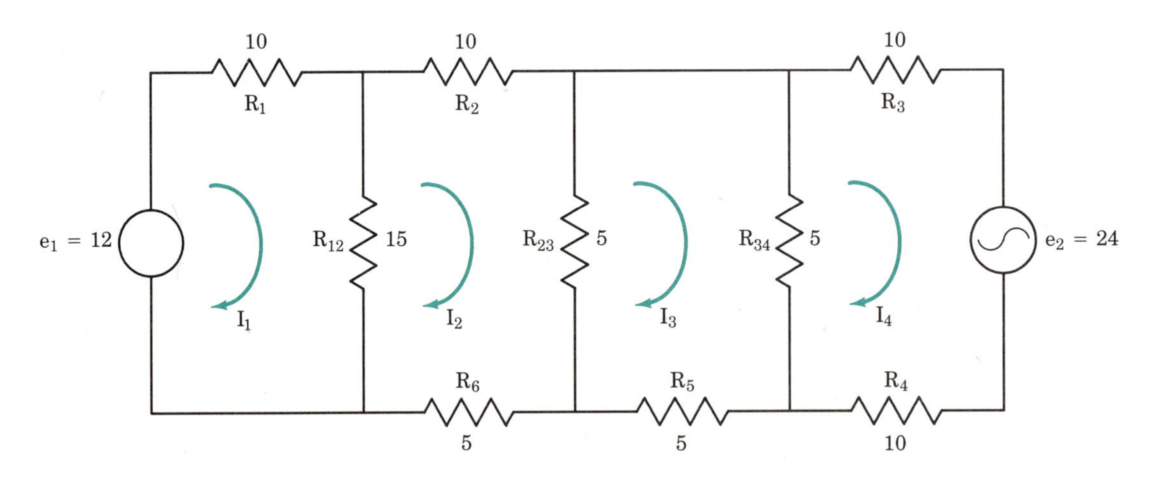

a. Determine the linear system for finding currents I_1, I_2, I_3, and I_4.

b. Solve the system for I_1, I_2, I_3, and I_4.

SECTION 1.6 STATISTICS: LEAST SQUARES MODEL (OPTIONAL)

Regression analysis is a statistical technique that can be used to develop an equation to estimate mathematically how certain variables are related. For example, suppose we had access to last year's sales

TABLE 1.1. Historical Sales Figures and Years of Experience for 10 Salespersons

Salesperson	*Years of Experience*	*Annual Sales ($1000s)*
1	1	80
2	3	97
3	4	92
4	4	102
5	6	103
6	8	111
7	10	119
8	10	123
9	11	117
10	13	136

records for 10 salespersons in a company and that we also knew the number of years of sales experience for each individual. A regression analysis of these data would provide a mathematical equation that could be used to predict the sales volume of an individual given the number of years of selling experience.

As a first step in determining if a relationship exists between two variables, we could plot, or graph, the available data for the two variables. Suppose that a sales manager has records containing data on annual sales volumes and years of experience. This information is summarized in Table 1.1. We then plot these data on a graph with years of selling experience on the horizontal axis and annual sales on the vertical axis. We now have a **scatter diagram**, as shown in Figure 1.4. A scatter diagram helps us visualize data and draw preliminary conclusions about the possibility of a relationship between the variables.

In regression analysis, statisticians commonly classify a variable as an **independent** or a **dependent variable**. This classification scheme is used to indicate which variable is doing the predicting or explaining (independent variable) and which variable is being predicted or explained (dependent variable). In our example, the years of selling experience is the independent variable. It is used to predict the sales volume, or dependent variable. The usual practice in drawing a scatter diagram is to plot the data with the independent variable on the horizontal axis and the dependent variable on the vertical axis.

The scatter diagram in Figure 1.4 allows us to draw some preliminary conclusions. In giving an overview of the data, it indicates that in this case there is a good chance that the variables are related. Note that the points corresponding to low years of experience also are low in annual sales; high years of experience points are relatively high in annual sales. In fact, it appears that the relationship between these two variables may be approximated by a straight line. We now show

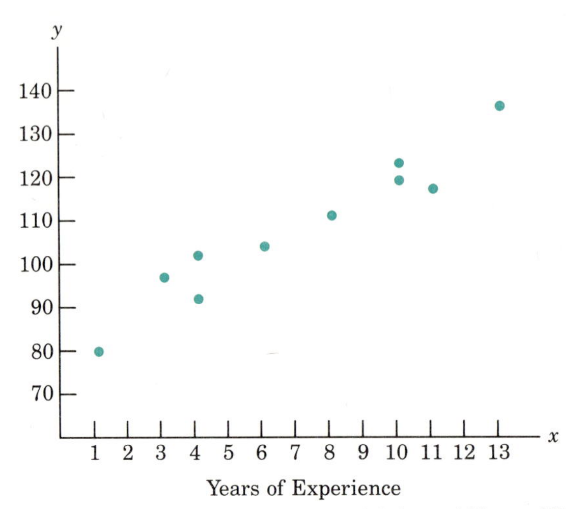

FIGURE 1.4. Scatter Diagram of Annual Sales and Years of Experience

how to develop such a straight line using an approach referred to as the **least squares method**.

There are several lines drawn for the scatter diagram in Figure 1.5. Each of these lines appears to fit the data. One of the problems in regression analysis is to find the straight line that best fits the data.

To solve the problem, we assume the equation of the best fitting line has the form

$$\hat{y} = b_0 + b_1 x$$

where

$\hat{y}$ = estimated value for the dependent variable

x = value of the independent variable

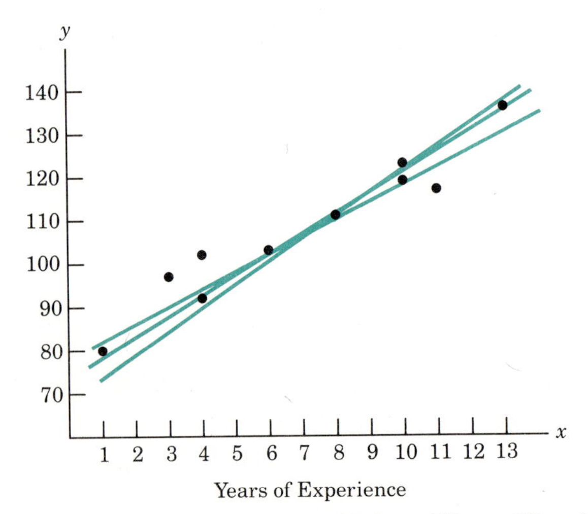

FIGURE 1.5. Scatter Diagram of Annual Sales and Years of Experience

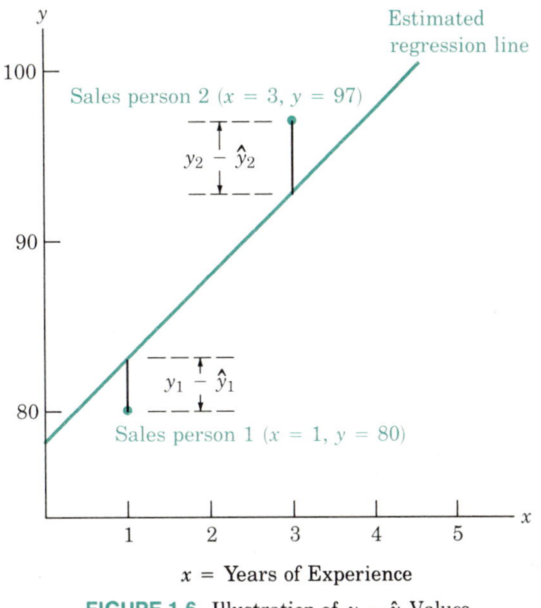

FIGURE 1.6. Illustration of $y_i - \hat{y}_i$ Values

From high school algebra we know

b_0 = intercept of the y-axis (value of $\hat{y}$ when $x = 0$)

b_1 = change in the dependent variable corresponding to the change in the independent variable (i.e., the slope of the line)

The least squares method chooses the line that minimizes the sum of squares of the differences between the observed values of the dependent variable (y_i) and the estimated values of the dependent variable ($\hat{y}_i$). This is what is meant by *best*. See Figure 1.6 for illustration of the $y_i - \hat{y}_i$ values.

Statisticians have shown that when the method of least squares is used to determine the line $\hat{y} = b_0 + b_1x$, then the best values of b_0 and b_1 are solutions of the following system of linear equations:

$$nb_0 + \left(\sum_{i=1}^{n} x_i\right)b_1 = \sum_{i=1}^{n} y_i \tag{1.6.1}$$

$$\left(\sum_{i=1}^{n} x_i\right)b_0 + \left(\sum_{i=1}^{n} x_i^2\right)b_1 = \sum_{i=1}^{n} x_i y_i \tag{1.6.2}$$

Example 1.6.1 Calculations necessary to develop the least squares line are listed in Table 1.2.

Substituting the appropriate values from Table 2 into the system of equations results in

$$10b_0 + 70b_1 = 1080$$

$$70b_0 + 632b_1 = 8128$$

TABLE 1.2. Calculations Necessary to Develop the Least Squares Estimated Regression Line

Salesperson (i)	x_i	y_i	$x_i y_i$	x_i^2
1	1	80	80	1
2	3	97	291	9
3	4	92	368	16
4	4	102	408	16
5	6	103	618	36
6	8	111	888	64
7	10	119	1190	100
8	10	123	1230	100
9	11	117	1287	121
10	13	136	1768	169
Totals	70	1080	8128	632
	Σx_i	Σy_i	$\Sigma x_i y_i$	Σx_i^2

Solving this system yields $b_0 = 80$ and $b_1 = 4$. Thus the best-fitting line found by the method of least squares is $\hat{y} = 80 + 4x$. ■

Example 1.6.2 Find the least squares line for the following data:

x_i	y_i	$x_i y_i$	x_i^2
0	1	0	0
1	4	4	1
2	3	6	4
3	2	6	9
4	4	16	16
5	6	30	25
15	20	62	55
Σx_i	Σy_i	$\Sigma x_i y_i$	Σx_i^2

Substituting the appropriate values from the given data into the system of equations results in

$$6b_0 + 15b_1 = 20$$

$$15b_0 + 55b_1 = 62$$

Solving this system yields $b_0 = \frac{34}{21}$ and $b_1 = \frac{24}{35}$. The best-fitting line found by the least squares method is $\hat{y} = \frac{34}{21} + \frac{24}{35}x$.

Thus if $x = 15$, then $\hat{y} = \frac{34}{21} + \left(\frac{24}{35}\right)15 = \frac{250}{21}$. The least squares line gives us an estimate for y given that $x = 15$. ■

The method of least squares is used to fit the best straight line to a set of data that has a linear relationship. In business, the data are often nonlinear. Many cases of nonlinear data can be best fitted with an exponential curve, $y = ae^{bx}$. This is a special case of the line, since taking logarithms of both sides of $y = ae^{bx}$ results in

$$\begin{aligned}\ln y &= \ln(ae^{bx})\\ &= \ln a + \ln(e^{bx})\\ &= \ln a + bx\end{aligned}$$

a linear function with slope b and vertical intercept $\ln a$. Thus to fit an exponential function to nonlinear data, (1) fit a straight line using

$$\ln \hat{y} = \ln a + bx$$

and (2) solve for $\hat{y}$,

$$\hat{y} = e^{\ln(a)+bx} = e^{\ln(a)}e^{bx} = ae^{bx}$$

This technique is illustrated with a short computational example.

Example 1.6.3 Find the best exponential function fit to the given data:

x	16	20	24	29	33
y	25	27	30	37	44

Solution We first find the least squares line for the data

x	16	20	24	29	33
$\ln y$	3.219	3.296	3.401	3.611	3.784

The last row consists of approximations of $\ln y$ for $y = 25, 27, 30, 37$, and 44 using a hand-held calculator.

In order to solve for b and $\ln a$, the following table is constructed:

x	$\ln y$	$x \ln y$	x^2
16	3.219	51.504	256
20	3.296	65.920	400
24	3.401	81.624	576
29	3.611	104.719	841
33	3.784	124.872	1089
122	17.311	428.639	3162
Σx_i	Σy_i	$\Sigma x_i y_i$	Σx_i^2

Using equations (1.6.1) and (1.6.2), since $n = 5$ we have

$$\begin{aligned}5 \ln a + 122b &= 17.311\\ 122 \ln a + 3162b &= 428.639\end{aligned}$$

Solving, we see that $\ln a = 2.639$ and $b = 0.03374$ are good approximations of the solution. The least square line for these values is

$$\ln \hat{y} = 2.639 + 0.03374x$$

and the corresponding exponential function is

$$\begin{aligned}\hat{y} &= e^{2.369 + 0.03374x} \\ &= e^{2.369}e^{0.03374x} \\ &= 14.00e^{0.03374x}\end{aligned}$$

■

SUMMARY

Term to know:

Scatter diagram

Method to know:

Least squares method

EXERCISE SET 1.6

1. The following data were collected regarding the starting monthly salary and the grade point average (GPA) for undergraduate students who had obtained a degree in business administration:

GPA	*Salary*
2.6	$ 900
3.4	$1200
3.6	$1600
3.2	$1100
3.5	$1400
2.9	$1000

   a. Develop a scatter diagram for this data with GPA on the horizontal axis.

   b. What does the scatter diagram developed in (a) suggest about the relationship between the two variables?

2. Tyler Realty collected the following data regarding the selling price of new homes and the size of the home measured in terms of square footage of living space:

Square Footage	*Selling Price*
2,500	$124,000
2,400	$108,000
1,800	$ 92,000
3,000	$146,000
2,300	$110,000

a. Develop a scatter diagram for this data with square footage on the horizontal axis.

b. What does the scatter diagram developed in (a) suggest about the relationship between the two variables?

3. The owner of a local grocery store varied the price of a 1-pound loaf of bread at random for six consecutive weeks. The following data show the price per loaf and the number of loaves sold for that week:

Price	*Loaves Sold*
$.60	220
$.62	200
$.58	280
$.60	250
$.64	190
$.62	240

a. Develop a scatter diagram for this data with price on the horizontal axis.

b. What does the scatter diagram developed in (a) indicate about the relationship between price and the number of loaves sold?

4. Using the data in Exercise 1, develop the least squares estimated regression line. Use Equations (1.6.1) and (1.6.2). For ease of calculation, you can treat the salary data in units of $100s, i.e., 9, 12, 16, 11, 14, and 10.

5. Using the estimated regression line developed in Exercise 4, predict the starting salary for a person with a 3.0 grade point average.

6. Using the data in Exercise 2, develop the best-fitting line according to the least squares method. For ease of calculation you can treat the square footage data in units of $1000s and the selling price data in units of $1000s.

7. Using the estimated regression line developed in Exercise 6, predict the selling price for a home with 2200 square feet of living space.

8. Using the data in Exercise 3, develop an estimated regression function that can be used to predict the number of units sold, given the price.

9. Using the estimated regression line developed in Exercise 8, predict the number of loaves sold for a price of $.63.

10. The system of equations (1.6.1) and (1.6.2) can be solved for b_0 and b_1. Show that

$$b_1 = \frac{n\Sigma x_i y_i - \Sigma x_i \Sigma y_i}{n\Sigma x_i^2 - (\Sigma x_i)^2}$$

$$b_0 = \bar{y} - b_1\bar{x}$$

where

x_i = value of independent variable for the ith observation

y_i = value of dependent variable for the ith observation

$\bar{x}$ = mean value for the independent variable $\left(\frac{1}{n}\Sigma x_i\right)$

$\bar{y}$ = mean value for the dependent variable $\left(\frac{1}{n}\Sigma y_i\right)$

n = total number of observations

11. Find a least squares exponential function that fits the data in each case.

a.

x	2	3	4	5	6
y	115	102	95	87	77

b.

x	5.6	7.8	8.2	9.5	10.3
y	200	312	470	520	612

12. A product was test marketed at six different prices and the total demand (y thousands) was estimated for a price of x dollars per unit. The data is given in the table:

Price (x)	30	32	34	36	38	40
Demand (y)	76	88	97	109	116	121

The marketing manager has decided to use an exponential function to fit the data best. Find the least squares exponential function that fits the data.

This exercise may best be solved using the accompanying software or a hand-held calculator.

13. Find the least square line for the following set of data.

x_i	y_i
0.22	1.18
1.31	4.27
2.54	2.99
3.72	2.45
4.21	4.63

SECTION 1.7 MATRIX ADDITION

There are other methods to solve linear systems besides the one developed in Section 1.2. One method makes further use of matrices (covered in this chapter) and another makes use of determinants (Chapter 5).

Matrices have other uses besides solving systems of linear equations. The matrix concept is important in model building and thus is used in many applications. We therefore develop some of the theory of matrices now. Some applications of matrices were developed in the last two sections of this chapter.

Matrices have an arithmetic of their own. Just as with ordinary numbers (under certain conditions, which we shall determine), matrices can be added, subtracted, and multiplied. Matrices cannot be divided in the ordinary sense, but (under certain conditions, which we shall determine), matrices can be inverted. Matrix inversion corresponds to the division process.

Recall (Section 1.1) that a matrix is a rectangular array of numbers usually enclosed by brackets or parentheses. As in Section 1.1, we use brackets. Thus, a general $m \times n$ matrix may be written as

$$A_{m\times n} = \begin{bmatrix} a_{11} & a_{12} & \cdots & a_{1n} \\ a_{21} & a_{22} & \cdots & a_{2n} \\ \vdots & \vdots & & \vdots \\ a_{m1} & a_{m2} & \cdots & a_{mn} \end{bmatrix}$$

It is convenient to abbreviate the general $m \times n$ matrix as

$$A = [a_{ij}]_{m\times n}$$

When the context is clear, we use $A = [a_{ij}]$.

Before developing the arithmetic of matrices, it is important that we know when two matrices are equal. Matrices are equal if and only if they have the same sizes and the same corresponding entries.

Definition 1.7.1 Two matrices $A = [a_{ij}]$ and $B = [b_{ij}]$ are **equal** provided they have the same size $(m \times n)$ and the entries in the corresponding positions are equal ($a_{ij} = b_{ij}$ for each i and j). Symbolically, $A = B$ if and only if $a_{ij} = b_{ij}$ for each i and j.

Example 1.7.2 a. If

$$\begin{bmatrix} a_{11} & a_{12} & a_{13} \\ a_{21} & a_{22} & a_{23} \end{bmatrix} = \begin{bmatrix} 0 & 1 & -1 \\ 2 & 3 & -4 \end{bmatrix}$$

then $a_{11} = 0$, $a_{12} = 1$, $a_{13} = -1$, $a_{21} = 2$, $a_{22} = 3$, and $a_{23} = -4$

b. The matrices

$$[1 \quad 2] \quad \text{and} \quad \begin{bmatrix} 2 \\ 1 \end{bmatrix}$$

are not equal because they do not have the same size whereas

$$\begin{bmatrix} 1 & 2 \\ 4 & 3 \end{bmatrix} \quad \text{and} \quad \begin{bmatrix} 1 & 2 \\ 3 & 4 \end{bmatrix}$$

are not equal because their (2, 1)-entries are different. ■

Example 1.7.3 If

$$\begin{bmatrix} x+y & x+z \\ y+z & x \end{bmatrix} = \begin{bmatrix} 2 & 4 \\ 4 & 1 \end{bmatrix}$$

find the values of x, y, and z.

Solution These 2×2 matrices are equal if and only if

$$\begin{aligned} x + y &= 2 \\ x + z &= 4 \\ y + z &= 4 \\ x &= 1 \end{aligned}$$

This is a 4×3 linear system of equations that can be solved without using the reduction method. From the last equation, $x = 1$, which means that y must equal 1 (from $x + y = 2$) and z must equal 3 (from $x + z = 4$). Substituting these values into the third equation, $y + z = 1 + 3 = 4$. Thus in order for the matrices to be equal, $x = 1$, $y = 1$, and $z = 3$. ■

Let us now define the operation of addition for matrices. We do not add matrices unless they have the same size; when adding, we add the corresponding entries.

Definition 1.7.4 Let $A = [a_{ij}]$ and $B = [b_{ij}]$ both be $m \times n$ matrices. Their **sum**, $A + B$, is the $m \times n$ matrix $C = [c_{ij}]$, where for each i and j,

$$c_{ij} = a_{ij} + b_{ij}$$

Example 1.7.5 Find $A + B$, where

$$A = \begin{bmatrix} 2 & 1 & 4 \\ 3 & -2 & 7 \end{bmatrix} \quad \text{and} \quad B = \begin{bmatrix} -3 & 2 & 1 \\ -4 & -6 & 2 \end{bmatrix}$$

Solution

$$A + B = \begin{bmatrix} 2 + (-3) & 1 + 2 & 4 + 1 \\ 3 + (-4) & -2 + (-6) & 7 + 2 \end{bmatrix} = \begin{bmatrix} -1 & 3 & 5 \\ -1 & -8 & 9 \end{bmatrix}$$ ■

Example 1.7.6 If

$$A = \begin{bmatrix} 2 & 1 & 4 \\ -7 & 0 & 3 \end{bmatrix} \quad \text{and} \quad B = \begin{bmatrix} 0 & -1 \\ 1 & 2 \end{bmatrix}$$

then $A + B$ is not defined. ■

REMARK The set of $m \times n$ matrices is said to be *closed under addition* because the sum of any two $m \times n$ matrices exists and is an $m \times n$ matrix.

We now develop the arithmetic properties of matrix addition. All the verifications reduce to the properties of ordinary addition of real numbers.

Theorem 1.7.7 *Matrix addition is associative: If A, B, and C are all $m \times n$ matrices, then*

$$(A + B) + C = A + (B + C)$$

Proof. By the definition of matrix addition, the general element of $(A + B) + C$ is $(a_{ij} + b_{ij}) + c_{ij}$, whereas the general element of $A + (B + C)$ is $a_{ij} + (b_{ij} + c_{ij})$ for $A = [a_{ij}]$, $B = [b_{ij}]$, and $C = [c_{ij}]$. Since the addition of real numbers is associative, we have $(a_{ij} + b_{ij}) + c_{ij} = a_{ij} + (b_{ij} + c_{ij})$ for each i and j, which means, by equality of matrices, that $(A + B) + C = A + (B + C)$.

Theorem 1.7.8 *Matrix addition is commutative: If A and B are both $m \times n$ matrices, then*

$$A + B = B + A$$

Proof. If $A = [a_{ij}]$ and $B = [b_{ij}]$, then by the definition of matrix addition, the general element of $A + B$ is $a_{ij} + b_{ij}$ and of $B + A$, $b_{ij} + a_{ij}$. Since the addition of real numbers is commutative, $a_{ij} + b_{ij} = b_{ij} + a_{ij}$ for each i and j, and hence $A + B = B + A$.

As a result of these two theorems, it is immaterial in adding a string of matrices how the sums are carried out. For example, $A + B + C + D = A + D + C + B = D + A + C + B$, and so on.

Definition 1.7.9 The $m \times n$ matrix denoted by 0 has all zero entries and is called the $m \times n$ **zero matrix**.

Theorem 1.7.10 *Additive identity: For any $m \times n$ matrix A,*

$$A + 0 = 0 + A = A$$

Verification is to be supplied in Exercise 20.

Definition 1.7.11 Let $A = [a_{ij}]$. The **negative** of A, denoted by $-A$, is the $m \times n$ matrix whose entries are $-a_{ij}$; that is, $-A = [-a_{ij}]$. The **difference** of A and B, denoted by $A - B$, is the sum $A + (-B)$. If $A = [a_{ij}]$ and $B = [b_{ij}]$, then $A - B = [a_{ij} - b_{ij}]$.

Theorem 1.7.12 *Additive inverse: For every $m \times n$ matrix A, there exists an $m \times n$ additive inverse matrix, denoted by $-A$, such that*

$$A + (-A) = (-A) + A = 0$$

where 0 is the $m \times n$ zero matrix.

Verification is to be supplied in Exercise 21.

Matrices of the same size may also be subtracted.

Example 1.7.13 Let

$$A = \begin{bmatrix} 2 & -1 \\ -1 & 3 \\ 4 & 2 \end{bmatrix} \quad \text{and} \quad B = \begin{bmatrix} -4 & 8 \\ 6 & 0 \\ 2 & 1 \end{bmatrix}$$

Then

$$A - B = \begin{bmatrix} 2 & -1 \\ -1 & 3 \\ 4 & 2 \end{bmatrix} - \begin{bmatrix} -4 & 8 \\ 6 & 0 \\ 2 & 1 \end{bmatrix}$$

$$= \begin{bmatrix} 2-(-4) & -1-8 \\ -1-6 & 3-0 \\ 4-2 & 2-1 \end{bmatrix} = \begin{bmatrix} 6 & -9 \\ -7 & 3 \\ 2 & 1 \end{bmatrix} \quad \blacksquare$$

Example 1.7.14 Find the solution of the matrix equation

$$X + \begin{bmatrix} 1 & 0 & 0 \\ 2 & 0 & 3 \end{bmatrix} = \begin{bmatrix} 0 & 1 & 1 \\ 2 & 0 & 2 \end{bmatrix}$$

Solution Applying the properties of this section, we have

$$X = \begin{bmatrix} 0 & 1 & 1 \\ 2 & 0 & 2 \end{bmatrix} - \begin{bmatrix} 1 & 0 & 0 \\ 2 & 0 & 3 \end{bmatrix}$$

$$= \begin{bmatrix} 0-1 & 1-0 & 1-0 \\ 2-2 & 0-0 & 2-3 \end{bmatrix} = \begin{bmatrix} -1 & 1 & 1 \\ 0 & 0 & -1 \end{bmatrix} \quad \blacksquare$$

SUMMARY

Terms to know:

Equality of matrices
Sum of matrices
Zero matrix
Negative of a matrix
Difference of two matrices

Theorems to know:

Matrix addition is associative.
Matrix addition is commutative.
There is an additive identity in the set of matrices.
For each matrix, there is an additive inverse.

EXERCISE SET 1.7

Perform the indicated operations, if possible.

1. $\begin{bmatrix} 2 & 1 & 4 \\ 8 & 2 & 9 \\ 6 & 2 & 1 \end{bmatrix} + \begin{bmatrix} 0 & 1 & 0 \\ 1 & 0 & 1 \\ 0 & 0 & 1 \end{bmatrix}$

2. $\begin{bmatrix} 1 & 0 & 0 \\ 1 & 0 & 0 \\ 0 & 1 & 0 \end{bmatrix} - \begin{bmatrix} 0 & 0 & 1 \\ 0 & 1 & 0 \\ 1 & 0 & 1 \end{bmatrix}$

3. $\left(\begin{bmatrix} 2 & 1 & 4 \\ 8 & 7 & 2 \end{bmatrix} + \begin{bmatrix} 7 & 9 & 6 \\ 2 & 1 & 4 \end{bmatrix}\right) + \begin{bmatrix} 0 & 1 & 1 \\ 1 & 0 & 0 \end{bmatrix}$

4. $\begin{bmatrix} 3 & 2 & -1 & 6 \\ 4 & 2 & 0 & 1 \\ -1 & 3 & 2 & 0 \end{bmatrix} + \begin{bmatrix} -3 & -2 & 1 & -6 \\ -4 & -2 & 0 & -1 \\ 1 & -3 & -2 & 0 \end{bmatrix}$

5. $\begin{bmatrix} 3 & 2 & -1 \\ 4 & 2 & 0 \\ -1 & 3 & 2 \end{bmatrix} - \begin{bmatrix} 3 & 2 & -1 \\ 4 & 2 & 0 \\ -1 & 3 & 2 \end{bmatrix}$

6. $\begin{bmatrix} 2 & 1 & 4 \\ 8 & 7 & 2 \end{bmatrix} + \left(\begin{bmatrix} 7 & 9 & 6 \\ 2 & 1 & 4 \end{bmatrix} + \begin{bmatrix} 0 & 1 & 1 \\ 1 & 0 & 0 \end{bmatrix}\right)$

7. $\begin{bmatrix} 1 & 1 & 3 \\ 4 & 0 & 2 \end{bmatrix} + \begin{bmatrix} 1 & 2 \\ 4 & 1 \end{bmatrix}$

8. $\begin{bmatrix} 1 & 0 & -1 \\ 0 & 2 & 1 \\ 0 & 0 & 1 \end{bmatrix} + \begin{bmatrix} 1 & 0 \\ 1 & 2 \\ 1 & 0 \end{bmatrix}$

Simplify each expression, where

$$A = \begin{bmatrix} 0 & -1 & 2 \\ -4 & 3 & 2 \end{bmatrix}, \quad B = \begin{bmatrix} 1 & -2 & 0 \\ 4 & -2 & 1 \end{bmatrix}, \text{ and } C = \begin{bmatrix} 5 & -7 & -1 \\ 6 & -4 & 2 \end{bmatrix}$$

9. $(A + B) + C$ and $A + (B + C)$
10. $A + A$
11. $B + A + C$
12. $A - B - C$ and $A - (B + C)$
13. $C - A + B$ and $C - (A - B)$
14. $A + (-A)$
15. $B + B + B$

Find the values of the unknowns.

16. $\begin{bmatrix} -3 & -2 \\ 4 & -7 \\ -6 & 5 \end{bmatrix} + \begin{bmatrix} x & y \\ r & s \\ u & v \end{bmatrix} = \begin{bmatrix} 0 & 0 \\ 0 & 0 \\ 0 & 0 \end{bmatrix}$

17. $\begin{bmatrix} \frac{2}{3}x + \frac{1}{2}y \\ -\frac{1}{6}x + y \end{bmatrix} = \begin{bmatrix} -1 \\ 1 \end{bmatrix}$

18. $X + \begin{bmatrix} -2 & 0 \\ 0 & 2 \end{bmatrix} = \begin{bmatrix} 5 & -12 \\ 11 & 7 \end{bmatrix}$

19. $\begin{bmatrix} 3 & 0 & -2 \\ 1 & -1 & 0 \\ 0 & 4 & 5 \end{bmatrix} - X = \begin{bmatrix} -1 & 2 & 0 \\ 0 & -4 & 3 \\ 7 & 2 & -1 \end{bmatrix}$

20. Let A and 0 be $m \times n$ matrices. Prove that $A + 0 = 0 + A = A$.

21. Let A be an $m \times n$ matrix. Prove that there exists an $m \times n$ matrix $-A$ such that
$$A + (-A) = (-A) + A = 0$$

SECTION 1.8 MATRIX MULTIPLICATION

There are many ways in which two matrices might be multiplied. However, there are two useful ways: the multiplication of a matrix by a number and the multiplication of a matrix by another matrix. We first define the product of a scalar (number) and a matrix.

Definition 1.8.1 Let A be an $m \times n$ matrix and let c be any scalar. The **scalar product** of c and A, denoted by cA, is an $m \times n$ matrix whose entries are ca_{ij}. In other words, a matrix can be multiplied by a scalar by multiplying every entry of the matrix by the scalar.

This definition is consistent with the rules for addition and subtraction. For example, let

$$A = \begin{bmatrix} 6 & -3 \\ 4 & -5 \end{bmatrix} \quad \text{and} \quad B = \begin{bmatrix} 5 & 7 \\ 0 & 1 \end{bmatrix}$$

Then

$$A + A = \begin{bmatrix} 6 & -3 \\ 4 & -5 \end{bmatrix} + \begin{bmatrix} 6 & -3 \\ 4 & -5 \end{bmatrix} = \begin{bmatrix} 12 & -6 \\ 8 & -10 \end{bmatrix}$$

$$2A = 2\begin{bmatrix} 6 & -3 \\ 4 & -5 \end{bmatrix} = \begin{bmatrix} (2)(6) & 2(-3) \\ (2)(4) & (2)(-5) \end{bmatrix} = \begin{bmatrix} 12 & -6 \\ 8 & -10 \end{bmatrix}$$

$$A - B = \begin{bmatrix} 6 & -3 \\ 4 & -5 \end{bmatrix} - \begin{bmatrix} 5 & 7 \\ 0 & 1 \end{bmatrix} = \begin{bmatrix} 1 & -10 \\ 4 & -6 \end{bmatrix}$$

$$A + (-1)B = \begin{bmatrix} 6 & -3 \\ 4 & -5 \end{bmatrix} + (-1)\begin{bmatrix} 5 & 7 \\ 0 & 1 \end{bmatrix}$$

$$= \begin{bmatrix} 6 & -3 \\ 4 & -5 \end{bmatrix} + \begin{bmatrix} -5 & -7 \\ 0 & -1 \end{bmatrix} = \begin{bmatrix} 1 & -10 \\ 4 & -6 \end{bmatrix}$$

Scalar multiplication satisfies the following properties for A and B, two $m \times n$ matrices, and a and b, two scalars:

1. $a(bA) = (ab)A$
2. $(a + b)A = aA + bA$
3. $a(A + B) = aA + aB$
4. $1A = A$
5. $-1A = -A$
6. $a0 = 0$, where 0 is the $m \times n$ zero matrix.
7. $0A = 0$, where the 0 on the left is the scalar 0.

The proofs of these properties are left for the exercises.

Example 1.8.2 Simplify $2(A + B) - 4(\frac{1}{2}B - C) + (C - A)$.

Solution Applying scalar multiplication and addition properties yields

$$2(A + B) - 4(\tfrac{1}{2}B - C) + (C - A)$$

$$= 2A + 2B - 2B + 4C + C - A = A + 5C$$ ■

Example 1.8.3 Let A and B be $m \times n$ matrices and a be a scalar different than zero. Solve

$$aX + A = B$$

for X.

Solution Since $aX + A = B$, then

$$aX = B - A$$

and

$$X = \left(\frac{1}{a}\right)(B - A)$$ ■

Multiplication of a matrix A by a scalar a multiplies every entry of A by a. This appears to be a reasonable definition. However, the operation of multiplying a matrix by a matrix may seem a little strange at first sight, but it is the natural definition for the applications we will consider. First, we define the product of an $m \times n$ matrix with an $n \times 1$ matrix.

Using matrix equality, the system

$$\begin{aligned} 2x_1 + 3x_2 + 4x_3 &= 1 \\ -x_1 + x_2 + 2x_3 &= 2 \\ -x_1 + x_2 + x_3 &= 3 \end{aligned}$$

can be written in the matrix form

$$\begin{bmatrix} 2x_1 + 3x_2 + 4x_3 \\ -x_1 + x_2 + 2x_3 \\ -x_1 + x_2 + x_3 \end{bmatrix} = \begin{bmatrix} 1 \\ 2 \\ 3 \end{bmatrix}$$

Observe that $2x_1 + 3x_2 + 4x_3 = (2)(x_1) + (3)(x_2) + (4)(x_3)$ is the sum of the products of the corresponding entries of the first row of the 3×3 coefficient matrix A and the 3×1 matrix X where

$$A = \begin{bmatrix} 2 & 3 & 4 \\ -1 & 1 & 2 \\ -1 & 1 & 1 \end{bmatrix} \quad \text{and} \quad X = \begin{bmatrix} x_1 \\ x_2 \\ x_3 \end{bmatrix}$$

Furthermore,

$$-x_1 + x_2 + 2x_3 = (-1)(x_1) + (1)(x_2) + (2)(x_3)$$

is the sum of the products of the corresponding entries of the second row of A with the corresponding entries of X; and

$$-x_1 + x_2 + x_3 = (-1)(x_1) + (1)(x_2) + (1)(x_3)$$

is the sum of the products of the corresponding entries of the third row of A and the corresponding entries of X. Thus we have the following definition.

Definition 1.8.4 The **product** of A, an $m \times n$ matrix, and X, an $n \times 1$ matrix, is

$$\begin{bmatrix} a_{11} & a_{12} & \cdots & a_{1n} \\ a_{21} & a_{22} & \cdots & a_{2n} \\ \vdots & \vdots & & \vdots \\ a_{m1} & a_{m2} & \cdots & a_{mn} \end{bmatrix} \begin{bmatrix} x_1 \\ x_2 \\ \vdots \\ x_n \end{bmatrix}$$

$$= \begin{bmatrix} a_{11}x_1 & + & a_{12}x_2 & + \cdots + & a_{1n}x_n \\ a_{12}x_1 & + & a_{22}x_2 & + \cdots + & a_{2n}x_n \\ \vdots & & \vdots & & \vdots \\ a_{m1}x_1 & + & a_{m2}x_2 & + \cdots + & a_{mn}x_n \end{bmatrix}$$

Example 1.8.5 The product

$$\begin{bmatrix} 3 & 2 & 4 & 1 \\ -1 & -2 & 5 & -3 \\ 7 & 9 & 8 & 6 \end{bmatrix} \begin{bmatrix} 2 \\ 3 \\ -3 \\ 2 \end{bmatrix}$$

$$= \begin{bmatrix} (3)(2) + (2)(3) + (4)(-3) + (1)(2) \\ (-1)(2) + (-2)(3) + (5)(-3) + (-3)(2) \\ (7)(2) + (9)(3) + (8)(-3) + (6)(2) \end{bmatrix} = \begin{bmatrix} 2 \\ -29 \\ 29 \end{bmatrix} \quad ■$$

Now consider the system

$$\begin{aligned} a_{11}x_1 + a_{12}x_2 &= y_1 \\ a_{21}x_1 + a_{22}x_2 &= y_2 \end{aligned} \tag{1.8.1}$$

Using the definition of multiplication of an $m \times n$ matrix and an $n \times 1$ matrix as well as the definition of matrix equality, system (1.8.1) can be rewritten as

$$\begin{bmatrix} a_{11} & a_{12} \\ a_{21} & a_{22} \end{bmatrix} \begin{bmatrix} x_1 \\ x_2 \end{bmatrix} = \begin{bmatrix} y_1 \\ y_2 \end{bmatrix} \tag{1.8.2}$$

One can view this linear system as a transformation (or mapping, or function) from the (x_1, x_2)-plane to the (y_1, y_2)-plane given by system (1.8.1) and characterized by the 2×2 coefficient matrix

$$A = \begin{bmatrix} a_{11} & a_{12} \\ a_{21} & a_{22} \end{bmatrix} \tag{1.8.3}$$

which is called the **matrix of the transformation**. A second transformation from the (y_1, y_2)-plane to the (z_1, z_2)-plane,

$$\begin{aligned} b_{11}y_1 + b_{12}y_2 &= z_1 \\ b_{21}y_1 + b_{22}y_2 &= z_2 \end{aligned} \quad \text{or} \quad \begin{bmatrix} b_{11} & b_{12} \\ b_{21} & b_{22} \end{bmatrix} \begin{bmatrix} y_1 \\ y_2 \end{bmatrix} = \begin{bmatrix} z_1 \\ z_2 \end{bmatrix} \tag{1.8.4}$$

is characterized by the coefficient matrix

$$B = \begin{bmatrix} b_{11} & b_{12} \\ b_{21} & b_{22} \end{bmatrix}$$

To obtain a direct connection between the (x_1, x_2)-plane and the (z_1, z_2)-plane, we can substitute the values of y_1 and y_2 from (1.8.1) into the appropriate places in (1.8.4) to obtain

$$\begin{aligned} b_{11}(a_{11}x_1 + a_{12}x_2) + b_{12}(a_{21}x_1 + a_{22}x_2) &= z_1 \\ b_{21}(a_{11}x_1 + a_{12}x_2) + b_{22}(a_{21}x_1 + a_{22}x_2) &= z_2 \end{aligned}$$

Simplifying yields

$$\begin{aligned} (b_{11}a_{11} + b_{12}a_{21})x_1 + (b_{11}a_{12} + b_{12}a_{22})x_2 &= z_1 \\ (b_{21}a_{11} + b_{22}a_{21})x_1 + (b_{21}a_{12} + b_{22}a_{22})x_2 &= z_2 \end{aligned}$$

with coefficient matrix

$$C = \begin{bmatrix} b_{11}a_{11} + b_{12}a_{21} & b_{11}a_{12} + b_{12}a_{22} \\ b_{21}a_{11} + b_{22}a_{21} & b_{21}a_{12} + b_{22}a_{22} \end{bmatrix}$$

which is obtained from transformation A followed by that of B, denoted by BA. Hence we can think of C as being equivalent to the product BA, that is,

$$\begin{bmatrix} b_{11} & b_{12} \\ b_{21} & b_{22} \end{bmatrix}\begin{bmatrix} a_{11} & a_{12} \\ a_{21} & a_{22} \end{bmatrix} = \begin{bmatrix} b_{11}a_{11} + b_{12}a_{21} & b_{11}a_{12} + b_{12}a_{22} \\ b_{21}a_{11} + b_{22}a_{21} & b_{21}a_{12} + b_{22}a_{22} \end{bmatrix}$$

The essence of this equation is generalized in the following definition.

Definition 1.8.6 Let $A = [a_{ij}]$ be an $m \times k$ matrix and $B = [b_{ij}]$ be a $k \times n$ matrix. These matrices can then be multiplied in the order AB, and their **product**

$$AB = C = [c_{ij}]$$

is the $m \times n$ matrix whose entries are defined for each ij by

$$c_{ij} = a_{i1}b_{1j} + a_{i2}b_{2j} + \cdots + a_{ik}b_{kj}$$

1. Observe that the (i, j)-entry of the product matrix C is obtained by multiplying each entry in the ith row of A by the corresponding entry of the jth column of B and summing the products:

$$i\text{th}\begin{bmatrix} a_{i1} & a_{i2} & \cdots & a_{ik} \end{bmatrix}\overset{j\text{th}}{\begin{bmatrix} b_{1j} \\ b_{2j} \\ \vdots \\ b_{kj} \end{bmatrix}} = \overset{j}{\begin{bmatrix} c_{ij} \end{bmatrix}}\, i$$

$$a_{i1}b_{1j} + a_{i2}b_{2j} + \cdots + a_{ik}b_{kj} = c_{ij}$$

2. Notice that the product AB is defined if and only if the number of columns of A is equal to the number of rows of B. Their product has the same number of rows as A and the same number of columns as B.

3. By definition of matrix multiplication, column j of C is just the product Ab_j of the matrix A with the $k \times 1$ matrix

$$b_j = \begin{bmatrix} b_{1j} \\ b_{2j} \\ \vdots \\ b_{kj} \end{bmatrix}$$

which is the jth column of B.

Example 1.8.7 Calculate the matrix product

$$\begin{bmatrix} 2 & 1 & 3 \\ 1 & 7 & 2 \end{bmatrix} \begin{bmatrix} 1 & 8 \\ 3 & 9 \\ 2 & 1 \end{bmatrix}$$

Solution Notice that the first matrix has three columns, whereas the second matrix has three rows; thus their product in this order exists. Since the first matrix has two rows and the second matrix has two columns, our product is a 2×2 matrix. Let the product be

$$\begin{bmatrix} c_{11} & c_{12} \\ c_{21} & c_{22} \end{bmatrix}$$

Then

$$\begin{aligned}
c_{11} &= a_{11}b_{11} + a_{12}b_{21} + a_{13}b_{31} \\
&= (2)(1) + (1)(3) + (3)(2) \\
&= 2 \quad + 3 \quad + 6 \quad = 11 \\
c_{12} &= a_{11}b_{12} + a_{12}b_{22} + a_{13}b_{32} \\
&= (2)(8) + (1)(9) + (3)(1) \\
&= 16 \quad + 9 \quad + 3 \quad = 28 \\
c_{21} &= a_{21}b_{11} + a_{22}b_{21} + a_{23}b_{31} \\
&= (1)(1) + (7)(3) + (2)(2) \\
&= 1 \quad + 21 \quad + 4 \quad = 26 \\
c_{22} &= a_{21}b_{12} + a_{22}b_{22} + a_{23}b_{32} \\
&= (1)(8) + (7)(9) + (2)(1) \\
&= 8 \quad + 63 \quad + 2 \quad = 73
\end{aligned}$$

Thus

$$\begin{bmatrix} 2 & 1 & 3 \\ 1 & 7 & 2 \end{bmatrix} \begin{bmatrix} 1 & 8 \\ 3 & 9 \\ 2 & 1 \end{bmatrix} = \begin{bmatrix} 11 & 28 \\ 26 & 73 \end{bmatrix}$$ ■

Example 1.8.8 Calculate all products of

$$A = \begin{bmatrix} 2 & 1 & 7 \\ 1 & 1 & 1 \end{bmatrix} \quad \text{and} \quad B = \begin{bmatrix} 1 & 1 \\ 1 & 9 \\ 2 & 3 \end{bmatrix}$$

Solution Notice that A is a 2×3 matrix and B is a 3×2 matrix, so AB can be calculated and is a 2×2 matrix

$$\begin{bmatrix} c_{11} & c_{12} \\ c_{21} & c_{22} \end{bmatrix}$$

Then

$$\begin{aligned} c_{11} &= a_{11}b_{11} + a_{12}b_{21} + a_{13}b_{31} \\ &= (2)(1) + (1)(1) + (7)(2) \\ &= 2 \quad + 1 \quad + 14 \quad = 17 \\ c_{12} &= a_{11}b_{12} + a_{12}b_{22} + a_{13}b_{32} \\ &= (2)(1) + (1)(9) + (7)(3) \\ &= 2 \quad + 9 \quad + 21 \quad = 32 \\ c_{21} &= a_{21}b_{11} + a_{22}b_{21} + a_{23}b_{31} \\ &= (1)(1) + (1)(1) + (1)(2) \\ &= 1 \quad + 1 \quad + 2 \quad = 4 \\ c_{22} &= a_{21}b_{12} + a_{22}b_{22} + a_{23}b_{32} \\ &= (1)(1) + (1)(9) + (1)(3) \\ &= 1 \quad + 9 \quad + 3 \quad = 13 \end{aligned}$$

Thus

$$\begin{bmatrix} 2 & 1 & 7 \\ 1 & 1 & 1 \end{bmatrix} \begin{bmatrix} 1 & 1 \\ 1 & 9 \\ 2 & 3 \end{bmatrix} = \begin{bmatrix} 17 & 32 \\ 4 & 13 \end{bmatrix}$$

On the other hand, BA can also be calculated, and the result is a 3×3 matrix. Consequently,

$$\begin{bmatrix} 1 & 1 \\ 1 & 9 \\ 2 & 3 \end{bmatrix} \begin{bmatrix} 2 & 1 & 7 \\ 1 & 1 & 1 \end{bmatrix} = \begin{bmatrix} d_{11} & d_{12} & d_{13} \\ d_{21} & d_{22} & d_{23} \\ d_{31} & d_{32} & d_{33} \end{bmatrix}$$

where

$$\begin{aligned}
d_{11} &= b_{11}a_{11} + b_{12}a_{21} \\
&= (1)(2) + (1)(1) = 3 \\
d_{12} &= b_{11}a_{12} + b_{12}a_{22} \\
&= (1)(1) + (1)(1) = 2 \\
d_{13} &= b_{11}a_{13} + b_{12}a_{23} \\
&= (1)(7) + (1)(1) = 8 \\
d_{21} &= b_{21}a_{11} + b_{22}a_{21} \\
&= (1)(2) + (9)(1) = 11 \\
d_{22} &= b_{21}a_{12} + b_{22}a_{22} \\
&= (1)(1) + (9)(1) = 10 \\
d_{23} &= b_{21}a_{13} + b_{22}a_{23} \\
&= (1)(7) + (9)(1) = 16 \\
d_{31} &= b_{31}a_{11} + b_{32}a_{21} \\
&= (2)(2) + (3)(1) = 7 \\
d_{32} &= b_{31}a_{12} + b_{32}a_{22} \\
&= (2)(1) + (3)(1) = 5 \\
d_{33} &= b_{31}a_{13} + b_{32}a_{23} \\
&= (2)(7) + (3)(1) = 17
\end{aligned}$$

We have

$$\begin{bmatrix} 1 & 1 \\ 1 & 9 \\ 2 & 3 \end{bmatrix} \begin{bmatrix} 2 & 1 & 7 \\ 1 & 1 & 1 \end{bmatrix} = \begin{bmatrix} 3 & 2 & 8 \\ 11 & 10 & 16 \\ 7 & 5 & 17 \end{bmatrix}$$

Notice that $AB \neq BA$:

$$\begin{bmatrix} 2 & 1 & 7 \\ 1 & 1 & 1 \end{bmatrix} \begin{bmatrix} 1 & 1 \\ 1 & 9 \\ 2 & 3 \end{bmatrix} = \begin{bmatrix} 17 & 32 \\ 4 & 13 \end{bmatrix}$$

$$\neq \begin{bmatrix} 3 & 2 & 8 \\ 11 & 10 & 16 \\ 7 & 5 & 17 \end{bmatrix} = \begin{bmatrix} 1 & 1 \\ 1 & 9 \\ 2 & 3 \end{bmatrix} \begin{bmatrix} 2 & 1 & 7 \\ 1 & 1 & 1 \end{bmatrix} \quad ■$$

The *order* in which matrices are multiplied can make a difference. In other words, *matrix multiplication is not commutative*, i.e., $AB \neq BA$ in general. Since the order of multiplication is important, we at times say matrix B is premultiplied by A or matrix A is postmultiplied by B.

Also, the product of two nonzero matrices may be zero.

Example 1.8.9 Let

$$A = \begin{bmatrix} 1 & 1 \\ 1 & 1 \end{bmatrix} \quad \text{and} \quad B = \begin{bmatrix} 2 & -2 \\ -2 & 2 \end{bmatrix}$$

Then

$$AB = \begin{bmatrix} 2 + (-2) & (-2) + 2 \\ 2 + (-2) & (-2) + 2 \end{bmatrix} = \begin{bmatrix} 0 & 0 \\ 0 & 0 \end{bmatrix}$$

■

Even though matrix multiplication is not commutative, other algebraic rules are valid. We present the following properties without proof.

1. *The product of two matrices is a matrix*: However, the product is not of the same order unless the matrices are square.
2. *Matrix multiplication is associative*: If the products exist:

$$A(BC) = (AB)C$$

3. *There is a multiplicative identity*:

$$AI_n = I_m A = A$$

where A is an $m \times n$ matrix and

$$I_k = \overbrace{\begin{bmatrix} 1 & 0 & 0 & \cdots & 0 \\ 0 & 1 & 0 & \cdots & 0 \\ 0 & 0 & 1 & \cdots & 0 \\ \vdots & \vdots & \vdots & & \vdots \\ 0 & 0 & 0 & \cdots & 1 \end{bmatrix}}^{k \text{ columns}} \; k \text{ rows}$$

4. *The distributive property holds:* Matrix multiplication is distributive with respect to addition if the products and sums exist:

$$(A + B)C = AC + BC \quad \text{[Right-hand distributive]}$$

$$A(B + C) = AB + AC \quad \text{[Left-hand distributive]}$$

In these properties, the order of the matrices must be such that multiplication and addition are defined in all cases. The proofs of (2),

(3), and (4) are left as exercises for 2×2 matrices. For higher-order matrices, the proofs use the same principle.

Finally, as with numbers, for an $n \times n$ matrix A, we define $A^2 = AA$ and, in general, $A^n = A(A^{n-1})$.

Example 1.8.10 Calculate A^3 where

$$A = \begin{bmatrix} 1 & 0 & 1 \\ 0 & 2 & 4 \\ 1 & 1 & 1 \end{bmatrix}$$

Solution First we calculate A^2, obtaining

$$\begin{bmatrix} 1\cdot 1 + 0\cdot 0 + 1\cdot 1 & 1\cdot 0 + 0\cdot 2 + 1\cdot 1 & 1\cdot 1 + 0\cdot 4 + 1\cdot 1 \\ 0\cdot 1 + 2\cdot 0 + 4\cdot 1 & 0\cdot 0 + 2\cdot 2 + 4\cdot 1 & 0\cdot 1 + 2\cdot 4 + 4\cdot 1 \\ 1\cdot 1 + 1\cdot 0 + 1\cdot 1 & 1\cdot 0 + 1\cdot 2 + 1\cdot 1 & 1\cdot 1 + 1\cdot 4 + 1\cdot 1 \end{bmatrix}$$

$$= \begin{bmatrix} 2 & 1 & 2 \\ 4 & 8 & 12 \\ 2 & 3 & 6 \end{bmatrix}$$

Then we calculate $A(A^2)$, obtaining

$$\begin{bmatrix} 1\cdot 2 + 0\cdot 4 + 1\cdot 2 & 1\cdot 1 + 0\cdot 8 + 1\cdot 3 & 1\cdot 2 + 0\cdot 12 + 1\cdot 6 \\ 0\cdot 2 + 2\cdot 4 + 4\cdot 2 & 0\cdot 1 + 2\cdot 8 + 4\cdot 3 & 0\cdot 2 + 2\cdot 12 + 4\cdot 6 \\ 1\cdot 2 + 1\cdot 4 + 1\cdot 2 & 1\cdot 1 + 1\cdot 8 + 1\cdot 3 & 1\cdot 2 + 1\cdot 12 + 1\cdot 6 \end{bmatrix}$$

$$= \begin{bmatrix} 4 & 4 & 8 \\ 16 & 28 & 48 \\ 8 & 12 & 20 \end{bmatrix}$$ ■

REMARK The system of linear equations

$$\begin{matrix} a_{11}x_1 + a_{12}x_2 + \cdots + a_{1n}x_n = b_1 \\ a_{21}x_1 + a_{22}x_2 + \cdots + a_{2n}x_n = b_2 \\ \vdots \qquad \vdots \qquad \qquad \vdots \\ a_{m1}x_1 + a_{m2}x_2 + \cdots + a_{mn}x_n = b_m \end{matrix}$$

can be compactly written as

$$AX = B$$

where

$$A = \begin{bmatrix} a_{11} & a_{12} & \cdots & a_{1n} \\ a_{21} & a_{22} & \cdots & a_{2n} \\ \vdots & \vdots & & \vdots \\ a_{m1} & a_{m2} & \cdots & a_{mn} \end{bmatrix}, \quad X = \begin{bmatrix} x_1 \\ x_2 \\ \vdots \\ x_n \end{bmatrix}, \quad \text{and} \quad B = \begin{bmatrix} b_1 \\ b_2 \\ \vdots \\ b_m \end{bmatrix}$$

21. Let

$$A = \begin{bmatrix} 0 & 1 & 0 & 1 \\ 1 & 0 & 0 & 0 \\ 0 & 0 & 0 & 1 \\ 1 & 0 & 1 & 0 \end{bmatrix}$$

Compute each of the following.

a. A^2

b. A^3

c. $A + A^2$

d. $A + A^2 + A^3$

22. Let A be an $m \times n$ matrix, and let S be the $1 \times m$ matrix having all entries equal to 1. Verify that SA is the $1 \times n$ matrix whose ith column is the sum of all entries lying in the ith column of A.

23. Let A be an $m \times n$ matrix and T be the $n \times 1$ matrix having all entries equal to 1. Verify that AT is the $m \times 1$ matrix whose ith row is the sum of all entries lying in the ith row of A.

24. a. Verify $A0 = 0$.

b. Verify $0A = 0$.

25. Prove each of the following for 2×2 matrices A, B, and C.

a. $A(BC) = (AB)C$

b. $A(B + C) = AB + AC$

c. $(B + C)A = BA + CA$

26. Let A be a square matrix of order n. Define its trace by

$$\text{tr}(A) = a_{11} + a_{22} + \cdots + a_{nn}$$

Verify that if A and B are square matrices of order n, each of the following is true.

a. For k a scalar, $\text{tr}(kA) = k\,\text{tr}(A)$

b. $\text{tr}(A + B) = \text{tr}(A) + \text{tr}(B)$

c. $\text{tr}(AB) = \text{tr}(BA)$

27. Under what conditions will multiplication of two matrices produce a product matrix that is each of the following?

a. 1×1

b. $m \times 1$ (column matrix)

c. $1 \times n$ (row matrix)

d. $n \times n$ (square matrix)

28. Verify the properties of scalar multiplication.

SECTION 1.9 THE INVERSE OF A MATRIX

Up to this point, we have discussed $m \times n$ matrices. We now restrict our discussion to $n \times n$ matrices, or **square matrices.**

The symbol I_n denotes the square matrice that has 1s in the diagonal entries (from top left to bottom right) and 0s in all other places. As we stated earlier, I_n is the identity matrix for multiplication—that is, $AI_n = I_nA = A$. We write I_n as I when the order n is obvious from the context of the problem.

Example 1.9.1 If

$$I = \begin{bmatrix} 1 & 0 \\ 0 & 1 \end{bmatrix}$$

for any 2×2 matrix

$$A = \begin{bmatrix} a_{11} & a_{12} \\ a_{21} & a_{22} \end{bmatrix}$$

$$AI = \begin{bmatrix} a_{11} \cdot 1 + a_{12} \cdot 0 & a_{11} \cdot 0 + a_{12} \cdot 1 \\ a_{21} \cdot 1 + a_{22} \cdot 0 & a_{21} \cdot 0 + a_{22} \cdot 1 \end{bmatrix} = \begin{bmatrix} a_{11} & a_{12} \\ a_{21} & a_{22} \end{bmatrix}$$

$$= A$$

■

The existence of identity matrices suggests that there may be matrices A and B such that $AB = I$ and $BA = I$. If so, A and B are *multiplicative inverses*: B is the inverse of A and A is the inverse of B. The symbol we use for multiplicative inverse of A is A^{-1}. We formalize this in the following definition.

Definition 1.9.2 Let A be a square matrix. The square matrix B (of the same order) is called an **inverse** of A provided

$$AB = I \quad \text{and} \quad BA = I$$

Example 1.9.3 The matrix

$$B = \begin{bmatrix} 5 & -6 \\ -4 & 5 \end{bmatrix}$$

is an inverse of

$$A = \begin{bmatrix} 5 & 6 \\ 4 & 5 \end{bmatrix}$$

since

$$AB = \begin{bmatrix} 5 & 6 \\ 4 & 5 \end{bmatrix} \begin{bmatrix} 5 & -6 \\ -4 & 5 \end{bmatrix} = \begin{bmatrix} 1 & 0 \\ 0 & 1 \end{bmatrix} = I$$

and

$$BA = \begin{bmatrix} 5 & -6 \\ -4 & 5 \end{bmatrix} \begin{bmatrix} 5 & 6 \\ 4 & 5 \end{bmatrix} = \begin{bmatrix} 1 & 0 \\ 0 & 1 \end{bmatrix} = I$$

■

In the preceding section we saw that the product of two nonzero matrices could be the zero matrix. So it should come as no surprise that some matrices have no inverses.

Example 1.9.5 Find an inverse of

$$A = \begin{bmatrix} 1 & -2 \\ -1 & 2 \end{bmatrix}$$

Solution Let

$$B = \begin{bmatrix} a & b \\ c & d \end{bmatrix}$$

be a multiplicative inverse of A. Then

$$\begin{bmatrix} 1 & -2 \\ -1 & 2 \end{bmatrix} \begin{bmatrix} a & b \\ c & d \end{bmatrix} = \begin{bmatrix} 1 & 0 \\ 0 & 1 \end{bmatrix}$$

Applying algebraic properties of matrices yields the systems

$$\begin{aligned} a - 2c &= 1 \quad \text{and} \quad & b - 2d &= 0 \\ -a + 2c &= 0 & -b + 2d &= 1 \end{aligned}$$

These systems are inconsistent. Thus an inverse of A does not exist, which means not every nonzero square matrix has an inverse. ■

A matrix that has an inverse is called a **nonsingular matrix**; a matrix that has no inverse is called a **singular matrix**. Before we develop a technique for finding an inverse of a nonsingular matrix, we first show that if a matrix has an inverse, that inverse is unique.

Theorem 1.9.6 *A nonsingular matrix has only one inverse.*

Proof. Let A be a nonsingular matrix and B be an inverse. This means

$$AB = I \quad \text{and} \quad BA = I$$

Suppose A has more than one inverse; let C be the second inverse.

Then,

$$AC = I \quad \text{and} \quad CA = I$$

Now

$$BA = I$$

$$(BA)C = IC$$

$$B(AC) = C$$

$$BI = C$$

$$B = C$$

Given the square matrix A, we now develop a technique to determine whether or not A^{-1} exists and to calculate it if it does exist.

Recall that the first column of any matrix B is

$$B\begin{bmatrix} 1 \\ 0 \\ \vdots \\ 0 \end{bmatrix}$$

In general, the ith column of any matrix B is Be_i, where e_i is the $n \times 1$ matrix with entries all 0 except for the ith row entry, which is 1.

The problem of finding $B = A^{-1}$ reduces to finding $n \times 1$ column matrices $x_1, x_2, \ldots, x_n$ for which each $Ax_i = e_i$. So solve

$$Ax = e_i$$

for each i *if possible*. (We'll see later that if any $Ax = e_i$ is unsolvable, A has no inverse.) Then construct $B = [x_1, x_2, \ldots, x_n]$ with the x_i as its columns.

It follows that $AB = I$. For the ith column of AB is

$$\begin{aligned} (AB)e_i &= A(Be_i) \\ &= Ax_i \qquad [\text{the } i\text{th column of } B \text{ is } Be_i = x_i] \\ &= e_i \qquad [\text{since } x_i \text{ is a solution of } Ax_i = e_i] \\ &= Ie_i \qquad [\text{the } i\text{th column of } I] \end{aligned}$$

Thus $AB = I$ since they have the same columns.

To solve the n equations $Ax = e_i$, we need to reduce the n matrices $[A|e_1], [A|e_2], \ldots, [A|e_n]$. But these reductions all go the same way and we can do this efficiently by reducing $[A|I] = [A|e_1, e_2, \ldots, e_n]$ to $[I|B]$ if possible. Then we mechanically read off $B = A^{-1}$.

Example 1.9.7 Find the inverse of

$$\begin{bmatrix} 2 & 1 \\ 3 & 2 \end{bmatrix}$$

Solution First construct the matrix

$$\left[\begin{array}{cc|cc} 2 & 1 & 1 & 0 \\ 3 & 2 & 0 & 1 \end{array}\right]$$

Now $(\frac{1}{2})$R1 yields

$$\left[\begin{array}{cc|cc} 1 & \frac{1}{2} & \frac{1}{2} & 0 \\ 3 & 2 & 0 & 1 \end{array}\right]$$

Next apply (-3)R1 to R2 to obtain

$$\left[\begin{array}{cc|cc} 1 & \frac{1}{2} & \frac{1}{2} & 0 \\ 0 & \frac{1}{2} & -\frac{3}{2} & 1 \end{array}\right]$$

Then $-$R2 to R1 yields

$$\left[\begin{array}{cc|cc} 1 & 0 & 2 & -1 \\ 0 & \frac{1}{2} & -\frac{3}{2} & 1 \end{array}\right]$$

Finally, (2)R2 yields

$$\left[\begin{array}{cc|cc} 1 & 0 & 2 & -1 \\ 0 & 1 & -3 & 2 \end{array}\right]$$

so

$$A^{-1} = \begin{bmatrix} 2 & -1 \\ -3 & 2 \end{bmatrix}$$

■

Example 1.9.8 Find the inverse of

$$\begin{bmatrix} 1 & 2 \\ 5 & 7 \end{bmatrix}$$

Solution First construct the matrix

$$\left[\begin{array}{cc|cc} 1 & 2 & 1 & 0 \\ 5 & 7 & 0 & 1 \end{array}\right]$$

(-5)R1 to R2 yields

$$\left[\begin{array}{cc|cc} 1 & 2 & 1 & 0 \\ 0 & -3 & -5 & 1 \end{array}\right]$$

$(-\frac{1}{3})$R2 yields

$$\left[\begin{array}{cc|cc} 1 & 2 & 1 & 0 \\ 0 & 1 & \frac{5}{3} & -\frac{1}{3} \end{array}\right]$$

Finally, (-2)R2 to R1 gives

$$\left[\begin{array}{cc|cc} 1 & 0 & -\frac{7}{3} & \frac{2}{3} \\ 0 & 1 & \frac{5}{3} & -\frac{1}{3} \end{array}\right]$$

Thus

$$A^{-1} = \begin{bmatrix} -\frac{7}{3} & \frac{2}{3} \\ \frac{5}{3} & -\frac{1}{3} \end{bmatrix}$$

To check:

$$\begin{bmatrix} 1 & 2 \\ 5 & 7 \end{bmatrix}\begin{bmatrix} -\frac{7}{3} & \frac{2}{3} \\ \frac{5}{3} & -\frac{1}{3} \end{bmatrix} = \begin{bmatrix} 1(-\frac{7}{3}) + 2(\frac{5}{3}) & 1(\frac{2}{3}) + 2(-\frac{1}{3}) \\ 5(-\frac{7}{3}) + 7(\frac{5}{3}) & 5(\frac{2}{3}) + 7(-\frac{1}{3}) \end{bmatrix}$$

$$= \begin{bmatrix} -\frac{7}{3} + \frac{10}{3} & \frac{2}{3} - \frac{2}{3} \\ -\frac{35}{3} + \frac{35}{3} & \frac{10}{3} - \frac{7}{3} \end{bmatrix} = \begin{bmatrix} 1 & 0 \\ 0 & 1 \end{bmatrix}$$

Recall that in the matrix B, the ith column is just the $n \times 1$ matrix x satisfying $Ax = e_i$. ■

In using this method, given the nonsingular matrix A we find the matrix B for which $AB = I$. That BA also equals I is the content of Theorem 1.9.11.

In general, determining inverses of matrices may be a fairly difficult task. However, there are some special matrices in which finding inverses is not so difficult.

1. Let

$$A = \begin{bmatrix} a & b \\ c & d \end{bmatrix}$$

Then

$$A^{-1} = \frac{1}{ad - bc}\begin{bmatrix} d & -b \\ -c & a \end{bmatrix}$$

provided $ad - bc \neq 0$. You are asked to verify this in Exercise 17.

2. The inverse of I is I—that is, $I^{-1} = I$.

The following are some useful facts about inverses of matrices.

Theorem 1.9.9 *If A is a matrix that has an inverse, then $(A^{-1})^{-1} = A$.*

The proof is left as an exercise.

Theorem 1.9.10 *If A and B are $n \times n$ matrices that have inverses, then so does AB, and $(AB)^{-1} = B^{-1}A^{-1}$. Moreover, if $A_1, A_2, \ldots, A_t$ are of the same order and invertible, then*

$$(A_1A_2 \cdots A_t)^{-1} = A_t^{-1}A_{t-1}^{-1} \cdots A_2^{-1}A_1^{-1}$$

Proof. Consider

$$(B^{-1}A^{-1})(AB) = B^{-1}(A^{-1}A)B = B^{-1}IB = B^{-1}B = I$$

and

$$(AB)(B^{-1}A^{-1}) = A(BB^{-1})A^{-1} = AIA^{-1} = AA^{-1} = I$$

Since the inverse of AB is unique, $(AB)^{-1} = B^{-1}A^{-1}$.

The proof of the final result is left as an exercise.

Recall from the definition that B is the inverse of A if both $AB = I$ and $BA = I$. Theorem 1.9.11 shows that we need only check that $AB = I$.

Theorem 1.9.11 *Suppose A and B are $n \times n$ matrices and $AB = I$; then $BA = I$. (Every left inverse is a right inverse.)*

The proof of Theorem 1.9.11 is too difficult to demonstrate at this time. It is proved in Chapter 4.

Theorem 1.9.12 *If A has an inverse, then the linear system $AX = B$, where A is an $n \times n$ matrix,*

$$X = \begin{bmatrix} x_1 \\ x_2 \\ \vdots \\ x_n \end{bmatrix}, \quad \text{and} \quad B = \begin{bmatrix} b_1 \\ b_2 \\ \vdots \\ b_n \end{bmatrix}$$

has the unique solution $X = A^{-1}B$.

Proof. Since $A(A^{-1}B) = (AA^{-1})B = IB = B$, $X = A^{-1}B$ is a solution of $AX = B$.

Let $AX = B$; then

$$A^{-1}(AX) = A^{-1}B$$

$$(A^{-1}A)X = A^{-1}B$$

$$IX = A^{-1}B$$

$$X = A^{-1}B$$

so $X = A^{-1}B$ is the unique solution to $AX = B$.

Corollary 1.9.13

Whenever A has an inverse, the homogeneous system $AX = 0$ has $X = 0$ as its only solution.

SUMMARY

Terms to know:

Square matrix
Inverse of a matrix
Nonsingular matrix
Singular matrix

Theorems to know:

The inverse of a matrix is unique.
$(A^{-1})^{-1} = A$
$(AB)^{-1} = B^{-1}A^{-1}$
If A and B are $n \times n$ matrices and $AB = I$, then $BA = I$.
If A is nonsingular and $AX = B$, then $X = A^{-1}B$.

Methods to know:

Determine whether or not a square matrix is nonsingular, and, if so, determine its inverse.

EXERCISE SET 1.9

Find the inverse of each of the following matrices, provided it exists.

1. $\begin{bmatrix} 7 & 9 \\ 1 & 3 \end{bmatrix}$

2. $\begin{bmatrix} 1 & 2 & 3 \\ 2 & 5 & 3 \\ 1 & 0 & 8 \end{bmatrix}$

3. $\begin{bmatrix} 0 & 1 \\ 1 & 0 \end{bmatrix}$

4. $\begin{bmatrix} 1 & 3 & 2 \\ 0 & 1 & 1 \\ 1 & 0 & 1 \end{bmatrix}$

5. $\begin{bmatrix} 1 & 3 & 2 \\ 0 & 1 & 1 \\ 0 & 0 & 1 \end{bmatrix}$

6. $\begin{bmatrix} 2 & 0 & 0 & 0 & 0 \\ 0 & -1 & 0 & 0 & 0 \\ 0 & 0 & 3 & 0 & 0 \\ 0 & 0 & 0 & -2 & 0 \\ 0 & 0 & 0 & 0 & 4 \end{bmatrix}$

7. $\begin{bmatrix} 2 & -3 & 4 & 1 \\ 0 & 1 & -2 & 1 \\ 0 & 0 & -2 & 4 \\ 0 & 0 & 0 & 3 \end{bmatrix}$

8. $\begin{bmatrix} 1 & -2 & 1 \\ -2 & 3 & 1 \\ 1 & 1 & -1 \end{bmatrix}$

9. $\begin{bmatrix} 1 & 1 & 0 & 0 & 0 \\ 1 & 1 & 1 & 0 & 0 \\ 0 & 1 & 1 & 1 & 0 \\ 0 & 0 & 0 & 1 & 1 \end{bmatrix}$

For what values of λ will the following matrices be invertible?

10. $\begin{bmatrix} 1-\lambda & 9 \\ 4 & 1-\lambda \end{bmatrix}$

11. $\begin{bmatrix} 1 & -16 \\ -4 & 1 \end{bmatrix} - \lambda \begin{bmatrix} 1 & 0 \\ 0 & 1 \end{bmatrix}$

Hint: Write as one matrix before trying to invert.

12. $\begin{bmatrix} 1-\lambda & 0 & 0 \\ -2 & 1-\lambda & 0 \\ 0 & -3 & 1-\lambda \end{bmatrix}$

13. a. Show by example that if A has an inverse and B has an inverse, it does not necessarily follow that $A + B$ has an inverse.

    b. Show that, in general, $(A + B)^{-1} \neq A^{-1} + B^{-1}$.

14. Show that $(A^{-1})^{-1} = A$.

15. Given

$$A = \begin{bmatrix} 3 & 4 \\ 2 & 3 \end{bmatrix}$$

Show $(A^{-1})^3 = (A^3)^{-1}$. This allows us to define $A^{-3} = (A^{-1})^3$ and, in general, $A^{-k} = (A^{-1})^k$, provided that A is nonsingular.

16. Given $AX + B = C$, prove that $X = A^{-1}C - A^{-1}B$, provided A is nonsingular.

17. Let

$$A = \begin{bmatrix} a & b \\ c & d \end{bmatrix}$$

and $ad - bc \neq 0$. Multiply to verify that

$$A^{-1} = \frac{1}{ad - bc} \begin{bmatrix} d & -b \\ -c & a \end{bmatrix}$$

18. If

$$A = \begin{bmatrix} 1 & -2 \\ 2 & 1 \end{bmatrix}$$

show that

$$A^{-1} = \begin{bmatrix} \frac{1}{5} & \frac{2}{5} \\ -\frac{2}{5} & \frac{1}{5} \end{bmatrix}$$

19. Let

$$D = \begin{bmatrix} a_{11} & \cdots & 0 \\ \vdots & & \vdots \\ 0 & \cdots & a_{nn} \end{bmatrix}$$

with $a_{ii} \neq 0$, $i = 1, 2, \ldots, n$. Verify

$$D^{-1} = \begin{bmatrix} \frac{1}{a_{11}} & \cdots & 0 \\ \vdots & & \vdots \\ 0 & \cdots & \frac{1}{a_{nn}} \end{bmatrix}$$

20. Let

$$D = \begin{bmatrix} 0 & 0 & 0 & 0 & a \\ 0 & 0 & 0 & b & 0 \\ 0 & 0 & c & 0 & 0 \\ 0 & d & 0 & 0 & 0 \\ e & 0 & 0 & 0 & 0 \end{bmatrix}$$

where a, b, c, d, and e are all nonzero. Verify that

$$D^{-1} = \begin{bmatrix} 0 & 0 & 0 & 0 & \frac{1}{e} \\ 0 & 0 & 0 & \frac{1}{d} & 0 \\ 0 & 0 & \frac{1}{c} & 0 & 0 \\ 0 & \frac{1}{b} & 0 & 0 & 0 \\ \frac{1}{a} & 0 & 0 & 0 & 0 \end{bmatrix}$$

21. If $AB = 0$ for a nonsingular matrix A, prove that $B = 0$.
22. If $A \neq I$ and $A^2 = A$, prove that A is singular.
23. If A_1, A_2, and A_3 are $n \times n$ invertible matrices, show that

$$(A_1 A_2 A_3)^{-1} = A_3^{-1} A_2^{-1} A_1^{-1}$$

24. If A_1, A_2, A_3, and A_4 are $n \times n$ invertible matrices, show that

$$(A_1 A_2 A_3 A_4)^{-1} = A_4^{-1} A_3^{-1} A_2^{-1} A_1^{-1}$$

25. Show that if $A_1, A_2, \ldots, A_n$ are of the same order and invertible, then

$$(A_1 A_2 \cdots A_n)^{-1} = A_n^{-1} A_{n-1}^{-1} \cdots A_2^{-1} A_1^{-1}$$

26. Show that $(AB)^{-1} = A^{-1}B^{-1}$ if and only if $AB = BA$, where A and B are invertible.
27. Show that if A is a square matrix of order n with a left inverse C and a right inverse D, then $C = D$.
28. If A and B are $n \times n$ square matrices and if $AB = 0$, then $A = 0$, or $B = 0$, or both A and B are singular.

Find the inverse of the following matrices. These exercises are best solved using the accompanying computer software or a hand-held calculator.

29. $\begin{bmatrix} 0.7 & -0.5 & 0.9 & -0.3 \\ -0.3 & 0.9 & -0.4 & 0.8 \\ -0.5 & -0.1 & 1.0 & 0.8 \\ 0.6 & -0.4 & -0.1 & 0.9 \end{bmatrix}$

30. $\begin{bmatrix} 1.1 & 1.9 & 1.1 \\ -0.9 & 3.1 & 0.9 \\ 2.1 & 3.9 & 2.1 \end{bmatrix}$

31. $\begin{bmatrix} 0.998 & 0.964 & 0.00 \\ 0.067 & 0.994 & 0.995 \\ 0.931 & 0.050 & 0.995 \end{bmatrix}$

SECTION 1.10 ELEMENTARY MATRICES AND THE *LU* DECOMPOSITION (OPTIONAL)

We have made use of elementary row transformations to solve systems of linear equations. These same row transformations can be accomplished by premultiplying a matrix by an elementary matrix. This idea has practical value when programming a computer to solve systems of equations and in determining the inverse of a matrix.

Definition 1.10.1 An **elementary matrix** is a matrix that can be obtained from an identity matrix by a single row operation.

To multiply row i of an $m \times n$ matrix A by a number k, we first multiply the ith row of the identity matrix by k, obtaining an elementary matrix E_1, and then form the product $E_1 A$.

Example 1.10.2 Given the matrix

$$A = \begin{bmatrix} 1 & -1 & 2 \\ 3 & 5 & 7 \\ 0 & -2 & 4 \end{bmatrix}$$

multiply the second row of A by k using elementary matrices.

Solution First, multiply the second row of

$$I = \begin{bmatrix} 1 & 0 & 0 \\ 0 & 1 & 0 \\ 0 & 0 & 1 \end{bmatrix}$$

by k to obtain

$$E_1 = \begin{bmatrix} 1 & 0 & 0 \\ 0 & k & 0 \\ 0 & 0 & 1 \end{bmatrix}$$

Now find the product

$$E_1A = \begin{bmatrix} 1 & 0 & 0 \\ 0 & k & 0 \\ 0 & 0 & 1 \end{bmatrix}\begin{bmatrix} 1 & -1 & 2 \\ 3 & 5 & 7 \\ 0 & -2 & 4 \end{bmatrix} = \begin{bmatrix} 1 & -1 & 2 \\ 3k & 5k & 7k \\ 0 & -2 & 4 \end{bmatrix}$$

which is A with row 2 multiplied by k. ■

Suppose we want to multiply row i of an $m \times n$ matrix by a nonzero number k and add the result to the jth row. We multiply the ith row of I by k and add it to the jth row, obtaining an elementary matrix E_2. We then form the product E_2A.

Example 1.10.3 Given the matrix

$$A = \begin{bmatrix} 1 & -1 & 2 \\ 3 & 5 & 7 \\ 0 & -2 & 4 \end{bmatrix}$$

multiply the first row by k and add it to the third row using an elementary matrix.

Solution Multiply the first row of I by k and add it to the third row to obtain

$$E_2 = \begin{bmatrix} 1 & 0 & 0 \\ 0 & 1 & 0 \\ k & 0 & 1 \end{bmatrix}$$

Now form the product

$$E_2A = \begin{bmatrix} 1 & 0 & 0 \\ 0 & 1 & 0 \\ k & 0 & 1 \end{bmatrix}\begin{bmatrix} 1 & -1 & 2 \\ 3 & 5 & 7 \\ 0 & -2 & 4 \end{bmatrix}$$

$$= \begin{bmatrix} 1 & -1 & 2 \\ 3 & 5 & 7 \\ 0 + k & -2 - k & 4 + 2k \end{bmatrix}$$

which is A with k row 1 added to row 3. ■

To interchange row i and row j of an $m \times n$ matrix, first interchange row i and row j of the identity matrix I, obtaining an elementary matrix E_3, and then form the product E_3A.

Example 1.10.4 Given the matrix

$$A = \begin{bmatrix} 1 & -1 & 2 \\ 3 & 5 & 7 \\ 0 & -2 & 4 \end{bmatrix}$$

interchange rows 1 and 3 using an elementary matrix. ■

Solution First, interchange rows 1 and 3 of the identity matrix I to obtain

$$E_3 = \begin{bmatrix} 0 & 0 & 1 \\ 0 & 1 & 0 \\ 1 & 0 & 0 \end{bmatrix}$$

Now form the product

$$E_3A = \begin{bmatrix} 0 & 0 & 1 \\ 0 & 1 & 0 \\ 1 & 0 & 0 \end{bmatrix}\begin{bmatrix} 1 & -1 & 2 \\ 3 & 5 & 7 \\ 0 & -2 & 4 \end{bmatrix} = \begin{bmatrix} 0 & -2 & 4 \\ 3 & 5 & 7 \\ 1 & -1 & 2 \end{bmatrix}$$

which is A with row 1 and row 3 interchanged. ■

We use the following notation.

1. E_1 denotes an elementary matrix that multiplies a row of a matrix by a nonzero number k.
2. E_2 denotes an elementary matrix that multiplies a row of a matrix by a number k and adds the result to another row.
3. E_3 denotes an elementary matrix that interchanges two rows of a matrix.

Theorem 1.10.5 *Given an $m \times n$ matrix A, an elementary row operation can be performed on A by the product EA, where E is an elementary matrix.*

Proof. The proof is rather tedious and essentially follows the methods used in the above examples.

Theorem 1.10.6 *Each elementary matrix has an inverse, which is also an elementary matrix.*

Proof. The inverse of E_1 is (since $k \neq 0$) the same as E_1 except that $1/k$ replaces k. If E_2 is the matrix obtained by adding k times row i to row j, then the inverse of E_2 is obtained by adding $-k$ times row i to row j. The inverse matrix of E_3 is E_3.

Example 1.10.7

$$E_1 = \begin{bmatrix} 1 & 0 & 0 \\ 0 & -2 & 0 \\ 0 & 0 & 1 \end{bmatrix} \text{ has inverse } E_1^{-1} = \begin{bmatrix} 1 & 0 & 0 \\ 0 & -\frac{1}{2} & 0 \\ 0 & 0 & 1 \end{bmatrix}$$

[here $k = -2$]

$$E_2 = \begin{bmatrix} 1 & 0 & 0 \\ 0 & 1 & 0 \\ 5 & 0 & 1 \end{bmatrix} \text{ has inverse } E_2^{-1} = \begin{bmatrix} 1 & 0 & 0 \\ 0 & 1 & 0 \\ -5 & 0 & 1 \end{bmatrix}$$

[here $k = 5$]

$$E_3 = \begin{bmatrix} 0 & 1 & 0 \\ 1 & 0 & 0 \\ 0 & 0 & 1 \end{bmatrix} \text{ has inverse } E_3^{-1} = \begin{bmatrix} 0 & 1 & 0 \\ 1 & 0 & 0 \\ 0 & 0 & 1 \end{bmatrix}$$

■

We may use elementary matrices to find the inverse of a matrix. Given an $n \times n$ matrix A, we may use elementary row operations applied to A to reduce A to the identity matrix I. If the elementary matrices corresponding to the elementary row operation are $E_1, E_2, \ldots, E_k$, then

$$E_k \cdots E_2 E_1 A = I$$

and so

$$A^{-1} = E_k \cdots E_2 E_1$$

and

$$A = E_1^{-1} E_2^{-1} \cdots E_k^{-1}.$$

Clearly, expressing A^{-1} as the product of elementary matrices appears to be a tedious process. However, expressing A^{-1} as such a product has great value when using computers. For example, if A is a 50×50 matrix, to characterize A we need to store $50^2 = 2500$ entries. Hence we also need 2500 entries to characterize A^{-1}. If we are lucky and need only a small number (say 25) of elementary row operations to change A into I, then we need only 25 elementary matrices to find A^{-1}. Each elementary matrix may be stored using at most three pieces of data as follows.

1. If E_1 denotes the elementary matrix that multiplies the ith row of a matrix by a nonzero number k, we need store only the number of the row (i) and the value of k.
2. If E_2 denotes the elementary matrix that multiplies the ith row of a matrix by a number k and adds the result to the jth row, we need store only i, j, and k.

3. If E_3 denotes the elementary matrix that interchanges rows i and j, we need store only i and j.

Since each elementary matrix requires at most three pieces of storage, the storing of A^{-1} requires at most $3e$ pieces of storage where e is the number of elementary row operations needed to change A into I. If $k = 25$, then A^{-1} can be stored using 75 pieces of storage rather than 2500, which is quite a savings in storage.

Definition 1.10.8 If A^{-1} is obtained from A by the elementary matrices $E_1, \ldots, E_k$, then

$$A^{-1} = E_k \cdots E_2 E_1$$

is called the **product form of the inverse.**

Example 1.10.9 Find the product form of the inverse of the matrix

$$A = \begin{bmatrix} 1 & 2 \\ 5 & 7 \end{bmatrix}$$

Solution In Example 1.9.8, we saw that upon applying the elementary row operations (-5)R1 to R2, $(-\frac{1}{3})$R2, and (-2)R2 to R1 to A, we obtained I. The product form of the inverse of A is

$$A^{-1} = \begin{bmatrix} 1 & -2 \\ 0 & 1 \end{bmatrix} \begin{bmatrix} 1 & 0 \\ 0 & -\frac{1}{3} \end{bmatrix} \begin{bmatrix} 1 & 0 \\ -5 & 1 \end{bmatrix}$$ ■

There are many important uses of elementary matrices. We can use elementary matrices to solve a linear system $AX = B$ of m equations in n unknowns. The method is illustrated using a simple example.

Solve

$$\begin{aligned} 3x_1 + x_2 &= 4 \\ 6x_1 + 5x_2 &= 3 \end{aligned} \qquad (1.10.1)$$

Applying row transformation (-2)R1 to R2 yields the system

$$\begin{aligned} 3x_1 + x_2 &= 4 \\ 3x_2 &= -5 \end{aligned} \qquad (1.10.2)$$

System (1.10.1) can be rewritten as

$$\begin{bmatrix} 3 & 1 \\ 6 & 5 \end{bmatrix} \begin{bmatrix} x_1 \\ x_2 \end{bmatrix} = \begin{bmatrix} 4 \\ 3 \end{bmatrix}$$

where

$$A = \begin{bmatrix} 3 & 1 \\ 6 & 5 \end{bmatrix} \tag{1.10.3}$$

and system (1.10.2) as

$$\begin{bmatrix} 3 & 1 \\ 0 & 3 \end{bmatrix}\begin{bmatrix} x_1 \\ x_2 \end{bmatrix} = \begin{bmatrix} 4 \\ -5 \end{bmatrix} \tag{1.10.4}$$

Corresponding to the given row transformation (-2)R1 to R2 is the elementary matrix

$$E_2 = \begin{bmatrix} 1 & 0 \\ -2 & 1 \end{bmatrix}$$

Premultiplying (1.10.3) by E_2 yields (1.10.4):

$$\begin{bmatrix} 1 & 0 \\ -2 & 1 \end{bmatrix}\begin{bmatrix} 3 & 1 \\ 6 & 5 \end{bmatrix}\begin{bmatrix} x_1 \\ x_2 \end{bmatrix} = \begin{bmatrix} 1 & 0 \\ -2 & 1 \end{bmatrix}\begin{bmatrix} 4 \\ 3 \end{bmatrix}$$

$$\begin{bmatrix} 3 & 1 \\ 0 & 3 \end{bmatrix}\begin{bmatrix} x_1 \\ x_2 \end{bmatrix} = \begin{bmatrix} 4 \\ -5 \end{bmatrix}$$

where

$$E_2 A = \begin{bmatrix} 3 & 1 \\ 0 & 3 \end{bmatrix}$$

Now,

$$A = \begin{bmatrix} 3 & 1 \\ 6 & 5 \end{bmatrix} \qquad E_2 = \begin{bmatrix} 1 & 0 \\ -2 & 1 \end{bmatrix}$$

and

$$E_2 A = \begin{bmatrix} 3 & 1 \\ 0 & 3 \end{bmatrix}$$

E_2 has an inverse, which is

$$E_2^{-1} = \begin{bmatrix} 1 & 0 \\ 2 & 1 \end{bmatrix}$$

Thus letting $U = \begin{bmatrix} 3 & 1 \\ 0 & 3 \end{bmatrix}$

$$A = E_2^{-1}\begin{bmatrix} 3 & 1 \\ 0 & 3 \end{bmatrix} = E_2^{-1}U$$

Observe that U is in row echelon form and U is an upper triangular matrix, that is, $u_{ij} = 0$ for $i < j$.

Also, E_2 is a lower triangular matrix: $e_{ij} = 0$ for $i > j$. Furthermore, E_2^{-1} is also lower triangular. Hence

$$A = LU = \begin{bmatrix} 1 & 0 \\ 2 & 1 \end{bmatrix} \begin{bmatrix} 3 & 1 \\ 0 & 3 \end{bmatrix}$$

the product of a lower triangular matrix, $L = E_2^{-1}$, with an upper triangular matrix, U. The factorization $A = LU$ is called an **LU decomposition** of A. This means

$$AX = B$$

can now be rewritten as

$$LUX = B$$

where

$$X = \begin{bmatrix} x_1 \\ x_2 \end{bmatrix} \quad \text{and} \quad B = \begin{bmatrix} 4 \\ 3 \end{bmatrix}$$

We can solve the system $AX = B$ by letting $Y = UX$ and first backsolving $LY = B$ for Y; then we backsolve $UX = Y$ for X. Since

$$A = E_2^{-1} \begin{bmatrix} 3 & 1 \\ 0 & 3 \end{bmatrix} = \begin{bmatrix} 1 & 0 \\ 2 & 1 \end{bmatrix} \begin{bmatrix} 3 & 1 \\ 0 & 3 \end{bmatrix}$$

then $AX = B$ can be rewritten as

$$\underset{L}{\begin{bmatrix} 1 & 0 \\ 2 & 1 \end{bmatrix}} \overbrace{\begin{bmatrix} 3 & 1 \\ 0 & 3 \end{bmatrix} \begin{bmatrix} x_1 \\ x_2 \end{bmatrix}}^{Y} = \underset{B}{\begin{bmatrix} 4 \\ 3 \end{bmatrix}}$$

We solve the system

$$\underset{L}{\begin{bmatrix} 1 & 0 \\ 2 & 1 \end{bmatrix}} \underset{Y}{\begin{bmatrix} y_1 \\ y_2 \end{bmatrix}} = \underset{B}{\begin{bmatrix} 4 \\ 3 \end{bmatrix}}$$

obtaining $y_1 = 4$ and $y_2 = -2y_1 + 3 = -8 + 3 = -5$. Continuing, we now solve

$$\underset{U}{\begin{bmatrix} 3 & 1 \\ 0 & 3 \end{bmatrix}} \underset{X}{\begin{bmatrix} x_1 \\ x_2 \end{bmatrix}} = \underset{Y}{\begin{bmatrix} 4 \\ -5 \end{bmatrix}}$$

obtaining

$$3x_2 = -5 \quad \text{or} \quad x_2 = -\tfrac{5}{3}$$

$$3x_1 + x_2 = 4 \quad \text{or} \quad 3x_1 + \left(-\tfrac{5}{3}\right) = 4 \text{ or } x_1 = \tfrac{17}{9}$$

We can always solve $AX = B$ by this method:

1. Factor A as LU.
2. Backsolve $LY = B$.
3. Backsolve $UX = Y$.

If we do no row interchanges, we can always factor A as LU. The matrix L is not as difficult to construct as it may appear. We *do not* have to invert and multiply the elementary matrices to find L. We merely make the following observation:

> To reduce the nonsingular matrix A to row echelon form, the only row operation we *must* use is multiply row i by k and add to row j where $i < j$. Then to find $L = [l_{ij}]$, we let $l_{ij} = -k$ for $i > j$, $l_{ii} = 1$ for all i and $l_{ij} = 0$ for $i < j$.

Notice that in reducing the nonsingular matrix A to row echelon form, the only row operation we use is *multiply row i by k and add to row j* for $j > i$.

Example 1.10.10 Solve

$$\begin{bmatrix} 3 & 4 & 2 \\ 1 & 5 & 8 \\ 1 & 1 & 1 \end{bmatrix}\begin{bmatrix} x_1 \\ x_2 \\ x_3 \end{bmatrix} = \begin{bmatrix} 1 \\ 4 \\ 7 \end{bmatrix}$$

Solution We reduce

$$A = \begin{bmatrix} 3 & 4 & 2 \\ 1 & 5 & 8 \\ 1 & 1 & 1 \end{bmatrix}$$

to U by row operations $(-\frac{1}{3})$R1 to R2, (so $l_{21} = \frac{1}{3}$), $(-\frac{1}{3})$R1 to R3 (so $l_{31} = \frac{1}{3}$), and finally add $(\frac{1}{11})$R2 to R3 (so $l_{32} = -\frac{1}{11}$) obtaining

$$U = \begin{bmatrix} 3 & 4 & 2 \\ 0 & \frac{11}{3} & \frac{22}{3} \\ 0 & 0 & 1 \end{bmatrix}$$

$$L = \begin{bmatrix} 1 & 0 & 0 \\ \frac{1}{3} & 1 & 0 \\ \frac{1}{3} & -\frac{1}{11} & 1 \end{bmatrix}$$ [You should verify that $LU = A$.]

where $l_{21} = \frac{1}{3}$ comes from $(-\frac{1}{3})$R1 to R2, $l_{31} = \frac{1}{3}$ comes from $(-\frac{1}{3})$R1 to R3, and $l_{32} = -\frac{1}{11}$ comes from $(\frac{1}{11})$R2 to R3.

Now we solve using the LU decomposition of A. First, we solve $LY = B$, or

$$y_1 = 1$$

$$\tfrac{1}{3}y_1 + y_2 = 4$$

$$\tfrac{1}{3}y_1 - \tfrac{1}{11}y_2 + y_3 = 7$$

obtaining $y_1 = 1$, $y_2 = \frac{11}{3}$, and $y_3 = 7$, the desired solution. Next we solve $UX = Y$, or

$$3x_1 + 4x_2 + 2x_3 = 1$$

$$\tfrac{11}{3}x_2 + \tfrac{22}{3}x_3 = \tfrac{11}{3}$$

$$x_3 = 7$$

obtaining $x_3 = 7$, $x_2 = -13$ and $x_1 = 13$. ■

There are several advantages to solving a system of equations using the LU decomposition.

1. Backsolving the LU factorization is faster than continuing to reduce A by row operations to the identity matrix (if possible).
2. L keeps track of the row operations used in solving the system of equations.
3. When programming a computer to solve systems of equations, backsolving is more numerically stable (introduces less error into the solution).

Note that if we must solve many systems of the form $AX = B$ with the same coefficient matrix A, we need *row reduce only A once* and then backsolve $LY = B$ and $UX = Y$ for each case.

In using the LU decomposition to solve $AX = B$, the backsolving of $LY = B$ is easily done because L is lower triangular. In obtaining U from A, we use only the elementary row operations $k\mathrm{R}i$ to $\mathrm{R}j$ where $i < j$. Thus $U = E_n E_{n-1} \cdots E_1 A$ where each E_i is lower triangular. The next two theorems show that L is lower triangular.

Theorem 1.10.11 *The inverse of a lower triangular matrix is lower triangular.*

Theorem 1.10.12 *The product of lower triangular matrices is lower triangular.*

Theorems 1.10.11 and 1.10.12 together show that L is lower triangular, since $L = E_1^{-1}E_2^{-1} \cdots E_n^{-1}$ (where $E_n E_{n-1} \cdots E_1 A = U$), and L is a product of the lower triangular matrices $E_1^{-1}, E_2^{-1}, \ldots, E_n^{-1}$. We prove these theorems in Section 1.11.

Before concluding this section, we give one more example of solving a system of equations using LU decomposition.

Example 1.10.13 Solve

$$\begin{bmatrix} 1 & 2 & 3 \\ 2 & 5 & 3 \\ 1 & 0 & 8 \end{bmatrix}\begin{bmatrix} x_1 \\ x_2 \\ x_3 \end{bmatrix} = \begin{bmatrix} -1 \\ 0 \\ 1 \end{bmatrix}$$

Solution We reduce

$$A = \begin{bmatrix} 1 & 2 & 3 \\ 2 & 5 & 3 \\ 1 & 0 & 8 \end{bmatrix}$$

to row echelon form (which is upper triangular) by applying, in succession, (-2)R1 to R2, $-$R1 to R3, and (2)R2 to R3 to obtain

$$U = \begin{bmatrix} 1 & 2 & 3 \\ 0 & 1 & -3 \\ 0 & 0 & -1 \end{bmatrix} \quad \text{and} \quad L = \begin{bmatrix} 1 & 0 & 0 \\ 2 & 1 & 0 \\ 1 & -2 & 1 \end{bmatrix}$$

where $l_{21} = 2$ comes from (-2)R1 to R2, $l_{31} = 1$ comes from $-$R1 to R3, and $l_{32} = -2$ comes from (2)R2 to R3. First we note that backsolving

$$\overset{L}{\begin{bmatrix} 1 & 0 & 0 \\ 2 & 1 & 0 \\ 1 & -2 & 1 \end{bmatrix}}\overset{Y}{\begin{bmatrix} y_1 \\ y_2 \\ y_3 \end{bmatrix}} = \overset{B}{\begin{bmatrix} -1 \\ 0 \\ 1 \end{bmatrix}}$$

yields $y_1 = -1$, $y_2 = 2$, and $y_3 = 6$. Backsolving

$$\overset{U}{\begin{bmatrix} 1 & 2 & 3 \\ 0 & 1 & -3 \\ 0 & 0 & -1 \end{bmatrix}}\overset{X}{\begin{bmatrix} x_1 \\ x_2 \\ x_3 \end{bmatrix}} = \overset{Y}{\begin{bmatrix} -1 \\ 2 \\ 6 \end{bmatrix}}$$

yields $x_3 = -6$, $x_2 = -16$, and $x_1 = 49$. ■

Throughout this chapter we have row reduced a matrix A to a matrix U in row echelon form. Clearly, the matrix U in row echelon form is not unique. We can obtain many such matrices by switching two rows at some stage of reducing A to U. However, if we consider matrices A that are invertible, then U is unique in some sense.

First, row reduce A to U, obtaining the LU decomposition of A. Note that L has only 1s on the diagonal. Since A is invertible, the diagonal entries of U are nonzero (we will verify this fact in the next section). Factor these diagonal entries from U to obtain a diagonal matrix D, a new upper triangular matrix U' with 1s on the diagonal, and $U = DU'$. In Example 1.10.10,

$$U = \begin{bmatrix} 3 & 4 & 2 \\ 0 & \frac{11}{3} & \frac{22}{3} \\ 0 & 0 & 1 \end{bmatrix}$$

$$= \begin{bmatrix} 3 & 0 & 0 \\ 0 & \frac{11}{3} & 0 \\ 0 & 0 & 1 \end{bmatrix} \begin{bmatrix} 1 & \frac{4}{3} & \frac{2}{3} \\ 0 & 1 & 2 \\ 0 & 0 & 1 \end{bmatrix}$$

Then

$$A = LDU'$$

which is called the ***LDU* decomposition** of A. In this case the matrices L, D, and U' are all unique.

Theorem 1.10.14 *If a matrix A is factored as $A = LDU'$, where L is a lower diagonal matrix with 1s on the diagonal, D is a diagonal matrix, all of whose diagonal entries are nonzero, and U' is an upper triangular matrix whose diagonal entries are all 1, and if $A = L_1D_1U_1'$, with L_1, D_1, and U_1' as above, then $L = L_1$, $D = D_1$, and $U' = U_1'$.*

Proof. Since $A = LDU' = L_1D_1U_1'$, we have that $L^{-1}L_1 = DU'U_1'^{-1}D_1^{-1}$. The left-hand side is a lower triangular matrix and the right-hand side is an upper triangular matrix. Thus both sides must be diagonal matrices. Since $L^{-1}L_1$ has only 1s on the diagonal, $L^{-1}L_1 = I$, and so $L = L_1$.

Next, $D^{-1}D_1 = U'U_1'^{-1}$, so $U'U_1'^{-1}$ is a diagonal matrix with 1s on the diagonal; therefore, $U'U_1'^{-1} = I$, or $U' = U_1'$. Thus $D^{-1}D_1 = I$, or $D = D_1$. Thus the assertion of the theorem is proved.

SUMMARY

Terms to know:

Elementary matrix

LU decomposition

LDU decomposition

Theorems to know:

Each elementary matrix has an inverse, which is also an elementary matrix.

Each matrix has an *LU* decomposition.

Methods to know:

Solve a system of linear equations with coefficient matrix A using the *LU* decomposition of A.

EXERCISE SET 1.10

Using the LU decomposition, solve the following systems of equations.

1. $\begin{bmatrix} 1 & 2 & 3 \\ 1 & 1 & 1 \\ -1 & 0 & 2 \end{bmatrix} X = \begin{bmatrix} 1 \\ 1 \\ 1 \end{bmatrix}$

2. $\begin{bmatrix} 1 & -1 & 1 \\ 0 & 1 & -1 \\ 1 & -1 & 1 \end{bmatrix} X = \begin{bmatrix} 4 \\ -1 \\ 3 \end{bmatrix}$

3. $\begin{bmatrix} -2 & 2 & -2 \\ 3 & -1 & -2 \\ 2 & -1 & -1 \end{bmatrix} X = \begin{bmatrix} 10 \\ -1 \\ 6 \end{bmatrix}$

4. $\begin{bmatrix} 2 & -3 \\ -6 & 9 \end{bmatrix} X = \begin{bmatrix} 1 \\ 4 \end{bmatrix}$

5. $\begin{bmatrix} 2 & 1 & 1 \\ 3 & -2 & -1 \\ 4 & -7 & 3 \end{bmatrix} X = \begin{bmatrix} 8 \\ 1 \\ 10 \end{bmatrix}$

6. $\begin{bmatrix} 1 & 1 & 1 \\ -2 & 5 & 2 \\ -7 & 7 & 1 \end{bmatrix} X = \begin{bmatrix} 0 \\ 0 \\ 0 \end{bmatrix}$

7. Prove that for any matrix B, $E_2 B$ is the matrix obtained from B by adding a constant k times row i to row j.

8. Prove that for any matrix B, $E_3 B$ is the matrix obtained from B by interchanging rows i and j of B.

9. If T is lower triangular and any $t_{ii} = 0$, prove that T has no inverse.

10. Solve the given systems of equations using the *LU* decomposition by reducing the coefficient matrix only once.

$$\begin{bmatrix} 1 & 3 & 1 \\ 0 & 1 & 2 \\ 1 & 0 & 1 \end{bmatrix} \begin{bmatrix} x_1 \\ x_2 \\ x_3 \end{bmatrix} = B$$

for each B.

a. $B = \begin{bmatrix} -1 \\ 1 \\ 1 \end{bmatrix}$ b. $B = \begin{bmatrix} 0 \\ -2 \\ 1 \end{bmatrix}$

c. $B = \begin{bmatrix} 4 \\ -3 \\ 2 \end{bmatrix}$ d. $B = \begin{bmatrix} -2 \\ 3 \\ 4 \end{bmatrix}$

11. Apply the operations a row i to row j and a column i to column j to I obtaining matrices F and G, respectively.

a. Does $F = G$?

b. Given the 3×3 matrix

$$A = \begin{bmatrix} 1 & 2 & 3 \\ 4 & 5 & 6 \\ 7 & 8 & 9 \end{bmatrix}$$

form the matrix G corresponding to -2 row 3 to column 1.

c. What is AG?

d. Perform the operation -2 column 3 to column 1 to A. Do the matrices obtained in (c) and (d) agree?

SECTION 1.11 TWO IMPORTANT CLASSES OF MATRICES

In this section we consider triangular matrices and matrix transposition. Later sections of this text demonstrate the usefulness of these two concepts.

Definition 1.11.1 An $m \times n$ matrix U is an **upper triangular matrix** if $u_{ij} = 0$ for $i > j$.

Example 1.11.2 The matrices

$$\begin{bmatrix} 1 & 2 & 4 \\ 0 & 5 & -1 \\ 0 & 0 & 2 \end{bmatrix} \quad \text{and} \quad \begin{bmatrix} 1 & 0 & 5 & 7 \\ 0 & 0 & 1 & 3 \\ 0 & 0 & 2 & 1 \end{bmatrix}$$

are upper triangular matrices, whereas

$$\begin{bmatrix} 1 & 0 & 1 \\ 1 & 1 & 2 \\ 0 & 0 & 3 \end{bmatrix}$$

is not upper triangular since the $(2, 1)$-entry is not zero. ■

Upper triangular matrices are those whose entries below the main diagonal are zero. Note also that a matrix in row echelon form is upper triangular. It is this property of matrices that makes backsolving possible.

Definition 1.11.3 An $m \times n$ matrix L is **lower triangular** if $l_{ij} = 0$ for $i < j$.
An $n \times n$ matrix D is a **diagonal matrix** if $d_{ij} = 0$ for $i \neq j$.
A matrix is a **triangular matrix** if it is either upper triangular or lower triangular.

Theorem 1.11.4 *A matrix is a diagonal matrix if and only if it is both upper and lower triangular.*

In Chapter 5 it is an easy exercise to prove the following result using determinants.

Theorem 1.11.5 *An upper triangular $n \times n$ matrix T is invertible if and only if its diagonal entries are all nonzero.*

Theorem 1.11.6 *If T is an invertible upper triangular matrix, then T^{-1} is upper triangular.*

Proof. Let $B = T^{-1}$. Since

$$1 = b_{11}t_{11} + b_{12} \cdot 0 + \cdots + b_{in} \cdot 0$$

$b_{11} = 1/t_{11}$. For any $i > 1$,

$$0 = b_{i1}t_{11} + b_{i2} \cdot 0 + \cdots + b_{in} \cdot 0$$

so $b_{i1} = 0$ for $i > 1$.

Next we consider $i > 2$; then

$$\begin{aligned} 0 &= b_{i1}t_{12} + b_{i2}t_{22} + b_{i3}t_{32} + \cdots + b_{in}t_{n2} \\ &= b_{i2}t_{22} \end{aligned}$$

Since $t_{22} \neq 0$, $b_{i2} = 0$ for $i > 2$.

Now suppose $b_{ik} = 0$ for $i > k$, $k = 1, 2, \ldots, p$; then for $i > p + 1$,

$$\begin{aligned} 0 &= b_{i1}t_{1(p+1)} + \cdots + b_{ip}t_{p(p+1)} + b_{i(p+1)}t_{(p+1)(p+1)} + \cdots + b_{in}t_{n(p+1)} \\ &= b_{i(p+1)}t_{(p+1)(p+1)} \end{aligned}$$

and so $b_{i(p+1)} = 0$ for $i > p + 1$. By induction, we are done.

We next consider another class of matrices.

Definition 1.11.7 Let $M = [m_{ij}]$ be an $m \times n$ matrix. The **transpose** of M, denoted by M^T, is the $n \times m$ matrix whose (i, j)-entry is m_{ji}.

Hence M^T is obtained from M by interchanging the rows and columns of M.

Example 1.11.8

a. $$\begin{bmatrix} 1 & 2 & 4 \\ 0 & 1 & 6 \\ 4 & 0 & 2 \end{bmatrix}^T = \begin{bmatrix} 1 & 0 & 4 \\ 2 & 1 & 0 \\ 4 & 6 & 2 \end{bmatrix}$$

b. $$\begin{bmatrix} 1 & 2 & 4 \\ 1 & 1 & -2 \end{bmatrix}^T = \begin{bmatrix} 1 & 1 \\ 2 & 1 \\ 4 & -2 \end{bmatrix}$$

c. $$\begin{bmatrix} 1 & 2 & 3 \\ 0 & 1 & 2 \\ 0 & 0 & -1 \end{bmatrix}^T = \begin{bmatrix} 1 & 0 & 0 \\ 2 & 1 & 0 \\ 3 & 2 & -1 \end{bmatrix}$$ ■

We leave it to the reader to prove the following result.

Theorem 1.11.9 *For any $m \times n$ matrices A and B, and real number r,*

a. $(A + B)^T = A^T + B^T$

b. $(rA)^T = rA^T$

c. $(A^T)^T = A$

Also, if A is an $m \times n$ matrix and B is an $n \times k$ matrix, we have the next result.

Theorem 1.11.10 $(AB)^T = B^T A^T$.

Proof. Let $AB = C$, $A^T = Y$, $B^T = Z$, and consider $C^T = X$. Then

$$\begin{aligned} x_{ij} = c_{ji} &= a_{j1}b_{1i} + a_{j2}b_{2i} + \cdots + a_{jn}b_{ni} \\ &= y_{1j}z_{i1} + y_{2j}z_{i2} + \cdots + y_{nj}z_{in} \\ &= z_{i1}y_{1j} + z_{i2}y_{2j} + \cdots + z_{in}y_{nj} \\ &= (i, j)\text{-entry of } ZY\ (= B^T A^T) \end{aligned}$$

We leave it as an exercise to prove the following theorem.

Theorem 1.11.11 *A matrix B is upper triangular if and only if B^T is lower triangular.*

Definition 1.11.12 A matrix S is a **symmetric** matrix if and only if $S^T = S$. A matrix A is **antisymmetric** if $A^T = -A$.

Example 1.11.13 The matrices

$$\begin{bmatrix} 1 & 2 & 3 \\ 2 & 1 & -1 \\ 3 & -1 & 4 \end{bmatrix} \quad \text{and} \quad \begin{bmatrix} 0 & -1 & \pi \\ -1 & x & 27 \\ \pi & 27 & 4 \end{bmatrix}$$

are symmetric, whereas

$$\begin{bmatrix} 1 & 0 & 4 \\ 1 & 2 & 3 \\ 4 & 3 & 5 \end{bmatrix}$$

is not since the $(1, 2)$-entry, 0, and the $(2, 1)$-entry, 1, are not equal. ■

In Chapter 6 we show that symmetric matrices have some very nice properties. However, until then we must be satisfied with the next result.

Theorem 1.11.14 *Let A be an $n \times n$ matrix. Then*

$$A = S + Y$$

for some symmetric matrix S and antisymmetric matrix Y.

Proof. Let $S = \frac{1}{2}(A + A^T)$ and $Y = \frac{1}{2}(A - A^T)$. S is symmetric since

$$\begin{aligned} S^T &= \left[\tfrac{1}{2}(A + A^T)\right]^T = \tfrac{1}{2}(A + A^T)^T \\ &= \tfrac{1}{2}\left[A^T + (A^T)^T\right] = \tfrac{1}{2}(A^T + A) \end{aligned}$$

Likewise, $Y^T = -Y$. Furthermore,

$$\begin{aligned} S + Y &= \tfrac{1}{2}(A + A^T) + \tfrac{1}{2}(A - A^T) \\ &= \tfrac{1}{2}A + \tfrac{1}{2}A^T + \tfrac{1}{2}A - \tfrac{1}{2}A^T \\ &= \tfrac{1}{2}A + \tfrac{1}{2}A + \tfrac{1}{2}A^T - \tfrac{1}{2}A^T \\ &= \left(\tfrac{1}{2} + \tfrac{1}{2}\right)A \\ &= 1A \\ &= A \end{aligned}$$

Example 1.11.15 Write each of the following matrices as the sum of a symmetric and an antisymmetric matrix:

a. $\begin{bmatrix} 1 & 3 & 0 \\ 2 & 1 & 4 \\ 0 & 1 & 5 \end{bmatrix}$

b. $\begin{bmatrix} -1 & \pi & e \\ \pi & 0 & 2 \\ e & 2 & 4 \end{bmatrix}$

Solution a. Let

$$A = \begin{bmatrix} 1 & 3 & 0 \\ 2 & 1 & 4 \\ 0 & 1 & 5 \end{bmatrix}$$

Then

$$A^T = \begin{bmatrix} 1 & 2 & 0 \\ 3 & 1 & 1 \\ 0 & 4 & 5 \end{bmatrix}$$

so

$$A + A^T = \begin{bmatrix} 1+1 & 3+2 & 0+0 \\ 2+3 & 1+1 & 4+1 \\ 0+0 & 1+4 & 5+5 \end{bmatrix} = \begin{bmatrix} 2 & 5 & 0 \\ 5 & 2 & 5 \\ 0 & 5 & 10 \end{bmatrix}$$

$$A - A^T = \begin{bmatrix} 1-1 & 3-2 & 0-0 \\ 2-3 & 1-1 & 4-1 \\ 0-0 & 1-4 & 5-5 \end{bmatrix} = \begin{bmatrix} 0 & 1 & 0 \\ -1 & 0 & 3 \\ 0 & -3 & 0 \end{bmatrix}$$

Thus

$$S = \tfrac{1}{2}\begin{bmatrix} 2 & 5 & 0 \\ 5 & 2 & 5 \\ 0 & 5 & 10 \end{bmatrix} = \begin{bmatrix} 1 & 2.5 & 0 \\ 2.5 & 1 & 2.5 \\ 0 & 2.5 & 5 \end{bmatrix}$$

whereas

$$Y = \tfrac{1}{2}\begin{bmatrix} 0 & 1 & 0 \\ -1 & 0 & 3 \\ 0 & -3 & 0 \end{bmatrix} = \begin{bmatrix} 0 & 0.5 & 0 \\ -0.5 & 0 & 1.5 \\ 0 & -1.5 & 0 \end{bmatrix}$$

b. Notice that

$$A = A^T = \begin{bmatrix} -1 & \pi & e \\ \pi & 0 & 2 \\ e & 2 & 4 \end{bmatrix}$$

is symmetric so

$$S = \tfrac{1}{2}(A + A^T) = \tfrac{1}{2}(A + A) = \tfrac{1}{2}(2A) = (\tfrac{1}{2} \cdot 2)A = A$$

whereas

$$Y = \tfrac{1}{2}(A - A^T) = \tfrac{1}{2}(A - A) = \tfrac{1}{2}0 = 0$$ ■

SUMMARY

Terms to know:

Upper triangular matrix
Lower triangular matrix
Diagonal matrix
Transpose of a matrix
Symmetric matrix
Antisymmetric matrix

Theorems to know:

An upper triangular $n \times n$ matrix T is invertible if and only if its diagonal entries are all nonzero.

The inverse of an invertible upper triangular matrix is upper triangular.

Let A be an $m \times n$ matrix, then A can be written as $A = S + Y$ where S is a symmetric matrix and Y is antisymmetric.

Methods to know:

Write an $m \times n$ matrix A as the sum of a symmetric matrix S and an antisymmetric matrix Y.

EXERCISE SET 1.11

Find the inverse of each matrix.

1. $\begin{bmatrix} 1 & 2 & 0 \\ 0 & 1 & 3 \\ 0 & 0 & 2 \end{bmatrix}$
2. $\begin{bmatrix} 1 & 4 & 0 \\ 0 & 1 & 0 \\ 0 & 0 & 5 \end{bmatrix}$
3. $\begin{bmatrix} 7 & 0 & 0 \\ 0 & 5 & 0 \\ 0 & 0 & -2 \end{bmatrix}$
4. $\begin{bmatrix} 1 & 0 & 0 \\ 4 & 2 & 0 \\ 1 & 3 & 2 \end{bmatrix}$
5. $\begin{bmatrix} 7 & 0 & 0 \\ -1 & -2 & 0 \\ 0 & 1 & 3 \end{bmatrix}$
6. $\begin{bmatrix} 4 & 7 & 0 \\ 0 & 1 & 0 \\ 0 & 0 & 1 \end{bmatrix}$
7. Given the matrices

$$A = \begin{bmatrix} 1 & 3 & 1 \\ 0 & 1 & 1 \end{bmatrix} \quad \text{and} \quad B = \begin{bmatrix} 1 & 1 \\ 0 & 2 \\ 0 & 0 \end{bmatrix}$$

   a. Calculate AB.

   b. Calculate BA.

8. Let A be an $m \times n$ upper triangular matrix and B be an $n \times k$ upper triangular matrix. Must the product AB be an upper triangular matrix?

Find the transpose of each matrix.

9. $\begin{bmatrix} 1 & 3 & 4 \\ -7 & 3 & 4 \end{bmatrix}$
10. $\begin{bmatrix} 4 & 1 & 3 & 2 \\ -7 & 6 & 4 & 1 \\ 5 & 3 & e & \pi \\ 4 & 6 & -1 & 9 \\ 2 & 1 & 3 & 0 \end{bmatrix}$
11. $\begin{bmatrix} 0 & 1 \\ 2 & 4 \\ 5 & -3 \\ 0 & 2 \end{bmatrix}$
12. $\begin{bmatrix} 0 & 4 & 2 \\ -4 & 0 & 3 \\ -2 & -3 & 0 \end{bmatrix}$
13. Let B be an $m \times n$ matrix and let A be the $m \times n$ matrix $A = BB^T$. Prove that A is symmetric. What about $C = B^TB$?
14. Let A and B be $m \times n$ matrices. Prove that $(A + B)^T = A^T + B^T$.
15. Prove that for any matrix A, $(A^T)^T = A$.

Write each given matrix as the sum of a symmetric matrix and an anti-symmetric matrix.

16. $\begin{bmatrix} 1 & 4 \\ 8 & 3 \end{bmatrix}$

17. $\begin{bmatrix} -1 & 0 & 3 \\ 4 & 1 & 2 \\ 0 & 1 & 5 \end{bmatrix}$

18. $\begin{bmatrix} 0 & 4 & 3 \\ -4 & 0 & 2 \\ -3 & 5 & 0 \end{bmatrix}$

19. $\begin{bmatrix} -6 & 2 & 5 \\ 1 & 4 & 3 \\ 0 & 0 & 0 \end{bmatrix}$

20. Let $A = S_1 + Y_1$, where S_1 is a symmetric matrix and Y_1 is an antisymmetric matrix. Prove that $S_1 = \frac{1}{2}(A + A^T)$ and $Y_1 = \frac{1}{2}(A - A^T)$.

    Hint: $A^T = S_1^T + Y_1^T = S_1 - Y_1$.

21. Let A be an $m \times n$ upper triangular matrix and B be an $n \times k$ upper triangular matrix. Prove that AB is an upper triangular matrix.

    Hint: Let $i > j$ and $AB = C$. Then

    $$c_{ij} = a_{i1}b_{1j} + a_{i2}b_{2j} + \cdots + a_{i(i-1)}b_{(i-1)j} + a_{ii}b_{ij} + \cdots + a_{in}b_{nj}$$

    Note that $a_{i1} = a_{i2} = \cdots = a_{i(i-1)} = 0$ (why?) and $b_{ij} = b_{(i+1)j} = \cdots = b_{nj} = 0$ (why?).

22. Let B be an anti-symmetric matrix. Prove that the diagonal elements of B are all zero.

CHAPTER TWO

VECTORS IN R^n

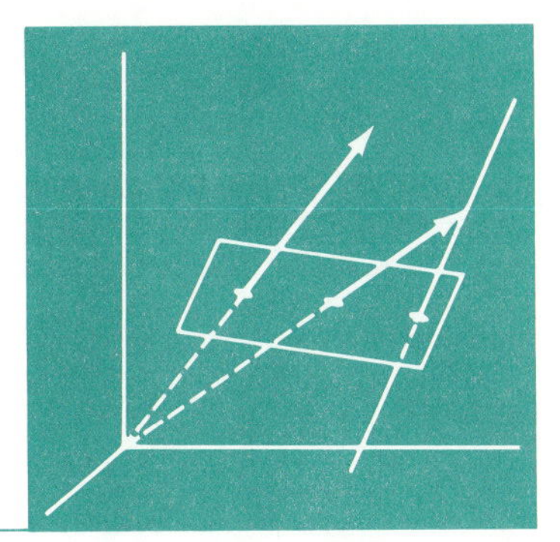

SECTION 2.1 VECTORS: AN INTRODUCTION

An $m \times n$ matrix is a rectangular array of real numbers consisting of m rows and n columns. A matrix with only one row, i.e., a $1 \times n$ matrix, is a **row vector**. A **column vector** is a matrix with only one column, i.e., an $m \times 1$ matrix. The entries of a vector are usually called **coordinates**, or **components**.

A **vector of n dimensions** has n components and is either a row vector or a column vector. The collection of all n-dimensional row (or column) vectors is denoted by R^n and is called **n-space** or **n-dimensional space**. Thus all n-dimensional row vectors $\mathbf{x} = [x_1, x_2, \ldots, x_n]$ constitute R^n or all n-dimensional column vectors

$$\mathbf{x} = \begin{bmatrix} x_1 \\ x_2 \\ \vdots \\ x_n \end{bmatrix}$$

constitute R^n. In the statements of definitions and theorems, vectors can be either row or column vectors. In this book, we choose to write column vectors for convenience.

Since vectors are matrices with one column, we have the following definition of the equality of two vectors.

Definition 2.1.1 Two vectors

$$\mathbf{x} = \begin{bmatrix} x_1 \\ x_2 \\ \vdots \\ x_m \end{bmatrix} \quad \text{and} \quad \mathbf{y} = \begin{bmatrix} y_1 \\ y_2 \\ \vdots \\ y_n \end{bmatrix}$$

are **equal** if and only if $m = n$ and $x_i = y_i$ for $1 \leqq i \leqq n$. That is, $\mathbf{x}$ and $\mathbf{y}$ in R^n are equal if their corresponding components are equal.

Example 2.1.2 $\begin{bmatrix} 3 \\ 4 \end{bmatrix} = \begin{bmatrix} r \\ r+1 \end{bmatrix}$ provided $r = 3$. ■

The one-dimensional vectors in R^1 are denoted by $[x_1]$. Since this space is the same as the set R of real numbers, we drop the brackets and superscript and regard R^1 as R. When dealing with vectors or matrices, it is customary to refer to real numbers as **scalars**.

In R^2, each vector

$$\begin{bmatrix} a \\ b \end{bmatrix}$$

may be visualized as the arrow with tail at the origin $(0, 0)$ and head at the point (a, b), as shown in Figure 2.1. The length of the arrow is $\sqrt{a^2 + b^2}$, as can be seen by applying the Pythagorean theorem to the right triangle with vertices $(0, 0)$, $(0, a)$ and (a, b).

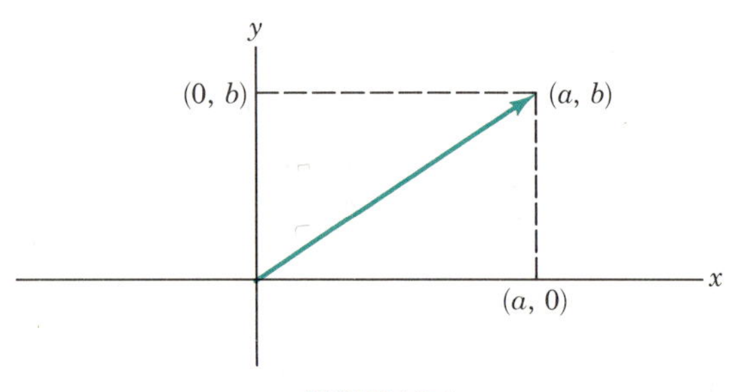

FIGURE 2.1

In R^3, each vector

$$\begin{bmatrix} a \\ b \\ c \end{bmatrix}$$

may be visualized as the arrow with tail at the origin and head at the point (a, b, c) as shown in Figure 2.2. The length of this arrow is $\sqrt{a^2 + b^2 + c^2}$.

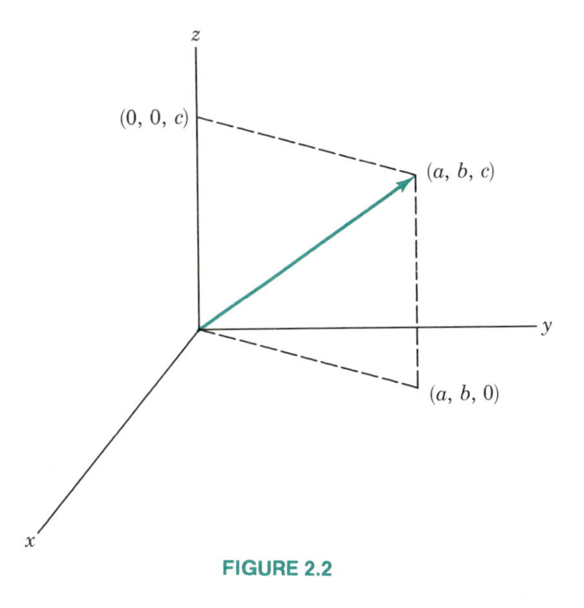

FIGURE 2.2

In R^2, the point (a, b) and the vector

with tail at $(0, 0)$ are closely related: The head of the vector

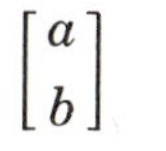

is (a, b), whereas to the point (a, b) is associated the vector

$$\begin{bmatrix} a \\ b \end{bmatrix}$$

which may be visualized geometrically as the arrow from $(0, 0)$ to (a, b).

Similarly, in R^3 the head of the vector

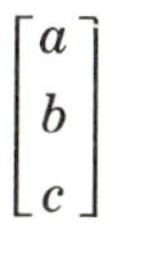

is (a, b, c), whereas to each point (a, b, c) is associated the vector

$$\begin{bmatrix} a \\ b \\ c \end{bmatrix}$$

which may be visualized geometrically as the arrow from $(0, 0, 0)$ to (a, b, c).

It is in this sense that points in R^2 and R^3 are regarded as vectors. In R^2 and R^3 we call $\mathbf{x}$ a vector when we want to emphasize the direction along the line from the origin to the point x rather than the location x, which is suggested by the word *point.*

We adopt the same convention for R^n even though vectors and points cannot be visualized; that is to say, **the words "vector" and "point" may and will be used interchangeably.**

We motivate our definition of distance between vectors in R^n by the geometry of vectors in R^2 and R^3.

In Figure 2.3, the line segment joining (x_1, x_2) and (y_1, y_2) is the hypoteneuse of the right triangle with vertices (x_1, x_2), (y_1, y_2), and (x_1, y_2). Two sides of this triangle have lengths $|x_1 - y_1|$ and $|x_2 - y_2|$, and the third side, by the Pythagorean theorem, has length

$$\sqrt{(x_1 - y_1)^2 + (x_2 - y_2)^2}$$

Geometrically, the length of the hypoteneuse is the distance between (x_1, x_2) and (y_1, y_2).

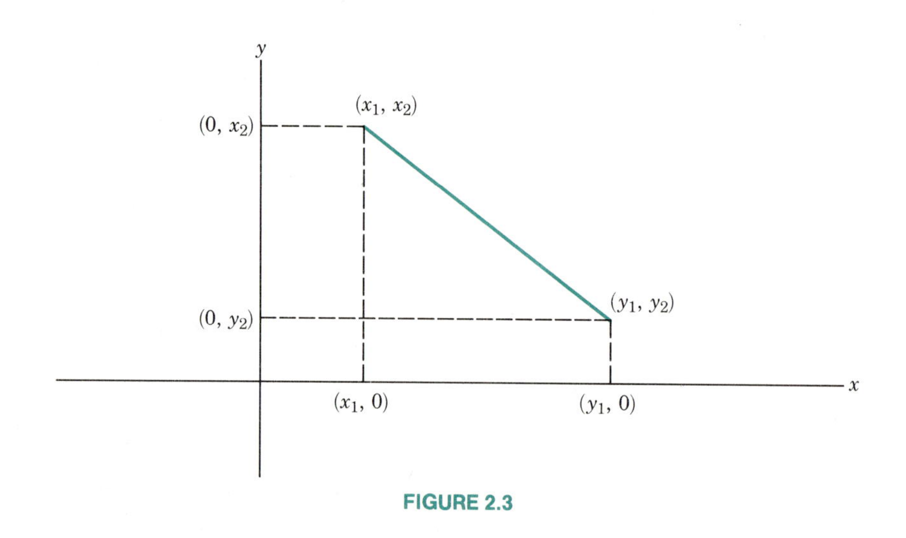

FIGURE 2.3

In R^3, the distance between the points (x_1, x_2, x_3) and (y_1, y_2, y_3) should be (geometrically) the length of the hypoteneuse of the right triangle shown in Figure 2.4, namely,

$$\sqrt{(x_1 - y_1)^2 + (x_2 - y_2)^2 + (x_3 - y_3)^2}$$

In R^n, we define the distance between two points in an analogous way to the formulas in R^2 and R^3, which were motivated by the geometry of R^2 in Figure 2.3 and R^3 in Figure 2.4.

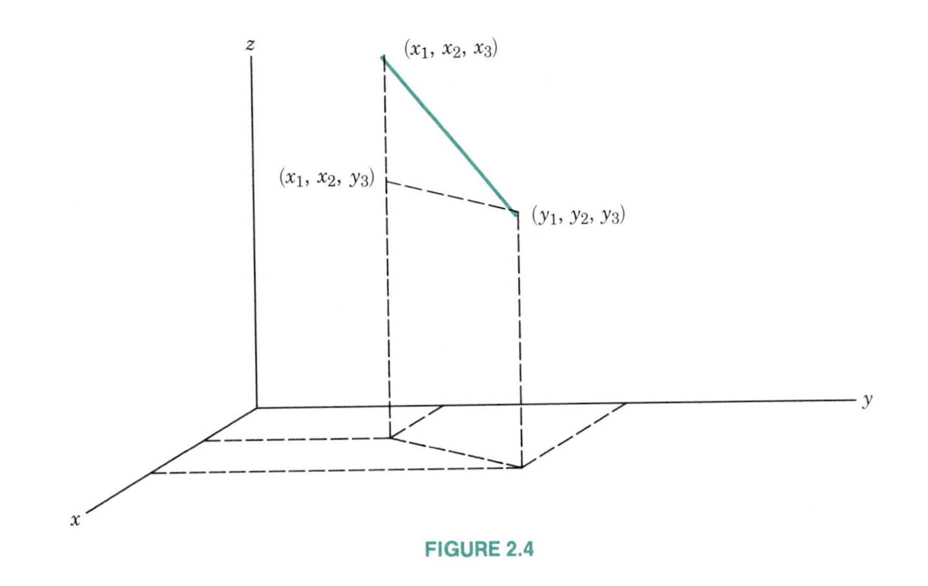

FIGURE 2.4

Definition 2.1.3 The **distance between** two points

$$\mathbf{x} = \begin{bmatrix} x_1 \\ \vdots \\ x_n \end{bmatrix} \quad \text{and} \quad \mathbf{y} = \begin{bmatrix} y_1 \\ \vdots \\ y_n \end{bmatrix}$$

in R^n, denoted by $d(\mathbf{x}, \mathbf{y})$, is defined to be the number

$$d(\mathbf{x}, \mathbf{y}) = \sqrt{(x_1 - y_1)^2 + (x_2 - y_2)^2 + \cdots + (x_n - y_n)^2}$$

According to this general distance formula for R^n, the distance between

$$\begin{bmatrix} 1 \\ 3 \\ 6 \end{bmatrix} \quad \text{and} \quad \begin{bmatrix} 2 \\ 6 \\ 4 \end{bmatrix}$$

in R^3 is

$$\sqrt{(1-2)^2 + (3-6)^2 + (6-4)^2} = \sqrt{14}$$

This is illustrated in Figure 2.5.

This agrees with our concept of distance in R^3 because, by the Pythagorean theorem, the distance between $(2, 6, 4)$ and $(1, 3, 4)$ is

$$\sqrt{(2-1)^2 + (6-3)^2 + (y-y)^2} = \sqrt{10}$$

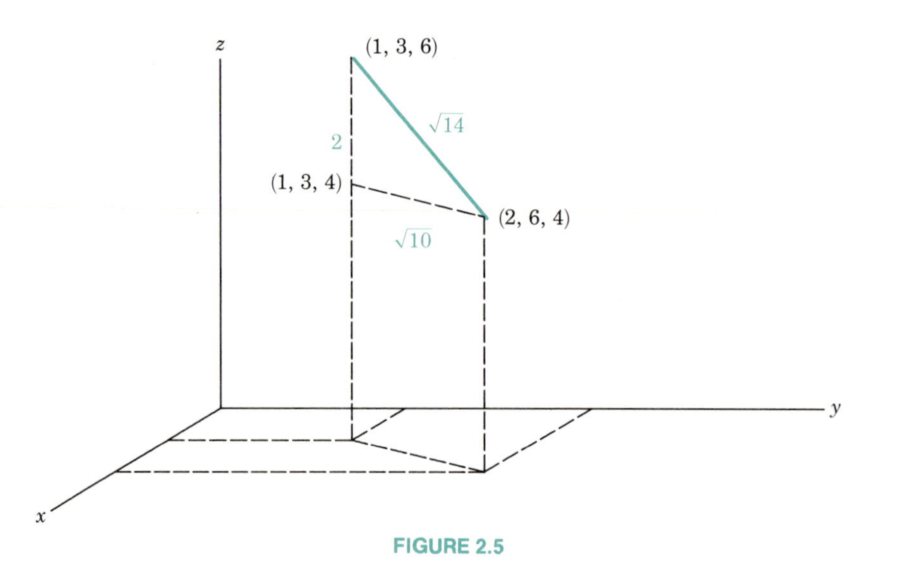

FIGURE 2.5

and the square of the distance between (1, 3, 6) and (2, 6, 4) is

$$(\sqrt{10})^2 + (6-4)^2 = 14$$

The distance formula for R^n may be used to calculate, in a routine way, distance between any two vectors $\mathbf{x}$ and $\mathbf{y}$ in R^n.

Example 2.1.4 Determine the distance between

$$\mathbf{x} = \begin{bmatrix} 1 \\ 2 \\ -1 \\ 3 \end{bmatrix} \quad \text{and} \quad \mathbf{y} = \begin{bmatrix} 0 \\ 1 \\ -2 \\ 1 \end{bmatrix}$$

in R^4.

Solution The distance is

$$\sqrt{(1-0)^2 + (2-1)^2 + (-1-(-2))^2 + (3-1)^2} = \sqrt{7}$$ ■

If a point x is regarded as a vector, then the length of the vector $\mathbf{x}$ is just the distance from x to the origin (denoted by $\mathbf{0}$). Since every coordinate of $\mathbf{0}$ is 0, it follows that the distance from

$$\mathbf{x} = \begin{bmatrix} x_1 \\ x_2 \\ \vdots \\ x_n \end{bmatrix}$$

to $\mathbf{0}$ is

$$\sqrt{x_1^2 + x_2^2 + \cdots + x_n^2}$$

For the record, we state the definition of the length of a vector in R^n.

Definition 2.1.5 The **length** of a vector

$$\mathbf{x} = \begin{bmatrix} x_1 \\ x_2 \\ \vdots \\ x_n \end{bmatrix}$$

denoted by $\|\mathbf{x}\|$, is given by

$$\|\mathbf{x}\| = \sqrt{x_1^2 + x_2^2 + \cdots + x_n^2}$$

The length of a vector $\mathbf{x}$ is sometimes called its **magnitude**, or **norm**. Note that $\|\mathbf{0}\| = 0$.

Example 2.1.6 Given

$$\mathbf{x} = \begin{bmatrix} 4 \\ x_2 \end{bmatrix}$$

determine a value of x_2 for which $\|\mathbf{x}\| = 5$.

Solution Since $\|\mathbf{x}\| = \sqrt{4^2 + x_2^2}$ and we want $\|\mathbf{x}\| = 5$, we solve $\sqrt{16 + x_2^2} = 5$:

$$\sqrt{16 + x_2^2} = 5$$

$$16 + x_2^2 = 25$$

$$x_2^2 = 9$$

$$x_2 = -3 \quad \text{or} \quad 3$$

We have two vectors, for which $\|\mathbf{x}\| = 5$, namely,

$$\begin{bmatrix} 4 \\ -3 \end{bmatrix} \quad \text{and} \quad \begin{bmatrix} 4 \\ 3 \end{bmatrix}$$

■

In R^1, the length of a vector is just the absolute value of a number. For if $\mathbf{x} = [x_1]$ is a point in R^1, then

$$\|\mathbf{x}\| = \sqrt{x_1^2} = |x_1|$$

Thus the length of a vector is a natural generalization of the absolute value of a number. The basic properties of absolute value are as follows.

1. $|x| > 0$ if, and only if, $x \neq 0$
2. $|rx| = |r|\,|x|$ for all $r \in R$
3. $|x + y| \leqq |x| + |y|$

It should not be too surprising, therefore, that analogous properties hold for the length of a vector.

1′. $\|\mathbf{x}\| > 0$ if, and only if, $\mathbf{x} \neq \mathbf{0}$

2′. $\|r\mathbf{x}\| = |r|\,\|\mathbf{x}\|$ for all $r \in R$

3′. $\|\mathbf{x} + \mathbf{y}\| \leqq \|\mathbf{x}\| + \|\mathbf{y}\|$

The first property of absolute value, $|x| > 0$ if $x \neq 0$, is easy to extend to length in R^n. Since

$$\|\mathbf{x}\| = \sqrt{x_1^2 + x_2^2 + \cdots + x_n^2}$$

it follows that $\|\mathbf{x}\| > 0$ if at least one of the coordinates of $\mathbf{x}$ is not zero. Since $\mathbf{x} \neq \mathbf{0}$ means at least one of the coordinates of $\mathbf{x}$ is not zero, then we have proved that

$$\|\mathbf{x}\| > 0 \quad \text{if, and only if,} \quad \mathbf{x} \neq \mathbf{0}$$

In order to get a generalization to R^n of the second property $|rx| = |r|\,|x|$, we introduce a definition.

Definition 2.1.7 If r is a number and $\mathbf{x}$ is a vector in R^n, the **scalar multiple** $r\mathbf{x}$ is defined to be the vector

$$r\mathbf{x} = \begin{bmatrix} rx_1 \\ rx_2 \\ \vdots \\ rx_n \end{bmatrix}$$

(Note that this definition agrees with that of multiplication of a matrix by a scalar.) With this definition of $r\mathbf{x}$, it follows that

$$\|r\mathbf{x}\| = \left\| r\begin{bmatrix} x_1 \\ x_2 \\ \vdots \\ x_n \end{bmatrix} \right\| = \left\| \begin{bmatrix} rx_1 \\ rx_2 \\ \vdots \\ rx_n \end{bmatrix} \right\| = \sqrt{(rx_1)^2 + (rx_2)^2 + \cdots + (rx_n)^2}$$

$$= \sqrt{r^2(x_1^2 + x_2^2 + \cdots + x_n)^2} = \sqrt{r^2}\,\|\mathbf{x}\|$$

$$= |r|\,\|\mathbf{x}\|$$

Example 2.1.8 Given

$$\mathbf{x} = \begin{bmatrix} 3 \\ 4 \end{bmatrix}$$

find a scalar r such that $\|r\mathbf{x}\| = 1$.

Solution We have

$$\|r\mathbf{x}\| = \left\| r\begin{bmatrix} 3 \\ 4 \end{bmatrix} \right\| = \left\| \begin{bmatrix} 3r \\ 4r \end{bmatrix} \right\| = \sqrt{(3r)^2 + (4r)^2}$$

$$= \sqrt{25r^2} = 5|r| \qquad \text{for all real numbers } r$$

We need r to satisfy the equation $5|r| = 1$ or $|r| = \frac{1}{5}$. There are two scalars, $r = \frac{1}{5}$ and $r = -\frac{1}{5}$, for which $\|r\mathbf{x}\| = 1$. ■

Geometrically, multiplying a vector by a scalar changes the length and/or direction of the vector, depending on the value of the scalar, as illustrated in Figure 2.6 (top of page 108) for vectors in R^2.

In particular, $(-1)\mathbf{x}$ has the same length as $\mathbf{x}$ but has the opposite direction from $\mathbf{x}$. We denote $(-1)\mathbf{x}$ as $-\mathbf{x}$, that is

$$(-1)\mathbf{x} = -\mathbf{x}.$$

The remaining property of absolute value, $|x + y| \leqq |x| + |y|$, is called the **triangular inequality**. To generalize the triangular inequality to vectors in R^n, it is necessary to define what is meant by the sum, $\mathbf{x} + \mathbf{y}$, of two vectors in R^n.

Definition 2.1.9 If

$$\mathbf{x} = \begin{bmatrix} x_1 \\ x_2 \\ \vdots \\ x_n \end{bmatrix} \quad \text{and} \quad \mathbf{y} = \begin{bmatrix} y_1 \\ y_2 \\ \vdots \\ y_n \end{bmatrix}$$

belong to R^n, then their **sum**, $\mathbf{x} + \mathbf{y}$, is defined to be the vector in R^n obtained by adding the corresponding entries:

$$\mathbf{x} + \mathbf{y} = \begin{bmatrix} x_1 \\ x_2 \\ \vdots \\ x_n \end{bmatrix} + \begin{bmatrix} y_1 \\ y_2 \\ \vdots \\ y_n \end{bmatrix} = \begin{bmatrix} x_1 + y_1 \\ x_2 + y_2 \\ \vdots \\ x_n + y_n \end{bmatrix}$$

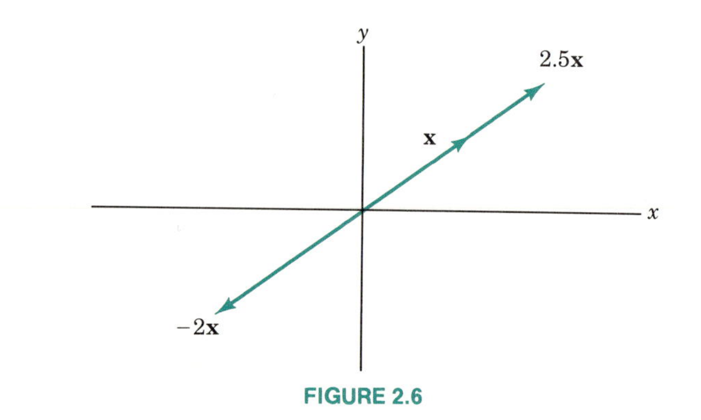

FIGURE 2.6

Note that for the sum of two vectors to be defined, they need to have the same dimension—i.e., belong to the same R^n. Again we note that this coincides with the definition of matrix addition.

Example 2.1.10

a. $$\begin{bmatrix} 7 \\ 9 \\ 3 \end{bmatrix} + \begin{bmatrix} 4 \\ 6 \\ 8 \end{bmatrix} = \begin{bmatrix} 7+4 \\ 9+6 \\ 3+8 \end{bmatrix} = \begin{bmatrix} 11 \\ 15 \\ 11 \end{bmatrix}$$

b. $\begin{bmatrix} 1 \\ 2 \end{bmatrix} + \begin{bmatrix} 7 \\ 9 \\ 3 \end{bmatrix}$ is not defined since $\begin{bmatrix} 1 \\ 2 \end{bmatrix} \in R^2$ and $\begin{bmatrix} 7 \\ 9 \\ 3 \end{bmatrix} \in R^3$

c. $$\begin{bmatrix} 2 \\ -1 \\ 5 \\ -3 \end{bmatrix} + \begin{bmatrix} -2 \\ 1 \\ -5 \\ 3 \end{bmatrix} = \begin{bmatrix} 2+(-2) \\ -1+(1) \\ 5+(-5) \\ -3+(3) \end{bmatrix} = \begin{bmatrix} 0 \\ 0 \\ 0 \\ 0 \end{bmatrix}$$

d. $$\begin{bmatrix} \frac{1}{2} \\ 0 \\ -3 \\ -\frac{4}{5} \\ 12 \end{bmatrix} + \begin{bmatrix} 0 \\ 0 \\ 0 \\ 0 \\ 0 \end{bmatrix} = \begin{bmatrix} \frac{1}{2}+0 \\ 0+0 \\ -3+0 \\ -\frac{4}{5}+0 \\ 12+0 \end{bmatrix} = \begin{bmatrix} \frac{1}{2} \\ 0 \\ -3 \\ -\frac{4}{5} \\ 12 \end{bmatrix}$$

e. $$-2\begin{bmatrix} -7 \\ 5 \\ 3 \\ -1 \end{bmatrix} + \frac{1}{2}\begin{bmatrix} 12 \\ -10 \\ 4 \\ -2 \end{bmatrix} = \begin{bmatrix} 14 \\ -10 \\ -6 \\ 2 \end{bmatrix} + \begin{bmatrix} 6 \\ -5 \\ 2 \\ -1 \end{bmatrix} = \begin{bmatrix} 20 \\ -15 \\ -4 \\ 1 \end{bmatrix}$$

f. $$6\begin{bmatrix} 3 \\ -2 \end{bmatrix} - 2\begin{bmatrix} -4 \\ 2 \end{bmatrix} = 6\begin{bmatrix} 3 \\ -2 \end{bmatrix} + (-2)\begin{bmatrix} -4 \\ 2 \end{bmatrix}$$
$$= \begin{bmatrix} 18 \\ -12 \end{bmatrix} + \begin{bmatrix} 8 \\ -4 \end{bmatrix} = \begin{bmatrix} 26 \\ -16 \end{bmatrix}$$ ■

Example 2.1.11 Given

$$\mathbf{x}_1 = \begin{bmatrix} 1 \\ 0 \\ -2 \end{bmatrix}, \quad \mathbf{x}_2 = \begin{bmatrix} 1 \\ 9 \\ 2 \end{bmatrix}, \quad \text{and} \quad \mathbf{x}_3 = \begin{bmatrix} -4 \\ 2 \\ 0 \end{bmatrix}$$

find $3\mathbf{x}_1 - 2\mathbf{x}_2 + (\frac{1}{2})\mathbf{x}_3$.

Solution

$$3\mathbf{x}_1 - 2\mathbf{x}_2 + (\tfrac{1}{2})\mathbf{x}_3 = 3\begin{bmatrix} 1 \\ 0 \\ -2 \end{bmatrix} - 2\begin{bmatrix} 1 \\ 9 \\ 2 \end{bmatrix} + \tfrac{1}{2}\begin{bmatrix} -4 \\ 2 \\ 0 \end{bmatrix}$$

$$= \begin{bmatrix} 3 \\ 0 \\ -6 \end{bmatrix} + \begin{bmatrix} -2 \\ -18 \\ -4 \end{bmatrix} + \begin{bmatrix} -2 \\ 1 \\ 0 \end{bmatrix}$$

$$= \begin{bmatrix} 3 + (-2) + (-2) \\ 0 + (-18) + 1 \\ -6 + (-4) + 0 \end{bmatrix} = \begin{bmatrix} -1 \\ -17 \\ -10 \end{bmatrix}$$

Example 2.1.12 Let $\mathbf{x}_1 = \begin{bmatrix} x_1 \\ y_1 \end{bmatrix}$, $\mathbf{x}_2 = \begin{bmatrix} x_2 \\ y_2 \end{bmatrix}$, $\mathbf{x}_3 = \begin{bmatrix} x_3 \\ y_3 \end{bmatrix}$ and $\mathbf{x}_4 = \begin{bmatrix} x_4 \\ y_4 \end{bmatrix}$. Then

$$a_1\mathbf{x}_1 + a_2\mathbf{x}_2 + a_3\mathbf{x}_3 + a_4\mathbf{x}_4 = a_1\begin{bmatrix} x_1 \\ y_1 \end{bmatrix} + a_2\begin{bmatrix} x_2 \\ y_2 \end{bmatrix} + a_3\begin{bmatrix} x_3 \\ y_3 \end{bmatrix} + a_4\begin{bmatrix} x_4 \\ y_4 \end{bmatrix}$$

$$= \begin{bmatrix} a_1x_1 \\ a_1y_1 \end{bmatrix} + \begin{bmatrix} a_2x_2 \\ a_2y_2 \end{bmatrix} + \begin{bmatrix} a_3x_3 \\ a_3y_3 \end{bmatrix} + \begin{bmatrix} a_4x_4 \\ a_4y_4 \end{bmatrix}$$

$$= \begin{bmatrix} a_1x_1 + a_2x_2 + a_3x_3 + a_4x_4 \\ a_1y_1 + a_2y_2 + a_3y_3 + a_4y_4 \end{bmatrix}$$

■

Example 2.1.13 Write

$$2\begin{bmatrix} x_1 \\ x_2 \\ x_3 \end{bmatrix} + 4\begin{bmatrix} y_1 \\ y_2 \\ y_3 \end{bmatrix} - 7\begin{bmatrix} z_1 \\ z_2 \\ z_3 \end{bmatrix} = \begin{bmatrix} 1 \\ 0 \\ 1 \end{bmatrix}$$

as a system of equations.

Solution

$$2\begin{bmatrix} x_1 \\ x_2 \\ x_3 \end{bmatrix} + 4\begin{bmatrix} y_1 \\ y_2 \\ y_3 \end{bmatrix} - 7\begin{bmatrix} z_1 \\ z_2 \\ z_3 \end{bmatrix} = \begin{bmatrix} 2x_1 \\ 2x_2 \\ 2x_3 \end{bmatrix} + \begin{bmatrix} 4y_1 \\ 4y_2 \\ 4y_3 \end{bmatrix} + \begin{bmatrix} -7z_1 \\ -7z_2 \\ -7z_3 \end{bmatrix}$$

$$= \begin{bmatrix} 2x_1 + 4y_1 - 7z_1 \\ 2x_2 + 4y_2 - 7z_2 \\ 2x_3 + 4y_3 - 7z_3 \end{bmatrix} = \begin{bmatrix} 1 \\ 0 \\ 1 \end{bmatrix}$$

Since corresponding entries are equal, we have the system

$$2x_1 + 4y_1 - 7z_1 = 1$$
$$2x_2 + 4y_2 - 7z_2 = 0$$
$$2x_3 + 4y_3 - 7z_3 = 1$$

Example 2.1.14 Let

$$\mathbf{x}_1 = \begin{bmatrix} 1 \\ -2 \end{bmatrix}, \quad \mathbf{x}_2 = \begin{bmatrix} 4 \\ 3 \end{bmatrix}, \quad \text{and} \quad \mathbf{x}_3 = \begin{bmatrix} -1 \\ 7 \end{bmatrix}$$

Write $a\mathbf{x}_1 + b\mathbf{x}_2 = \mathbf{x}_3$ as a system of equations. Solve the resulting system for a and b.

Solution

$$a\mathbf{x}_1 + b\mathbf{x}_2 = a\begin{bmatrix} 1 \\ -2 \end{bmatrix} + b\begin{bmatrix} 4 \\ 3 \end{bmatrix} = \begin{bmatrix} a \\ -2a \end{bmatrix} + \begin{bmatrix} 4b \\ 3b \end{bmatrix}$$
$$= \begin{bmatrix} a + 4b \\ -2a + 3b \end{bmatrix} = \begin{bmatrix} -1 \\ 7 \end{bmatrix}$$

Since corresponding entries are equal, we have

$$\begin{aligned} a + 4b &= -1 \\ -2a + 3b &= 7 \end{aligned} \quad \text{or} \quad \begin{bmatrix} 1 & 4 \\ -2 & 3 \end{bmatrix}\begin{bmatrix} a \\ b \end{bmatrix} = \begin{bmatrix} -1 \\ 7 \end{bmatrix}$$

The corresponding augmented matrix is

$$\left[\begin{array}{cc|c} 1 & 4 & -1 \\ -2 & 3 & 7 \end{array}\right]$$

Applying (2)R1 to R2 yields

$$\left[\begin{array}{cc|c} 1 & 4 & -1 \\ 0 & 11 & 5 \end{array}\right]$$

or

$$\begin{aligned} a + 4b &= -1 \\ 11b &= 5 \end{aligned}$$

Backsolving, the last equation yields $b = \frac{5}{11}$, so $a + 4b = a + 4(\frac{5}{11}) = a + \frac{20}{11} = -1$, or $a = -\frac{31}{11}$. Thus

$$-\tfrac{31}{11}\begin{bmatrix} 1 \\ -2 \end{bmatrix} + \tfrac{5}{11}\begin{bmatrix} 4 \\ 3 \end{bmatrix} = \begin{bmatrix} -1 \\ 7 \end{bmatrix}$$

Check:

$$-\tfrac{31}{11}\begin{bmatrix} 1 \\ -2 \end{bmatrix} + \tfrac{5}{11}\begin{bmatrix} 4 \\ 3 \end{bmatrix} = \begin{bmatrix} -\frac{31}{11} \\ \frac{62}{11} \end{bmatrix} + \begin{bmatrix} \frac{20}{11} \\ \frac{15}{11} \end{bmatrix} = \begin{bmatrix} -\frac{31}{11} + \frac{20}{11} \\ \frac{62}{11} + \frac{15}{11} \end{bmatrix}$$
$$= \begin{bmatrix} -\frac{11}{11} \\ \frac{77}{11} \end{bmatrix} = \begin{bmatrix} -1 \\ 7 \end{bmatrix}$$

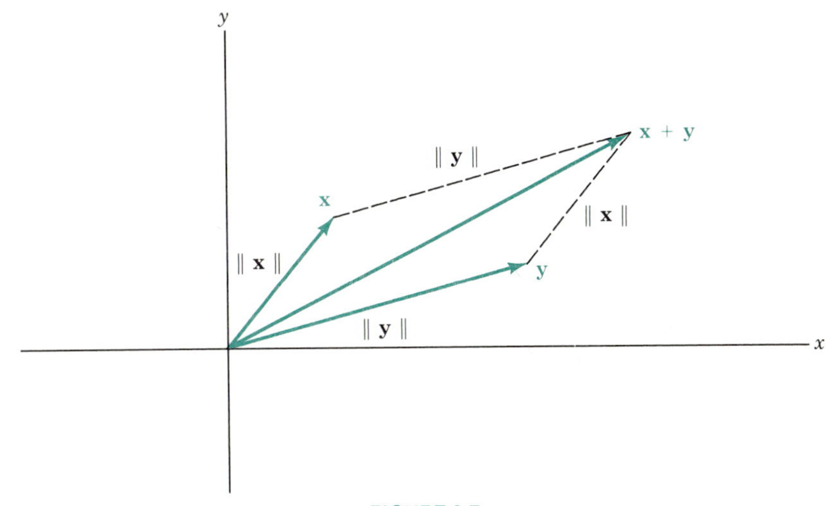

FIGURE 2.7

The sum of two vectors in R^2 and R^3 has a useful geometric interpretation. The diagram in R^2 (Figure 2.7) illustrates where the $\mathbf{x} + \mathbf{y}$ is located relative to $\mathbf{x}$ and $\mathbf{y}$. In fact, since the distance from $\mathbf{x}$ to $\mathbf{x} + \mathbf{y}$ is $\|\mathbf{y}\|$ and the distance from $\mathbf{y}$ to $\mathbf{x} + \mathbf{y}$ is $\|\mathbf{x}\|$, it follows that the four points $\mathbf{0}$, $\mathbf{x}$, $\mathbf{y}$, and $\mathbf{x} + \mathbf{y}$ are the vertices of a parallelogram, since if the opposite sides of a quadrilateral are equal, the quadrilateral is a parallelogram. The same is true in R^3, as is illustrated in Figure 2.8, because the two vectors $\mathbf{x}$ and $\mathbf{y}$ in R^3 determine a plane and the four points, $\mathbf{0}$, $\mathbf{x}$, $\mathbf{y}$, and $\mathbf{x} + \mathbf{y}$ are on this plane. This geometric interpretation of vectors in R^2 and R^3 explains why vector addition in R^2 is said to follow the **parallelogram law**.

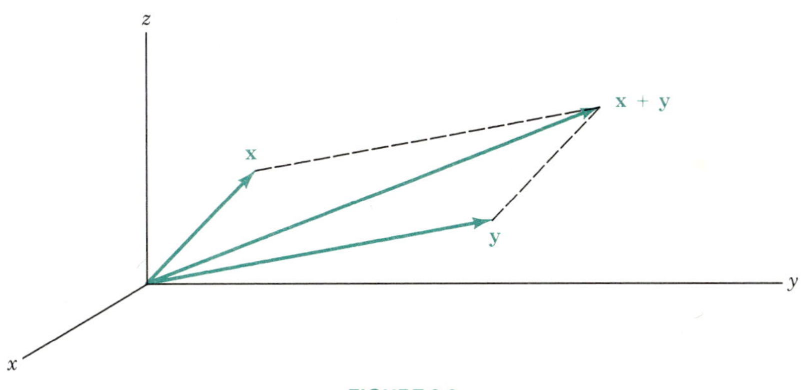

FIGURE 2.8

From Figure 2.9 (on top of page 112) in R^2 and the parallelogram law, it follows that $\|\mathbf{x} + \mathbf{y}\| \leq \|\mathbf{x}\| + \|\mathbf{y}\|$ because the length of one side of a triangle cannot be greater than the sum of lengths of the other two sides. This is a geometric proof of the triangular inequality. In the next section, the triangular inequality is shown to hold in R^n, for any n.

In real number addition, $x - y$ is that number which, when added to y, yields x. In R^n, we define $\mathbf{x} - \mathbf{y}$ similarly.

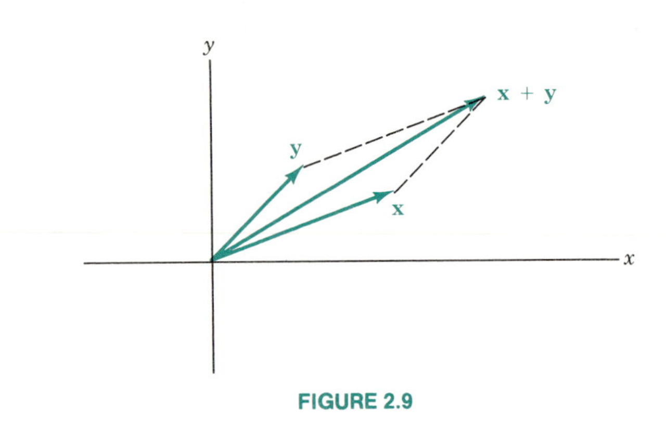

FIGURE 2.9

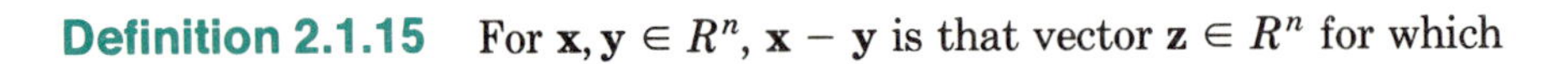

Definition 2.1.15 For $\mathbf{x}, \mathbf{y} \in R^n$, $\mathbf{x} - \mathbf{y}$ is that vector $\mathbf{z} \in R^n$ for which

$$\mathbf{z} + \mathbf{y} = \mathbf{y} + \mathbf{z} = \mathbf{x}.$$

To illustrate this geometrically in R^2, let $\mathbf{z} = \mathbf{x} - \mathbf{y}$. By the parallelogram law of vector addition, $\mathbf{x}$ is the diagonal of a parallelogram with sides $\mathbf{y}$ and $\mathbf{z}$. We wish to determine $\mathbf{z}$. Recalling our geometry, we see that the line segment from the head of $\mathbf{y}$ to the head of $\mathbf{x}$ has the same length as $\mathbf{z} = \mathbf{x} - \mathbf{y}$ and is parallel to the line through $\mathbf{0}$ determined by $\mathbf{z}$. See Figure 2.10. (Figure 2.10 illustrates why some texts interpret $\mathbf{x} - \mathbf{y}$ geometrically as the line segment from the head of $\mathbf{y}$ to the head of $\mathbf{x}$.)

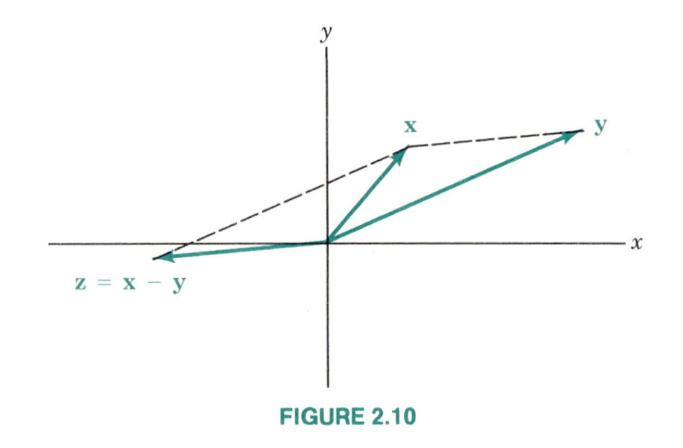

FIGURE 2.10

Theorem 2.1.16 $\mathbf{x} - \mathbf{y} = \mathbf{x} + (-\mathbf{y})$

Proof. First $\mathbf{x} = (\mathbf{x} - \mathbf{y}) + \mathbf{y}$ by definition. But

$$\begin{aligned}\mathbf{x} &= \mathbf{x} + \mathbf{0}\\ &= \mathbf{x} + [(-\mathbf{y}) + \mathbf{y}]\\ &= [\mathbf{x} + (-\mathbf{y})] + \mathbf{y}\end{aligned}$$

so

$$(\mathbf{x} - \mathbf{y}) + \mathbf{y} = [\mathbf{x} + (-\mathbf{y})] + \mathbf{y} = \mathbf{x}$$

or

$$\mathbf{x} - \mathbf{y} = \mathbf{x} + (-\mathbf{y})$$

Corollary
If

$$\mathbf{x} = \begin{bmatrix} x_1 \\ \vdots \\ x_n \end{bmatrix}$$

and

$$\mathbf{y} = \begin{bmatrix} y_1 \\ \vdots \\ y_n \end{bmatrix}$$

then

$$\mathbf{x} - \mathbf{y} = \begin{bmatrix} x_1 - y_1 \\ x_2 - y_2 \\ \vdots \\ x_n - y_n \end{bmatrix}$$

In establishing the three basic properties of length in R^n, it was useful to define the sum $\mathbf{x} + \mathbf{y}$ and the scalar multiple $r\mathbf{x}$. For future reference, the basic rules that govern these operations on vectors are given in Theorem 2.1.17. These statements follow from the fact that they are all properties of matrices.

Theorem 2.1.17 *The vectors in R^n satisfy the following properties:*

A1. For $\mathbf{x}_1, \mathbf{x}_2 \in R^n$, $\mathbf{x}_1 + \mathbf{x}_2 = \mathbf{x}_2 + \mathbf{x}_1$.

A2. For $\mathbf{x}_1, \mathbf{x}_2, \mathbf{x}_3 \in R^n$, $(\mathbf{x}_1 + \mathbf{x}_2) + \mathbf{x}_3 = \mathbf{x}_1 + (\mathbf{x}_2 + \mathbf{x}_3)$.

A3. The vector

$$\mathbf{0} = \begin{bmatrix} 0 \\ 0 \\ \vdots \\ 0 \end{bmatrix} \in R^n$$

satisfies $\mathbf{x} + \mathbf{0} = \mathbf{0} + \mathbf{x} = \mathbf{x}$ for all $\mathbf{x} \in R^n$.

A4. For each

$$\mathbf{x} = \begin{bmatrix} x_1 \\ x_2 \\ \vdots \\ x_n \end{bmatrix} \in R^n$$

the vector

$$-\mathbf{x} = \begin{bmatrix} -x_1 \\ -x_2 \\ \vdots \\ -x_n \end{bmatrix} \in R^n$$

satisfies

$$\mathbf{x} + (-\mathbf{x}) = (-\mathbf{x}) + \mathbf{x} = \mathbf{0}$$

M1. For all $a_1, a_2 \in R$ *and all* $\mathbf{x} \in R^n$,

$$(a_1 a_2)\mathbf{x} = a_1(a_2\mathbf{x})$$

M2. For all $a_1, a_2 \in R$ *and all* $\mathbf{x} \in R^n$,

$$(a_1 + a_2)\mathbf{x} = a_1\mathbf{x} + a_2\mathbf{x}$$

M3. For all $\mathbf{x} \in R^n$,

$$1\mathbf{x} = \mathbf{x}$$

M4. For all $a \in R$ *and* $\mathbf{x}_1, \mathbf{x}_2 \in R^n$

$$a(\mathbf{x}_1 + \mathbf{x}_2) = a\mathbf{x}_1 + a\mathbf{x}_2$$

We verify M1 and leave the verification of the remaining parts as exercises.

Proof. M1. Let

$$\mathbf{x} = \begin{bmatrix} x_1 \\ x_2 \\ \vdots \\ x_n \end{bmatrix}$$

Then

$$(a_1a_2)\mathbf{x} = \begin{bmatrix} (a_1a_2)x_1 \\ (a_1a_2)x_2 \\ \vdots \\ (a_1a_2)x_n \end{bmatrix}$$

But for all $a_1, a_2, y \in R$, $(a_1a_2)y = a_1(a_2y)$. Therefore,

$$\begin{bmatrix} (a_1a_2)x_1 \\ (a_1a_2)x_2 \\ \vdots \\ (a_1a_2)x_n \end{bmatrix} = \begin{bmatrix} a_1(a_2x_1) \\ a_1(a_2x_2) \\ \vdots \\ a_1(a_2x_n) \end{bmatrix} = a_1 \begin{bmatrix} a_2x_1 \\ a_2x_2 \\ \vdots \\ a_2x_n \end{bmatrix} = a_1(a_2\mathbf{x})$$

Hence $(a_1a_2)\mathbf{x} = a_1(a_2\mathbf{x})$, as we wished to show.

SUMMARY

Terms to know:

Row vector
Column vector
Coordinates
Components
Vector
R^n
n-space
Distance between two vectors
Length of a vector
Scalar multiple
Triangular inequality
Sum of vectors
Parallelogram law

Theorems to know:

1. For $\mathbf{x}_1, \mathbf{x}_2, \mathbf{x}_3 \in R^n$:
   A1. $\mathbf{x}_1 + \mathbf{x}_2 = \mathbf{x}_2 + \mathbf{x}_1$
   A2. $(\mathbf{x}_1 + \mathbf{x}_2) + \mathbf{x}_3 = \mathbf{x}_1 + (\mathbf{x}_2 + \mathbf{x}_3)$
   A3. $\mathbf{x} + \mathbf{0} = \mathbf{0} + \mathbf{x} = \mathbf{0}$
2. For $a_1, a_2 \in R$ and $\mathbf{x} \in R^n$:
   M1. $(a_1a_2)\mathbf{x} = a_1(a_2\mathbf{x})$
   M2. $(a_1 + a_2)\mathbf{x} = a_1\mathbf{x} + a_2\mathbf{x}$
   M3. $1\mathbf{x} = \mathbf{x}$
   M4. $a(\mathbf{x}_1 + \mathbf{x}_2) = a\mathbf{x}_1 + a\mathbf{x}_2$

EXERCISE SET 2.1

Perform the indicated calculations.

1. $\begin{bmatrix} 7 \\ 9 \\ 3 \end{bmatrix} + \begin{bmatrix} 2 \\ 4 \\ 8 \end{bmatrix}$ 2. $9\begin{bmatrix} 1 \\ 3 \end{bmatrix}$ 3. $3\begin{bmatrix} 7 \\ 9 \\ 6 \end{bmatrix} + 2\begin{bmatrix} 0 \\ 7 \\ 0 \end{bmatrix}$ 4. $8\begin{bmatrix} 2 \\ 1 \\ 4 \end{bmatrix} + \begin{bmatrix} 3 \\ 4 \\ 7 \end{bmatrix}$

5. $0\begin{bmatrix} 1 \\ -1 \\ 0 \\ 1 \end{bmatrix} + (-2)\begin{bmatrix} 2 \\ -1 \\ 0 \\ -1 \end{bmatrix}$ 6. $2\begin{bmatrix} 3 \\ 1 \\ 0 \\ 2 \end{bmatrix} + (-1)\begin{bmatrix} 5 \\ -1 \\ -2 \\ 3 \end{bmatrix} + 6\begin{bmatrix} 0 \\ 1 \\ 0 \\ 2 \end{bmatrix}$

7. Let $\mathbf{x}_1 = \begin{bmatrix} x_1 \\ y_1 \\ z_1 \end{bmatrix}$, $\mathbf{x}_2 = \begin{bmatrix} x_2 \\ y_2 \\ z_2 \end{bmatrix}$, and $\mathbf{x}_3 = \begin{bmatrix} x_3 \\ y_3 \\ z_3 \end{bmatrix}$. Find $a_1\mathbf{x}_1 + a_2\mathbf{x}_2 + a_3\mathbf{x}_3$.

8. Write $\begin{bmatrix} x_1 \\ x_2 \\ x_3 \end{bmatrix} - \begin{bmatrix} y_1 \\ y_2 \\ y_3 \end{bmatrix} + 2\begin{bmatrix} z_1 \\ z_2 \\ z_3 \end{bmatrix} = \begin{bmatrix} -1 \\ 1 \\ 0 \end{bmatrix}$ as a system of equations.

9. Let

$$\mathbf{x}_1 = \begin{bmatrix} 1 \\ 0 \\ -1 \end{bmatrix}, \quad \mathbf{x}_2 = \begin{bmatrix} 2 \\ -1 \\ -1 \end{bmatrix}, \quad \mathbf{x}_3 = \begin{bmatrix} 3 \\ -2 \\ 2 \end{bmatrix}, \quad \text{and} \quad \mathbf{x}_4 = \begin{bmatrix} 0 \\ 1 \\ 1 \end{bmatrix}$$

Write $a\mathbf{x}_1 + b\mathbf{x}_2 + c\mathbf{x}_3 = \mathbf{x}_4$ as a system of equations and solve this system for a, b, and c. Check your answer.

10. Let

$$\mathbf{x}_1 = \begin{bmatrix} 1 \\ 0 \\ 0 \\ 1 \end{bmatrix}, \quad \mathbf{x}_2 = \begin{bmatrix} 0 \\ 1 \\ 0 \\ 0 \end{bmatrix}, \quad \mathbf{x}_3 = \begin{bmatrix} 2 \\ 0 \\ 0 \\ 0 \end{bmatrix}, \quad \mathbf{x}_4 = \begin{bmatrix} 1 \\ 0 \\ 3 \\ 0 \end{bmatrix}, \quad \text{and} \quad \mathbf{x}_5 = \begin{bmatrix} 0 \\ 0 \\ 0 \\ 0 \end{bmatrix}$$

Write $a_1\mathbf{x}_1 + a_2\mathbf{x}_2 + a_3\mathbf{x}_3 + a_4\mathbf{x}_4 = \mathbf{x}_5$ as a system of equations and solve this system. Check your answer.

For Exercises 11–19, let

$$\mathbf{x}_1 = \begin{bmatrix} x_1 \\ y_1 \\ z_1 \end{bmatrix}, \quad \mathbf{x}_2 = \begin{bmatrix} x_2 \\ y_2 \\ z_2 \end{bmatrix}, \quad \text{and} \quad \mathbf{x}_3 = \begin{bmatrix} x_3 \\ y_3 \\ z_3 \end{bmatrix}$$

11. Verify that $\mathbf{x}_1 + (\mathbf{x}_2 + \mathbf{x}_3) = (\mathbf{x}_1 + \mathbf{x}_2) + \mathbf{x}_3$. This says *vector addition is associative.*

12. Let

$$\mathbf{0} = \begin{bmatrix} 0 \\ 0 \\ 0 \end{bmatrix}$$

$\mathbf{0}$ is the zero vector for R^3. Verify that $\mathbf{x}_1 + \mathbf{0} = \mathbf{x}_1$.

13. Verify that there exists a vector $\mathbf{v}$ in R^3 such that $\mathbf{x}_1 + \mathbf{v} = \mathbf{0}$.
14. Verify that $\mathbf{x}_1 + \mathbf{x}_2 = \mathbf{x}_2 + \mathbf{x}_1$. This says *vector addition is commutative.*
15. Verify that for any scalar b, $b(\mathbf{x}_1 + \mathbf{x}_2) = b\mathbf{x}_1 + b\mathbf{x}_2$.
16. Verify that for scalars b_1 and b_2, $(b_1 + b_2)\mathbf{x}_1 = b_1\mathbf{x}_1 + b_2\mathbf{x}_1$.
17. Verify that for the scalar 1 that $1\mathbf{x}_1 = \mathbf{x}_1$.
18. Verify $0\mathbf{x}_1 = \mathbf{0}$.
19. Verify $\mathbf{x}_1 - \mathbf{x}_2 = \mathbf{x}_1 + (-1)\mathbf{x}_2$.
20. Let

$$\mathbf{x}_1 = \begin{bmatrix} 7 \\ 4 \\ -3 \\ 2 \end{bmatrix} \quad \text{and} \quad \mathbf{x}_2 = \begin{bmatrix} -4 \\ 3 \\ -2 \\ 1 \end{bmatrix}$$

Find $\mathbf{x}_1 - \mathbf{x}_2$ and $\mathbf{x}_2 - \mathbf{x}_1$. Does $\mathbf{x}_1 - \mathbf{x}_2 = \mathbf{x}_2 - \mathbf{x}_1$?

21. Give a geometric interpretation to the inequality

$$\|\mathbf{x} - \mathbf{y}\| \leqq \|\mathbf{x} - \mathbf{z}\| + \|\mathbf{z} - \mathbf{y}\|$$

in R^2. Use the triangular inequality to prove this inequality in R^n.

Hint: $\mathbf{x} - \mathbf{y} = (\mathbf{x} - \mathbf{z}) + (\mathbf{z} - \mathbf{y})$.

22. If $\|\mathbf{x}\| = 0$, show that $\mathbf{x} = \mathbf{0}$.

SECTION 2.2 SCALAR PRODUCT

In this section we discuss one method by which we can multiply two vectors in R^n.

Definition 2.2.1 If

$$\mathbf{x} = \begin{bmatrix} x_1 \\ x_2 \\ \vdots \\ x_n \end{bmatrix} \quad \text{and} \quad \mathbf{y} = \begin{bmatrix} y_1 \\ y_2 \\ \vdots \\ y_n \end{bmatrix}$$

are two vectors in R^n, their **scalar product** is defined to be the number

$$\mathbf{x} \cdot \mathbf{y} = x_1y_1 + x_2y_2 + \cdots + x_ny_n$$

The scalar product $\mathbf{x} \cdot \mathbf{y}$ of two vectors $\mathbf{x}$ and $\mathbf{y}$ is also called the **dot product** or **inner product** of $\mathbf{x}$ and $\mathbf{y}$.

Notice that the inner product is the product

$$[x_1 \quad x_2 \quad \cdots \quad x_n]\begin{bmatrix} y_1 \\ y_2 \\ \vdots \\ y_n \end{bmatrix}$$

of the matrices x^T and y. Note also that $\sqrt{\mathbf{x} \cdot \mathbf{x}} = \|\mathbf{x}\| \geq 0$.

Example 2.2.2 a. Let

$$\mathbf{x} = \begin{bmatrix} 1 \\ 4 \\ 7 \end{bmatrix} \quad \text{and} \quad \mathbf{y} = \begin{bmatrix} 3 \\ -1 \\ 2 \end{bmatrix}$$

then $\mathbf{x} \cdot \mathbf{y} = (1)(3) + (4)(-1) + (7)(2) = 13$

b. If

$$\mathbf{x} = \begin{bmatrix} 1 \\ 2 \\ -1 \\ 1 \end{bmatrix} \quad \text{and} \quad \mathbf{y} = \begin{bmatrix} 1 \\ 0 \\ 3 \\ -6 \end{bmatrix}$$

then $\mathbf{x} \cdot \mathbf{y} = (1)(1) + (2)(0) + (-1)(3) + (1)(-6) = -8$ ■

The next theorem shows how the scalar product behaves with respect to vector addition and scalar multiplication.

Theorem 2.2.3 *The scalar product satisfies the following properties:*

1. *For any* $\mathbf{x}, \mathbf{y} \in R^n$, $\mathbf{x} \cdot \mathbf{y} = \mathbf{y} \cdot \mathbf{x}$.
2. *For any* $\mathbf{x}, \mathbf{y}, \mathbf{z} \in R^n$, $(\mathbf{x} + \mathbf{y}) \cdot \mathbf{z} = \mathbf{x} \cdot \mathbf{z} + \mathbf{y} \cdot \mathbf{z}$.
3. *For any* $\mathbf{x}, \mathbf{y} \in R^n$, $a \in R$, $(a\mathbf{x}) \cdot \mathbf{y} = a(\mathbf{x} \cdot \mathbf{y})$.
4. *For any* $\mathbf{x} \in R^n$, $\mathbf{x} \cdot \mathbf{x} \geqq 0$.
5. $\mathbf{x} \cdot \mathbf{x} = 0$ *if, and only if,* $\mathbf{x} = \mathbf{0}$.

We prove (1) and (3). You are asked to prove the remaining properties in the exercise set.

Proof. We begin with (1). Let

$$\mathbf{x} = \begin{bmatrix} x_1 \\ x_2 \\ \vdots \\ x_n \end{bmatrix} \quad \text{and} \quad \mathbf{y} = \begin{bmatrix} y_1 \\ y_2 \\ \vdots \\ y_n \end{bmatrix}$$

Then

$$\mathbf{x} \cdot \mathbf{y} = x_1y_1 + x_2y_2 + \cdots + x_ny_n = y_1x_1 + y_2x_2 + \cdots + y_nx_n = \mathbf{y} \cdot \mathbf{x}$$

since $ab = ba$ for all $a, b \in R$.

We next do (3). Let

$$\mathbf{x} = \begin{bmatrix} x_1 \\ x_2 \\ \vdots \\ x_n \end{bmatrix} \quad \text{and} \quad \mathbf{y} = \begin{bmatrix} y_1 \\ y_2 \\ \vdots \\ y_n \end{bmatrix}$$

Then

$$a\mathbf{x} = \begin{bmatrix} ax_1 \\ ax_2 \\ \vdots \\ ax_n \end{bmatrix}$$

and

$$\begin{aligned} (a\mathbf{x}) \cdot \mathbf{y} &= (ax_1)y_1 + (ax_2)y_2 + \cdots + (ax_n)y_n \\ &= a(x_1y_1) + a(x_2y_2) + \cdots + a(x_ny_n) \\ &= a(x_1y_1 + x_2y_2 + \cdots + x_ny_n) \\ &= a(\mathbf{x} \cdot \mathbf{y}) \end{aligned}$$

Figure 2.11 together with the law of cosines from trigonometry illustrate the following relation in R^2:

$$\|\mathbf{x} - \mathbf{y}\|^2 = \|\mathbf{x}\|^2 + \|\mathbf{y}\|^2 - 2\|\mathbf{x}\|\,\|\mathbf{y}\|\cos\theta$$

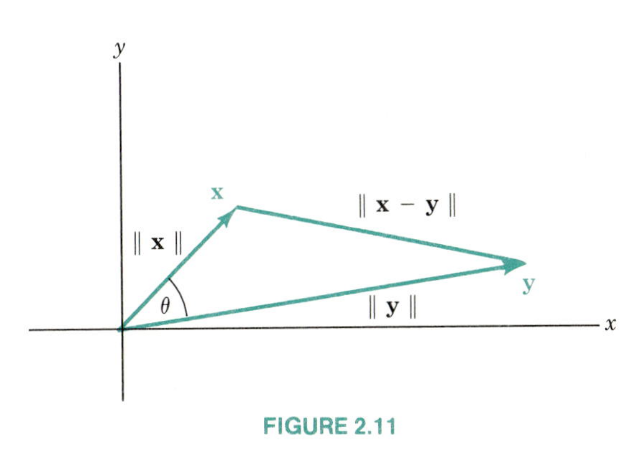

FIGURE 2.11

where θ is the angle between the two vectors

$$\mathbf{x} = \begin{bmatrix} x_1 \\ x_2 \end{bmatrix} \quad \text{and} \quad \mathbf{y} = \begin{bmatrix} y_1 \\ y_2 \end{bmatrix}$$

Necessarily, $0 \le \theta \le \pi$. On the other hand

$$\begin{aligned}
\|\mathbf{x} - \mathbf{y}\|^2 = \left\| \begin{bmatrix} x_1 - y_1 \\ x_2 - y_2 \end{bmatrix} \right\|^2 &= (x_1 - y_1)^2 + (x_2 - y_2)^2 \\
&= x_1^2 - 2x_1y_1 + y_1^2 + x_2^2 - 2x_2y_2 + y_2^2 \\
&= \left(x_1^2 + x_2^2\right) - 2(x_1y_1 + x_2y_2) + \left(y_1^2 + y_2^2\right) \\
&= \|\mathbf{x}\|^2 - 2(x_1y_1 + x_2y_2) + \|\mathbf{y}\|^2
\end{aligned}$$

It then follows that

$$x_1y_1 + x_2y_2 = \|\mathbf{x}\|\,\|\mathbf{y}\|\cos\theta$$

or

$$\mathbf{x} \cdot \mathbf{y} = \|\mathbf{x}\|\,\|\mathbf{y}\|\cos\theta$$

It is this formula from R^2 that suggests how to define the notion of an angle between two vectors in R^n.

Definition 2.2.4 If

$$\mathbf{x} = \begin{bmatrix} x_1 \\ x_2 \\ \vdots \\ x_n \end{bmatrix} \quad \text{and} \quad \mathbf{y} = \begin{bmatrix} y_1 \\ y_2 \\ \vdots \\ y_n \end{bmatrix}$$

are two vectors in R^n, neither of which is equal to $\mathbf{0}$, then the **angle θ between x and y** is defined to be that angle $0 \le \theta \le \pi$ for which

$$\cos\theta = \frac{\mathbf{x} \cdot \mathbf{y}}{\|\mathbf{x}\|\,\|\mathbf{y}\|}$$

Example 2.2.5 If $\mathbf{x} = \begin{bmatrix} 1 \\ 2 \end{bmatrix}$ and $\mathbf{y} = \begin{bmatrix} -1 \\ 3 \end{bmatrix}$, then

$$\cos\theta = \frac{\mathbf{x} \cdot \mathbf{y}}{\|\mathbf{x}\|\,\|\mathbf{y}\|} = \frac{(1)(-1) + (2)(3)}{\sqrt{1^2 + 2^2}\sqrt{(-1)^2 + 3^2}} = \frac{5}{\sqrt{5}\sqrt{10}} = \frac{\sqrt{2}}{2}$$

and so $\theta = \pi/4$. ■

In R^2, since $|\cos\theta| \leqq 1$, then

$$|\cos\theta| = \frac{|\mathbf{x}\cdot\mathbf{y}|}{\|\mathbf{x}\|\,\|\mathbf{y}\|} \leqq 1$$

or

$$|\mathbf{x}\cdot\mathbf{y}| \leqq \|\mathbf{x}\|\,\|\mathbf{y}\|$$

This inequality is called the **Cauchy–Schwartz inequality**.

In R^n, $n \geq 3$, we verify the Cauchy–Schwartz inequality as follows: First, if $\mathbf{x} = \mathbf{0}$, then both sides of the inequality are 0 and the inequality holds.

If $\mathbf{x} \neq \mathbf{0}$, then

$$0 \leq (t\mathbf{x} + \mathbf{y})\cdot(t\mathbf{x} + \mathbf{y}) = t^2(\mathbf{x}\cdot\mathbf{x}) + 2t(\mathbf{x}\cdot\mathbf{y}) + \mathbf{y}\cdot\mathbf{y}$$

Thus the quadratic equation $t^2(\mathbf{x}\cdot\mathbf{x}) + 2t(\mathbf{x}\cdot\mathbf{y}) + (\mathbf{y}\cdot\mathbf{y}) = 0$ either has no real roots or has a repeated root. Thus its discriminant must satisfy the following sequence of inequalities.

$$[2(\mathbf{x}\cdot\mathbf{y})]^2 - 4(\mathbf{x}\cdot\mathbf{x})(\mathbf{y}\cdot\mathbf{y}) \leq 0$$

$$4(\mathbf{x}\cdot\mathbf{y})^2 \leq 4(\mathbf{x}\cdot\mathbf{x})(\mathbf{y}\cdot\mathbf{y})$$

$$(\mathbf{x}\cdot\mathbf{y})^2 \leq (\mathbf{x}\cdot\mathbf{x})(\mathbf{y}\cdot\mathbf{y})$$

Finally, taking square roots of both sides, we have proved the following theorem.

Theorem 2.2.6 *In R^n, $n \geq 3$, for any two vectors $\mathbf{x}$ and $\mathbf{y}$,*

$$(\mathbf{x}\cdot\mathbf{y})^2 \leq (\mathbf{x}\cdot\mathbf{x})(\mathbf{y}\cdot\mathbf{y}), \quad \text{or} \quad |\mathbf{x}\cdot\mathbf{y}| \leq \|\mathbf{x}\|\,\|\mathbf{y}\|$$

Theorem 2.2.6 tells us that for any pair of nonzero vectors in R^n,

$$0 \leq \frac{(\mathbf{x}\cdot\mathbf{y})^2}{(\mathbf{x}\cdot\mathbf{x})(\mathbf{y}\cdot\mathbf{y})} \leq 1$$

Thus, upon taking square roots, we have

$$-1 \leq \frac{(\mathbf{x}\cdot\mathbf{y})}{\|\mathbf{x}\|\,\|\mathbf{y}\|} \leq 1$$

Hence, there is a unique angle θ between 0 and π for which

$$\cos\theta = \frac{(\mathbf{x}\cdot\mathbf{y})}{\|\mathbf{x}\|\,\|\mathbf{y}\|}$$

Having established the Cauchy–Schwartz inequality, we have the necessary tools to verify the **triangular inequality**.

$$\|\mathbf{x} + \mathbf{y}\| \leqq \|\mathbf{x}\| + \|\mathbf{y}\|$$

The strategy is to examine $\|\mathbf{x} + \mathbf{y}\|^2$ and then take square roots:

$$\begin{aligned}
\|\mathbf{x} + \mathbf{y}\|^2 &= (\mathbf{x} + \mathbf{y}) \cdot (\mathbf{x} + \mathbf{y}) \\
&= \mathbf{x} \cdot \mathbf{x} + \mathbf{x} \cdot \mathbf{y} + \mathbf{y} \cdot \mathbf{x} + \mathbf{y} \cdot \mathbf{y} \\
&= \|\mathbf{x}\|^2 + 2\mathbf{x} \cdot \mathbf{y} + \|\mathbf{y}\|^2 \\
&\leqq \|\mathbf{x}\|^2 + 2\|\mathbf{x}\|\,\|\mathbf{y}\| + \|\mathbf{y}\|^2 \quad \text{[Cauchy–Schwartz inequality]} \\
&= (\|\mathbf{x}\| + \|\mathbf{y}\|)^2
\end{aligned}$$

and so

$$\|\mathbf{x} + \mathbf{y}\| \leqq \|\mathbf{x}\| + \|\mathbf{y}\|$$

The relation

$$\mathbf{x} \cdot \mathbf{y} = \|\mathbf{x}\|\,\|\mathbf{y}\| \cos \theta$$

also gives us a simple test for deciding if two vectors are perpendicular. In order for two vectors $\mathbf{x}$ and $\mathbf{y}$ to be perpendicular, the cosine of the angle between them should be equal to 0 because the angle between them ought to be an odd integer multiple of $\pi/2$. This motivates the following definition, which uses the word *orthogonal* instead of the word *perpendicular*.

Definition 2.2.7 Two vectors $\mathbf{x}$ and $\mathbf{y}$ in R^n are **orthogonal** (i.e., perpendicular) if $\mathbf{x} \cdot \mathbf{y} = 0$.

Example 2.2.8 Since

$$\begin{bmatrix} 1 \\ 0 \\ -1 \\ 1 \end{bmatrix} \cdot \begin{bmatrix} 1 \\ 1 \\ 1 \\ 0 \end{bmatrix} = (1)(1) + (0)(1) + (-1)(1) + (1)(0) = 0$$

the two vectors

$$\begin{bmatrix} 1 \\ 0 \\ -1 \\ 1 \end{bmatrix} \quad \text{and} \quad \begin{bmatrix} 1 \\ 1 \\ 1 \\ 0 \end{bmatrix}$$

in R^4 are orthogonal. ■

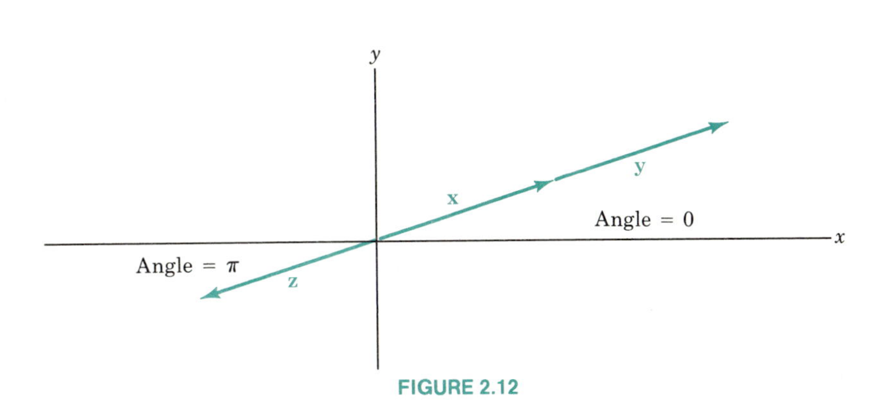

FIGURE 2.12

In R^2 and R^3, two vectors **x** and **y** are parallel if the cosine of the angle between them is equal to either 1 or -1, since the angle between them is either 0 or π (see Figure 2.12). This means that **x** and **y** ought to be parallel if

$$\mathbf{x} \cdot \mathbf{y} = \|\mathbf{x}\|\,\|\mathbf{y}\| \quad \text{or} \quad \mathbf{x} \cdot \mathbf{y} = -\|\mathbf{x}\|\,\|\mathbf{y}\|$$

We leave it as an exercise to show that

$$|\mathbf{x} \cdot \mathbf{y}| = \|\mathbf{x}\|\,\|\mathbf{y}\|$$

if and only if $\mathbf{x} = \lambda\mathbf{y}$ for some nonzero λ.

Definition 2.2.9 Two nonzero vectors **x** and **y** in R^n are defined to be **parallel** if there is a number λ such that $\mathbf{x} = \lambda\mathbf{y}$ (or $\mathbf{y} = \lambda\mathbf{x}$). If $\lambda > 0$ and $\mathbf{x} = \lambda\mathbf{y}$, **x** and **y** are said to have the **same direction**; if $\lambda < 0$ and $\mathbf{x} = \lambda\mathbf{y}$, **x** and **y** are said to have **opposite directions.**

Example 2.2.10 The vectors

$$\begin{bmatrix} 2 \\ -1 \\ 1 \end{bmatrix} \quad \text{and} \quad \begin{bmatrix} -2 \\ 1 \\ -1 \end{bmatrix}$$

are parallel but have opposite directions since

$$\begin{bmatrix} 2 \\ -1 \\ 1 \end{bmatrix} = -1 \begin{bmatrix} 2 \\ 1 \\ -1 \end{bmatrix}$$

The vectors

$$\begin{bmatrix} 1 \\ 2 \\ -1 \\ 3 \end{bmatrix} \quad \text{and} \quad \begin{bmatrix} 4 \\ 8 \\ -4 \\ 12 \end{bmatrix}$$

are parallel and have the same direction since

$$\begin{bmatrix} 4 \\ 8 \\ -4 \\ 12 \end{bmatrix} = 4 \begin{bmatrix} 1 \\ 2 \\ -1 \\ 3 \end{bmatrix}$$

■

Example 2.2.11 Let

$$\mathbf{x} = \begin{bmatrix} 1 \\ 3 \\ 1 \end{bmatrix}$$

Find a vector $\mathbf{y}$ in R^3 orthogonal to $\mathbf{x}$.

Solution Let

$$\mathbf{y} = \begin{bmatrix} y_1 \\ y_2 \\ y_3 \end{bmatrix}$$

Then for $\mathbf{x}$ and $\mathbf{y}$ to be orthogonal,

$$\mathbf{x} \cdot \mathbf{y} = (1)(y_1) + (3)(y_2) + (1)(y_3) = y_1 + 3y_2 + y_3 = 0$$

Hence, $y_1 = -3y_2 - y_3$, and so

$$\mathbf{y} = \begin{bmatrix} -3y_2 - y_3 \\ y_2 \\ y_3 \end{bmatrix} = y_2 \begin{bmatrix} -3 \\ 1 \\ 0 \end{bmatrix} + y_3 \begin{bmatrix} -1 \\ 0 \\ 1 \end{bmatrix}$$

for y_2 and y_3 arbitrarily chosen.

Choosing $y_2 = 1$ and $y_3 = 0$, we obtain

$$\mathbf{y} = \begin{bmatrix} -3 \\ 1 \\ 0 \end{bmatrix}$$

Of course, any value for y_2 and another value for y_3 also yields a vector orthogonal to $\mathbf{x}$. ■

SUMMARY

Terms to know:

Scalar product, dot product, inner product

Cosine of the angle between vectors

Cauchy–Schwartz inequality

Orthogonal vectors
Parallel vectors
Triangular inequality

Theorems to know:

For any $\mathbf{x}, \mathbf{y}, \mathbf{z} \in R^n$ and $a \in R$, we have

1. $\mathbf{x} \cdot \mathbf{y} = \mathbf{y} \cdot \mathbf{x}$
2. $(\mathbf{x} + \mathbf{y}) \cdot \mathbf{z} = \mathbf{x} \cdot \mathbf{z} + \mathbf{y} \cdot \mathbf{z}$
3. $(a\mathbf{x}) \cdot \mathbf{y} = a(\mathbf{x} \cdot \mathbf{y})$
4. $\mathbf{x} \cdot \mathbf{x} \geq 0$
5. $\mathbf{x} \cdot \mathbf{x} = 0$ if and only if $\mathbf{x} = \mathbf{0}$
6. $|\mathbf{x} \cdot \mathbf{y}| < \|\mathbf{x}\| \, \|\mathbf{y}\|$

EXERCISE SET 2.2

1. Find the lengths of the following vectors.

a. $\begin{bmatrix} 1 \\ 1 \end{bmatrix}$ b. $\begin{bmatrix} 1 \\ -1 \\ \frac{1}{2} \end{bmatrix}$ c. $\begin{bmatrix} 0 \\ -1 \\ 0 \end{bmatrix}$ d. $\begin{bmatrix} -1 \\ 3 \\ -1 \\ -2 \end{bmatrix}$

e. $\begin{bmatrix} -1 \\ 0 \\ 1 \end{bmatrix}$ f. $\begin{bmatrix} 1 \\ 2 \\ 3 \\ 1 \end{bmatrix}$ g. $\begin{bmatrix} 1 \\ -2 \\ 3 \\ -4 \\ \sqrt{6} \end{bmatrix}$

2. Find the scalar product of the following pairs of vectors.

a. $\begin{bmatrix} -1 \\ 1 \end{bmatrix}, \begin{bmatrix} 1 \\ 2 \end{bmatrix}$ b. $\begin{bmatrix} 1 \\ 0 \\ 1 \end{bmatrix}, \begin{bmatrix} -1 \\ 1 \\ 3 \end{bmatrix}$ c. $\begin{bmatrix} 1 \\ 0 \\ 1 \end{bmatrix}, \begin{bmatrix} 1 \\ 1 \\ 1 \end{bmatrix}$

d. $\begin{bmatrix} 1 \\ 1 \\ 1 \end{bmatrix}, \begin{bmatrix} -1 \\ -1 \\ 1 \end{bmatrix}$ e. $\begin{bmatrix} 0 \\ 1 \\ 3 \end{bmatrix}, \begin{bmatrix} 3 \\ 1 \\ 0 \end{bmatrix}$ f. $\begin{bmatrix} 1 \\ 1 \\ 1 \\ 1 \end{bmatrix}, \begin{bmatrix} -1 \\ 1 \\ -1 \\ 1 \end{bmatrix}$

3. Given that

$$\mathbf{x} = \begin{bmatrix} 1 \\ 0 \\ 0 \end{bmatrix} \quad \text{and} \quad \mathbf{y} = \begin{bmatrix} 0 \\ 1 \\ 1 \end{bmatrix}$$

find numbers r and s so that

$$r\mathbf{x} + s\mathbf{y} = \begin{bmatrix} 2 \\ 1 \\ 1 \end{bmatrix}$$

Can you also find numbers r and s so that

$$r\mathbf{x} + s\mathbf{y} = \begin{bmatrix} -1 \\ 1 \\ 0 \end{bmatrix}$$

4. Find the cosine of the angle between two vectors

$$\begin{bmatrix} 1 \\ 2 \\ 4 \end{bmatrix} \text{ and } \begin{bmatrix} -1 \\ 3 \\ 5 \end{bmatrix}$$

in 3-space.

5. Find a vector $\mathbf{b}$ orthogonal to the vector $\mathbf{a}$.

a. $\mathbf{a} = \begin{bmatrix} 1 \\ 1 \end{bmatrix}$ b. $\mathbf{a} = \begin{bmatrix} 1 \\ -1 \\ 3 \end{bmatrix}$ c. $\mathbf{a} = \begin{bmatrix} -1 \\ \frac{1}{2} \end{bmatrix}$

d. $\mathbf{a} = \begin{bmatrix} 1 \\ 1 \\ 1 \\ 1 \end{bmatrix}$ e. $\mathbf{a} = \begin{bmatrix} 1 \\ 3 \\ 3 \\ 1 \end{bmatrix}$ f. $\mathbf{a} = \begin{bmatrix} 1 \\ -1 \\ 3 \\ 1 \end{bmatrix}$

6. Prove that for $\mathbf{x}, \mathbf{y}, \mathbf{z} \in R^n$,

$$\mathbf{x} \cdot (\mathbf{y} + \mathbf{z}) = \mathbf{x} \cdot \mathbf{y} + \mathbf{x} \cdot \mathbf{z}$$

7. Find vectors $\mathbf{x}$, $\mathbf{y}$, and $\mathbf{z}$ for which $(\mathbf{x} \cdot \mathbf{y})\mathbf{z} \neq \mathbf{x}(\mathbf{y} \cdot \mathbf{z})$

8. Show that for $\mathbf{x} \neq \mathbf{0}$ and $\mathbf{y} \neq \mathbf{0}$,

$$|\mathbf{x} \cdot \mathbf{y}| = \|\mathbf{x}\| \, \|\mathbf{y}\|$$

if and only if there are numbers r and s, not both zero, such that $r\mathbf{x} = s\mathbf{y}$.

9. Verify the Cauchy–Schwartz inequality for each pair of vectors.

a. $\begin{bmatrix} 5 \\ -7 \\ 1 \end{bmatrix}$ and $\begin{bmatrix} 6 \\ 3 \\ 2 \end{bmatrix}$ b. $\begin{bmatrix} -3 \\ 5 \\ 2 \end{bmatrix}$ and $\begin{bmatrix} 6 \\ -10 \\ -4 \end{bmatrix}$

10. Prove that $\mathbf{0}$ is orthogonal to any vector $\mathbf{x}$.

11. Prove that if $\mathbf{a}$ is orthogonal to every vector $\mathbf{v} \in R^n$, then $\mathbf{a} = \mathbf{0}$.

Hint: Look at $\mathbf{a} \cdot \mathbf{a}$.

12. Let $\mathbf{x} \in R^n$ such that $\mathbf{x} \neq \mathbf{0}$. Explain why it is not always possible for us to conclude from the relation $\mathbf{x} \cdot \mathbf{y} = 0$ that $\mathbf{y} = \mathbf{0}$.
13. Let $\mathbf{x} \in R^n$ and $\mathbf{x} \cdot \mathbf{y} = 0$ for every vector $\mathbf{y} \in R^n$. Is it true that $\mathbf{x} = \mathbf{0}$? Why?
14. Prove that two nonzero vectors $\mathbf{x}$ and $\mathbf{y}$ in R^n are orthogonal if and only if $\|\mathbf{x} + \mathbf{y}\| = \|\mathbf{x} - \mathbf{y}\|$.
15. Prove that each of the vectors

$$\begin{bmatrix} 1 \\ 0 \\ 0 \end{bmatrix} \quad \begin{bmatrix} 0 \\ 1 \\ 0 \end{bmatrix} \quad \begin{bmatrix} 0 \\ 0 \\ 1 \end{bmatrix}$$

in R^3 is orthogonal to the other two.
16. Prove $(r\mathbf{x}) \cdot (s\mathbf{y}) = (rs)(\mathbf{x} \cdot \mathbf{y})$.
17. If $\|\mathbf{x}\| = \|\mathbf{y}\|$, where $\mathbf{x}$ and $\mathbf{y}$ are in R^n, prove that $\mathbf{x} + \mathbf{y}$ is orthogonal to $\mathbf{x} - \mathbf{y}$.
18. If $\mathbf{x}$ and $\mathbf{y}$ are in R^n, prove that $\|\mathbf{x} + \mathbf{y}\|^2 - \|\mathbf{x} - \mathbf{y}\|^2 = 4\mathbf{x} \cdot \mathbf{y}$.
19. Let $\mathbf{x}$ and $\mathbf{y}$ be nonzero vectors in R^n such that $\mathbf{x} \cdot \mathbf{y} = 0$. Prove that $\|\mathbf{x} + r\mathbf{y}\| \geqq \|\mathbf{x}\|$.
20. If, for every number r, $\|\mathbf{x} + r\mathbf{y}\| \geqq \|\mathbf{x}\|$, prove that $\mathbf{x} \cdot \mathbf{y} = 0$.
21. A function F that assigns to every pair $\mathbf{x}, \mathbf{y}$ of vectors in R^n a number $F(\mathbf{x}, \mathbf{y})$ such that

$$\begin{aligned} F(\mathbf{x}, \mathbf{y}) &= F(\mathbf{y}, \mathbf{x}) \\ F(\mathbf{x}, \mathbf{y} + \mathbf{z}) &= F(\mathbf{x}, \mathbf{y}) + F(\mathbf{x}, \mathbf{z}) \\ F(r\mathbf{x}, \mathbf{y}) &= rF(\mathbf{x}, \mathbf{y}) \\ F(\mathbf{x}, \mathbf{x}) &> 0 \text{ if } \mathbf{x} \neq \mathbf{0} \end{aligned}$$

is called an **inner product** on R^n.

a. If F is an inner product on R^n, show that $F(\mathbf{x}, r\mathbf{y}) = rF(\mathbf{x}, \mathbf{y})$.

b. Is $\mathbf{x} \cdot \mathbf{y}$ an inner product on R^n? Why?

c. Is

$$F(\mathbf{x}, \mathbf{y}) = \sum_{i=1}^{n} (x_i + y_i)^2 - \sum_{i=1}^{n} x_i^2 - \sum_{i=1}^{n} y_i^2$$

an inner product on R^n? Why?

SECTION 2.3 LINES AND PLANES IN R^n

The diagram in Figure 2.13 (on top of page 128) shows a straight line L in R^2 passing through point $\mathbf{a}$ and a nonzero vector $\mathbf{p}$ parallel to the line L.

According to the parallelogram law for vector addition, a point $\alpha(t)$ is on line L if and only if it has the form $\alpha(t) = \mathbf{a} + t\mathbf{p}$. This special situation in R^2 suggests how to define a line in R^n for any n.

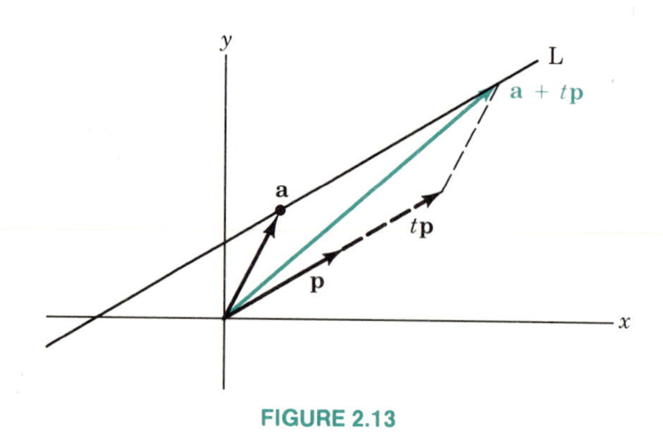

FIGURE 2.13

Definition 2.3.1 Given a vector **a** in R^n and a nonzero vector **p** in R^n, then

$$\boldsymbol{\alpha}(t) = \boldsymbol{\alpha} + t\mathbf{p}$$

is the **vector equation of a line** through the point **a** in the direction **p**.

Example 2.3.2 (a) The line in R^2 that goes through

$$\begin{bmatrix} 0 \\ 1 \end{bmatrix}$$

in the direction

$$\begin{bmatrix} 1 \\ 1 \end{bmatrix}$$

has the equation

$$\boldsymbol{\alpha}(t) = \begin{bmatrix} 0 \\ 1 \end{bmatrix} + t\begin{bmatrix} 1 \\ 1 \end{bmatrix} = \begin{bmatrix} t \\ 1 + t \end{bmatrix}$$

(b) The line in R^3 that goes through the point

$$\begin{bmatrix} 1 \\ 0 \\ 0 \end{bmatrix}$$

in the direction of

$$\begin{bmatrix} -1 \\ 3 \\ 1 \end{bmatrix}$$

has the equation

$$\boldsymbol{\alpha}(t) = \begin{bmatrix} 1 \\ 0 \\ 0 \end{bmatrix} + t\begin{bmatrix} -1 \\ 3 \\ 1 \end{bmatrix} = \begin{bmatrix} 1 - t \\ 3t \\ t \end{bmatrix}$$

■

In the plane R^2, the graph of $y = mx + b$ is a straight line that goes through the point $(0, b)$ and has slope m. To see how this description of a line in R^2 compares with the description of a line in R^2 in terms of vectors, note that, for every real number t, the point $(t, mt + b)$ is on the line. But since

$$\begin{bmatrix} t \\ mt + b \end{bmatrix} = \begin{bmatrix} 0 \\ b \end{bmatrix} + t\begin{bmatrix} 1 \\ m \end{bmatrix}$$

it follows that

$$\boldsymbol{\alpha}(t) = \begin{bmatrix} 0 \\ b \end{bmatrix} + t\begin{bmatrix} 1 \\ m \end{bmatrix}$$

is the vector equation of the line through $(0, b)$ in the direction $(1, m)$. Hence the graph of $y = mx + b$ and the vector equation

$$\boldsymbol{\alpha}(t) = \begin{bmatrix} 0 \\ b \end{bmatrix} + t\begin{bmatrix} 1 \\ m \end{bmatrix}$$

describes the same line in the plane R^2.

Example 2.3.3 The line in R^2 through $(0, 1)$ in the direction $(1, -2)$ is both the graph of $y = -2x + 1$ and the graph of

$$\boldsymbol{\alpha}(t) = \begin{bmatrix} 0 \\ 1 \end{bmatrix} + t\begin{bmatrix} 1 \\ -2 \end{bmatrix}$$ ■

It is clear that two points **a** and **b** in R^n should determine a line in R^n. The equation

$$\boldsymbol{\alpha}(t) = \mathbf{a} + t(\mathbf{b} - \mathbf{a})$$

describes the line through **a** in the direction $\mathbf{b} - \mathbf{a}$. But since $\boldsymbol{\alpha}(1) = \mathbf{b}$, this line goes through both of the points **a** and **b**. Therefore,

$$\boldsymbol{\alpha}(t) = \mathbf{a} + t(\mathbf{b} - \mathbf{a})$$

is the *equation of the line through the two points* **a** and **b** (see Figure 2.14).

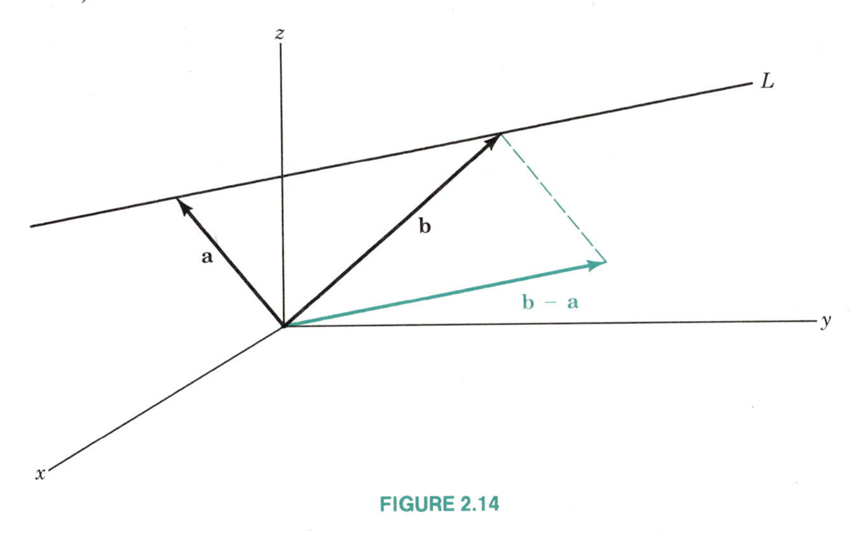

FIGURE 2.14

Example 2.3.4 The equation of the line in R^4 that goes through the two points

$$\begin{bmatrix} 1 \\ 2 \\ -1 \\ 0 \end{bmatrix} \quad \text{and} \quad \begin{bmatrix} 2 \\ 2 \\ 2 \\ 2 \end{bmatrix}$$

is

$$\alpha(t) = \begin{bmatrix} 1 \\ 2 \\ -1 \\ 0 \end{bmatrix} + t \begin{bmatrix} 2 - 1 \\ 2 - 2 \\ 2 - (-1) \\ 2 - 0 \end{bmatrix} = \begin{bmatrix} 1 \\ 2 \\ -1 \\ 0 \end{bmatrix} + t \begin{bmatrix} 1 \\ 0 \\ 3 \\ 2 \end{bmatrix}$$

■

Example 2.3.5 Show that the vector equations

$$\alpha_1(t) = \begin{bmatrix} 1 \\ 2 \\ 1 \end{bmatrix} + t \begin{bmatrix} 5 \\ 6 \\ -1 \end{bmatrix} \quad \text{and} \quad \alpha_2(s) = \begin{bmatrix} 6 \\ 8 \\ 0 \end{bmatrix} + s \begin{bmatrix} -5 \\ -6 \\ 1 \end{bmatrix}$$

describe the same line.

Solution For $t = 0$,

$$\alpha_1(0) = \begin{bmatrix} 1 \\ 2 \\ 1 \end{bmatrix}$$

and for $t = 2$,

$$\alpha_1(2) = \begin{bmatrix} 11 \\ 14 \\ -1 \end{bmatrix}$$

Thus

$$\begin{bmatrix} 1 \\ 2 \\ 1 \end{bmatrix} \quad \text{and} \quad \begin{bmatrix} 11 \\ 14 \\ -1 \end{bmatrix}$$

lie on $\alpha_1(t)$. If we can show these two points lie on $\alpha_2(t)$, we are done, since two points determine a unique straight line. To do this, we solve

$$\alpha_2(s) = \begin{bmatrix} 6 \\ 8 \\ 0 \end{bmatrix} + s \begin{bmatrix} -5 \\ -6 \\ 1 \end{bmatrix} = \begin{bmatrix} 1 \\ 2 \\ 1 \end{bmatrix}$$

for s, obtaining $s = 1$. To check, note that

$$\boldsymbol{\alpha}_2(1) = \begin{bmatrix} 1 \\ 2 \\ 1 \end{bmatrix}$$

We also solve

$$\boldsymbol{\alpha}_2(s) = \begin{bmatrix} 6 \\ 8 \\ 0 \end{bmatrix} + s \begin{bmatrix} -5 \\ -6 \\ 1 \end{bmatrix} = \begin{bmatrix} 11 \\ 14 \\ -1 \end{bmatrix}$$

obtaining $s = -1$. As a check,

$$\boldsymbol{\alpha}_2(-1) = \begin{bmatrix} 11 \\ 14 \\ -1 \end{bmatrix}$$

Thus, both points lie on $\boldsymbol{\alpha}_2(s)$. ■

We now find an equation for the plane passing through a given point **a** and perpendicular to a given vector **N**. Figure 2.15 shows that **x** is on the plane through **a** if the vector **N** (called the **normal vector**) is perpendicular to **x** − **a**. We have drawn both **N** and **x** − **a** with their tails at **a** to help us visualize the situation. Thus

$$\mathbf{N} \cdot (\mathbf{x} - \mathbf{a}) = 0$$

is an equation for the plane that goes through the point **a** and is

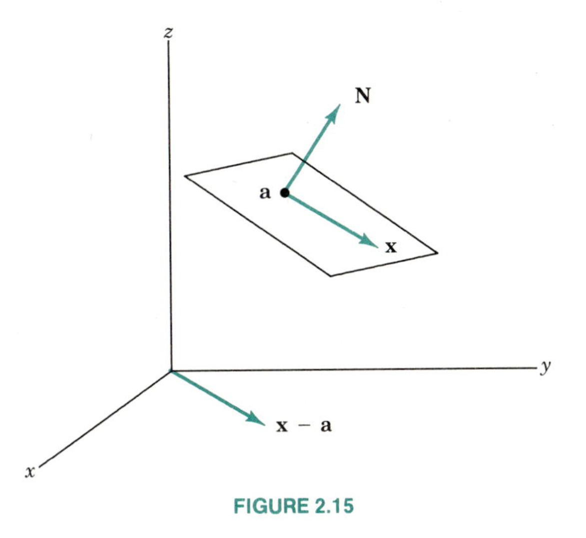

FIGURE 2.15

perpendicular to **N**, in the sense that **x** is on this plane if and only if $\mathbf{N} \cdot (\mathbf{x} - \mathbf{a}) = 0$.

If

$$\mathbf{N} = \begin{bmatrix} N_1 \\ N_2 \\ N_3 \end{bmatrix}, \quad \mathbf{a} = \begin{bmatrix} a_1 \\ a_2 \\ a_3 \end{bmatrix} \quad \text{and} \quad \mathbf{x} = \begin{bmatrix} x \\ y \\ z \end{bmatrix}$$

then our equation for the plane through **a** which is perpendicular to **N** can be written as

$$N_1(x - a_1) + N_2(y - a_2) + N_3(z - a_3) = 0$$

or

$$N_1 x + N_2 y + N_3 z = d$$

where

$$d = \mathbf{N} \cdot \mathbf{a}$$

Example 2.3.6 The equation of the plane determined by the point

$$\begin{bmatrix} 1 \\ 2 \\ -3 \end{bmatrix}$$

and normal vector

$$\mathbf{N} = \begin{bmatrix} -2 \\ 4 \\ \frac{1}{2} \end{bmatrix}$$

is given by

$$\mathbf{N} \cdot (\mathbf{x} - \mathbf{a}) = \begin{bmatrix} -2 \\ 4 \\ \frac{1}{2} \end{bmatrix} \cdot \begin{bmatrix} x - 1 \\ y - 2 \\ z + 3 \end{bmatrix}$$

$$= -2(x - 1) + 4(y - 2) + \tfrac{1}{2}(z + 3) = 0$$

or, upon simplifying,

$$-2x + 4y + \tfrac{1}{2}z = \tfrac{9}{2}$$ ■

This idea can be carried over to n-space quite easily.

Definition 2.3.7 If $\mathbf{a}$ is a point in n-space and $\mathbf{N}$ is a nonzero vector in n-space, then the set of all vectors $\mathbf{x}$ in n-space that satisfy the equation

$$\mathbf{N} \cdot (\mathbf{x} - \mathbf{a}) = 0$$

or

$$\mathbf{N} \cdot \mathbf{x} = d$$

where $d = \mathbf{N} \cdot \mathbf{a}$ is defined to be the **hyperplane through a with normal vector N.**

Example 2.3.8 Find the equation of the hyperplane in R^4 that goes through

$$\begin{bmatrix} 1 \\ 0 \\ 2 \\ 1 \end{bmatrix}$$

and is perpendicular to

$$\begin{bmatrix} -1 \\ 0 \\ 1 \\ -1 \end{bmatrix}$$

Solution For

$$\mathbf{x} = \begin{bmatrix} x \\ y \\ z \\ w \end{bmatrix}$$

the equation is

$$\mathbf{N} \cdot (\mathbf{x} - \mathbf{a}) = \begin{bmatrix} -1 \\ 0 \\ 1 \\ -1 \end{bmatrix} \cdot \begin{bmatrix} x - 1 \\ y \\ z - 2 \\ w - 1 \end{bmatrix} = 0$$

or

$$-1(x-1)+0(y)+1(z-2)+(-1)(w-1)=0$$

$$-x+z-w=0$$ ■

Example 2.3.9 A normal vector for the hyperplane $x+y-z=1$ in R^3 is

$$\begin{bmatrix} 1 \\ 1 \\ -1 \end{bmatrix}$$ ■

SUMMARY

Terms to know:

Vector equation of a line
Normal vector to a plane (hyperplane)
Hyperplane through **a** with normal vector **N**

EXERCISE SET 2.3

1. Write the vector equation of the line that passes through the given vector in the specified direction.

a. $\begin{bmatrix} 0 \\ 1 \\ 0 \end{bmatrix}$ in the direction $\begin{bmatrix} 1 \\ 1 \\ 1 \end{bmatrix}$

b. $\begin{bmatrix} 1 \\ 1 \\ 1 \end{bmatrix}$ in the direction $\begin{bmatrix} 0 \\ 1 \\ 0 \end{bmatrix}$

c. $\begin{bmatrix} 2 \\ 2 \\ -1 \end{bmatrix}$ having the same direction as $\begin{bmatrix} 1 \\ 1 \\ 1 \end{bmatrix}$

d. $\begin{bmatrix} 2 \\ -1 \\ 3 \\ 4 \end{bmatrix}$ in the direction $\begin{bmatrix} 1 \\ 0 \\ 1 \\ 0 \end{bmatrix}$

2. Find the vector equation of the line passing through the given points.

a. $\begin{bmatrix} 1 \\ 3 \end{bmatrix}, \begin{bmatrix} -1 \\ 4 \end{bmatrix}$ b. $\begin{bmatrix} 0 \\ 1 \end{bmatrix}, \begin{bmatrix} 1 \\ -1 \end{bmatrix}$

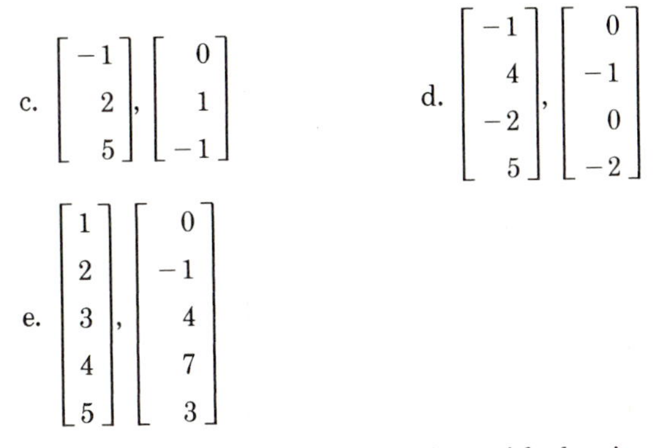

c. $\begin{bmatrix} -1 \\ 2 \\ 5 \end{bmatrix}, \begin{bmatrix} 0 \\ 1 \\ -1 \end{bmatrix}$ d. $\begin{bmatrix} -1 \\ 4 \\ -2 \\ 5 \end{bmatrix}, \begin{bmatrix} 0 \\ -1 \\ 0 \\ -2 \end{bmatrix}$

e. $\begin{bmatrix} 1 \\ 2 \\ 3 \\ 4 \\ 5 \end{bmatrix}, \begin{bmatrix} 0 \\ -1 \\ 4 \\ 7 \\ 3 \end{bmatrix}$

3. Find the vector equation of the lines with the given equations.

   a. $y = 4x - 2$ b. $y = -x$

   c. $y = (\frac{1}{2})x - 4$ d. $3x + 2y = 7$

   e. $y = 2$

4. Is it true that $\boldsymbol{\alpha}(t) = \mathbf{a} + t\mathbf{p}$ and $\boldsymbol{\beta}(t) = \mathbf{a} + 5t\mathbf{p}$ are equations of the same line? Why? What about $\boldsymbol{\psi}(t) = \mathbf{a} - t\mathbf{p}$?

5. Given two lines $\boldsymbol{\alpha}(t) = \mathbf{a} + t\mathbf{p}$ and $\boldsymbol{\beta}(t) = \mathbf{c} + t\mathbf{d}$, the angle between the two lines is the angle between $\mathbf{p}$ and $\mathbf{d}$. Find the cosine of the angle between the two lines whose equations are given.

   a. $\boldsymbol{\alpha}(t) = \begin{bmatrix} 1 \\ 1 \\ -1 \\ 2 \end{bmatrix} + \begin{bmatrix} 1 \\ 0 \\ 1 \\ -1 \end{bmatrix} t$ and $\boldsymbol{\beta}(t) = \begin{bmatrix} 0 \\ -3 \\ 2 \\ 1 \end{bmatrix} + \begin{bmatrix} 0 \\ 1 \\ 2 \\ -1 \end{bmatrix} t$

   b. $\boldsymbol{\alpha}(t) = \begin{bmatrix} 1 \\ 1 \\ -1 \end{bmatrix} + \begin{bmatrix} 1 \\ 0 \\ 1 \end{bmatrix} t$ and $\boldsymbol{\beta}(t) = \begin{bmatrix} 0 \\ -3 \\ 2 \end{bmatrix} + \begin{bmatrix} 0 \\ 1 \\ 2 \end{bmatrix} t$

6. Let $x + 2y - 2z = 1$ and $x + y + 3z = -2$ be equations of two planes. Write the equation of the line of intersection of the two planes.

7. Write an equation of a line perpendicular to the plane $x + y + z = 3$.

8. Show that every point on the line described by

$$\boldsymbol{\alpha}(t) = \begin{bmatrix} 1 \\ -1 \\ -1 \end{bmatrix} + \begin{bmatrix} 2 \\ -1 \\ -\frac{3}{2} \end{bmatrix} t$$

   is also a point in the plane described by $x - y + 2z = 0$.

SECTION 2.4 THE CROSS PRODUCT AND PLANES IN R^3

Another way to describe a plane is to give any three noncollinear points lying in the plane. However, from this information we must determine a vector $\mathbf{N}$ normal to the plane. This is possible (only in R^3) using an operation called the **cross product of two vectors.**

Definition 2.4.1 If

$$\mathbf{u} = \begin{bmatrix} u_1 \\ u_2 \\ u_3 \end{bmatrix} \quad \text{and} \quad \mathbf{v} = \begin{bmatrix} v_1 \\ v_2 \\ v_3 \end{bmatrix}$$

are vectors in R^3, the **cross product** $\mathbf{u} \times \mathbf{v}$ is the vector

$$\begin{bmatrix} u_2v_3 - u_3v_2 \\ u_3v_1 - u_1v_3 \\ u_1v_2 - u_2v_1 \end{bmatrix}$$

REMARK Unlike the scalar product, the cross product $\mathbf{u} \times \mathbf{v}$ is a vector.

Theorem 2.4.2 $\mathbf{u} \times \mathbf{v}$ *is perpendicular to both* $\mathbf{u}$ *and* $\mathbf{v}$.

Proof. We need to show that $(\mathbf{u} \times \mathbf{v}) \cdot \mathbf{v} = 0$:

$$(\mathbf{u} \times \mathbf{v}) \cdot \mathbf{v} = \begin{bmatrix} u_2v_3 - u_3v_2 \\ u_3v_1 - u_1v_3 \\ u_1v_2 - u_2v_1 \end{bmatrix} \cdot \begin{bmatrix} v_1 \\ v_2 \\ v_3 \end{bmatrix}$$

$$= u_2v_3v_1 - u_3v_2v_1 + u_3v_1v_2 - u_1v_3v_2 + u_1v_2v_3 - u_2v_1v_3$$

$$= 0$$

A similar calculation shows that $(\mathbf{u} \times \mathbf{v}) \cdot \mathbf{u} = 0$.

Example 2.4.3 Calculate

$$\begin{bmatrix} 1 \\ 3 \\ -1 \end{bmatrix} \times \begin{bmatrix} \frac{1}{2} \\ 0 \\ \frac{1}{3} \end{bmatrix}$$

and verify that it is perpendicular to each of

$$\begin{bmatrix} 1 \\ 3 \\ -1 \end{bmatrix} \quad \text{and} \quad \begin{bmatrix} \frac{1}{2} \\ 0 \\ \frac{1}{3} \end{bmatrix}$$

Solution The cross product is

$$\begin{bmatrix} 3(\frac{1}{3}) - (-1)0 \\ (-1)(\frac{1}{2}) - (1)(\frac{1}{3}) \\ (1)(0) - 3(\frac{1}{2}) \end{bmatrix} = \begin{bmatrix} 1 \\ -\frac{5}{6} \\ -\frac{3}{2} \end{bmatrix}$$

and

$$\begin{bmatrix} 1 \\ -\frac{5}{6} \\ -\frac{3}{2} \end{bmatrix} \cdot \begin{bmatrix} 1 \\ 3 \\ -1 \end{bmatrix} = 1(1) + (-\tfrac{5}{6})(3) + (-\tfrac{3}{2})(-1) = 0$$

Likewise,

$$\begin{bmatrix} 1 \\ -\frac{5}{6} \\ -\frac{3}{2} \end{bmatrix} \cdot \begin{bmatrix} \frac{1}{2} \\ 0 \\ \frac{1}{3} \end{bmatrix} = 1(\tfrac{1}{2}) + (-\tfrac{5}{6})(0) + (-\tfrac{3}{2})(\tfrac{1}{3}) = 0$$

■

We now return to our discussion of planes. Suppose we are given three noncollinear points on a plane: $P = (p_1, p_2, p_3)$, $Q = (q_1, q_2, q_3)$, and $R = (r_1, r_2, r_3)$.

1. Form the vectors

$$\mathbf{u} = \begin{bmatrix} q_1 - p_1 \\ q_2 - p_2 \\ q_3 - p_3 \end{bmatrix} \quad \text{and} \quad \mathbf{v} = \begin{bmatrix} r_1 - p_1 \\ r_2 - p_2 \\ r_3 - p_3 \end{bmatrix}$$

2. Form $\mathbf{N} = \mathbf{u} \times \mathbf{v}$, a normal to the plane determined by $\mathbf{u}$ and $\mathbf{v}$.
3. Form the equation of the plane.

This is illustrated geometrically in Figure 2.16.

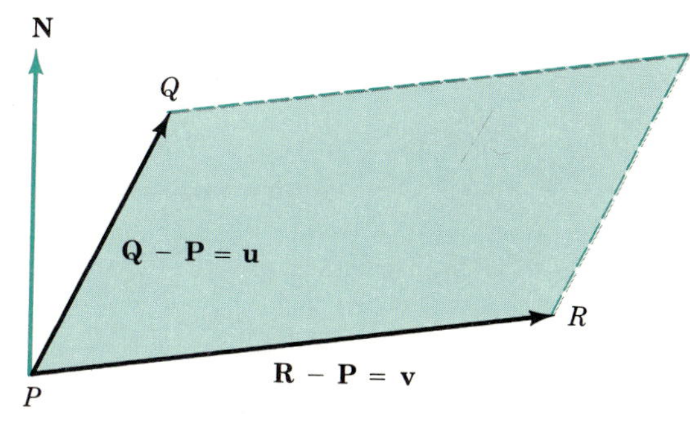

FIGURE 2.16

Example 2.4.4 Find the equation of the plane containing $(0, 1, 3)$, $(4, -1, -3)$, $(2, -1, 7)$.

Solution Let

$$\mathbf{u} = \begin{bmatrix} 4-0 \\ -1-1 \\ -3-3 \end{bmatrix} = \begin{bmatrix} 4 \\ -2 \\ -6 \end{bmatrix} \quad \text{and} \quad \mathbf{v} = \begin{bmatrix} 2-0 \\ -1-1 \\ 7-3 \end{bmatrix} = \begin{bmatrix} 2 \\ -2 \\ 4 \end{bmatrix}$$

A normal to this plane is

$$\mathbf{N} = \begin{bmatrix} 4 \\ -2 \\ -6 \end{bmatrix} \times \begin{bmatrix} 2 \\ -2 \\ 4 \end{bmatrix} = \begin{bmatrix} -2(4) - (-2)(-6) \\ -6(2) - 4(4) \\ 4(-2) - (2)(-2) \end{bmatrix} = \begin{bmatrix} -20 \\ -28 \\ -4 \end{bmatrix}$$

An equation of this plane is, using the point $(0, 1, 3)$,

$$-20(x-0) - 28(y-1) - 4(z-3) = 0$$

or, using the point $(4, -1, -3)$,

$$-20(x-4) - 28[y-(-1)] - 4[z-(-3)] = 0$$

or, using the point $(2, -1, 7)$

$$-20(x-2) - 28[y-(-1)] - 4(z-7) = 0$$

Each of these equations simplifies to $20x + 28y + 4z = 40$. ■

Other books, especially in engineering and physics, use the notation

$$x\mathbf{i} + y\mathbf{j}$$

to stand for a vector in R^2 whose first coordinate is x and whose second coordinate is y. In this alternate notation,

$$\mathbf{i} = \begin{bmatrix} 1 \\ 0 \end{bmatrix} \quad \text{and} \quad \mathbf{j} = \begin{bmatrix} 0 \\ 1 \end{bmatrix}$$

The notation used in this book

$$\begin{bmatrix} x \\ y \end{bmatrix}$$

is consistent with the notation $x\mathbf{i} + y\mathbf{j}$, which follows from the laws of vector addition and scalar multiplication:

$$\begin{aligned} x\mathbf{i} + y\mathbf{j} &= x\begin{bmatrix} 1 \\ 0 \end{bmatrix} + y\begin{bmatrix} 0 \\ 1 \end{bmatrix} \\ &= \begin{bmatrix} x \\ 0 \end{bmatrix} + \begin{bmatrix} 0 \\ y \end{bmatrix} \\ &= \begin{bmatrix} x \\ y \end{bmatrix} \end{aligned}$$

Similarly, the notation

$$x\mathbf{i} + y\mathbf{j} + z\mathbf{k}$$

is sometimes used to stand for vectors in R^3 whose first coordinate is x, whose second coordinate is y, and whose third coordinate is z. In this alternate notation,

$$\mathbf{i} = \begin{bmatrix} 1 \\ 0 \\ 0 \end{bmatrix}, \quad \mathbf{j} = \begin{bmatrix} 0 \\ 1 \\ 0 \end{bmatrix}, \quad \text{and} \quad \mathbf{k} = \begin{bmatrix} 0 \\ 0 \\ 1 \end{bmatrix}$$

Just as in R^2, it is easy to see that

$$\begin{bmatrix} x \\ y \\ z \end{bmatrix}$$

is consistent with the notation $x\mathbf{i} + y\mathbf{j} + z\mathbf{k}$, i.e.,

$$\begin{bmatrix} x \\ y \\ z \end{bmatrix} = x\mathbf{i} + y\mathbf{j} + z\mathbf{k}$$

We rewrite the definition of the cross product in terms of **i**, **j**, and **k** as

$$\begin{aligned} \mathbf{u} \times \mathbf{v} &= (u_1\mathbf{i} + u_2\mathbf{j} + u_3\mathbf{k}) \times (v_1\mathbf{i} + v_2\mathbf{j} + v_3\mathbf{k}) \\ &= (u_2v_3 - u_3v_2)\mathbf{i} - (u_1v_3 - u_3v_1)\mathbf{j} + (u_1v_2 - u_2v_1)\mathbf{k} \end{aligned}$$

In order to remember this formula more easily, we let

$$\begin{vmatrix} a & b \\ c & d \end{vmatrix} = ad - bc$$

and so

$$\mathbf{u} \times \mathbf{v} = \begin{vmatrix} u_2 & u_3 \\ v_2 & v_3 \end{vmatrix}\mathbf{i} - \begin{vmatrix} u_1 & u_3 \\ v_1 & v_3 \end{vmatrix}\mathbf{j} + \begin{vmatrix} u_1 & u_2 \\ v_1 & v_2 \end{vmatrix}\mathbf{k}$$

Expanding the third-order determinant by the first row gives

$$\mathbf{u} \times \mathbf{v} = \begin{vmatrix} \mathbf{i} & \mathbf{j} & \mathbf{k} \\ u_1 & u_2 & u_3 \\ v_1 & v_2 & v_3 \end{vmatrix}$$

(We discuss such determinants in Chapter 5.)

Example 2.4.5 Find $\mathbf{u} \times \mathbf{v}$ and $\mathbf{v} \times \mathbf{u}$, where $\mathbf{u} = \mathbf{i} - \mathbf{j} + \mathbf{k}$ and $\mathbf{v} = -2\mathbf{i} + 3\mathbf{j} + \mathbf{k}$.

Solution

$$\mathbf{u} \times \mathbf{v} = \begin{vmatrix} \mathbf{i} & \mathbf{j} & \mathbf{k} \\ 1 & -1 & 1 \\ -2 & 3 & 1 \end{vmatrix} = \begin{vmatrix} -1 & 1 \\ 3 & 1 \end{vmatrix}\mathbf{i} - \begin{vmatrix} 1 & 1 \\ -2 & 1 \end{vmatrix}\mathbf{j} + \begin{vmatrix} 1 & -1 \\ -2 & 3 \end{vmatrix}\mathbf{k}$$

$$= [(-1)(1) - (3)(1)]\mathbf{i} - [(1)(1) - (-2)(1)]\mathbf{j}$$

$$+ [(1)(3) - (-2)(-1)]\mathbf{k}$$

$$= -4\mathbf{i} - 3\mathbf{j} + \mathbf{k}$$

$$\mathbf{v} \times \mathbf{u} = \begin{vmatrix} \mathbf{i} & \mathbf{j} & \mathbf{k} \\ -2 & 3 & 1 \\ 1 & -1 & 1 \end{vmatrix} = \begin{vmatrix} 3 & 1 \\ -1 & 1 \end{vmatrix}\mathbf{i} - \begin{vmatrix} -2 & 1 \\ 1 & 1 \end{vmatrix}\mathbf{j} + \begin{vmatrix} -2 & 3 \\ 1 & -1 \end{vmatrix}\mathbf{k}$$

$$= [(3)(1) - (-1)(1)]\mathbf{i} - [(-2)(1) - (1)(1)]\mathbf{j}$$

$$+ [(-2)(-1) - (1)(3)]\mathbf{k}$$

$$= 4\mathbf{i} + 3\mathbf{j} - \mathbf{k}$$

■

Notice that $\mathbf{u} \times \mathbf{v} = -(\mathbf{v} \times \mathbf{u})$, which is one of the algebraic properties of the cross-product operation listed in the following theorem.

Theorem 2.4.6 *For* $\mathbf{a}$ *and* $\mathbf{b}$ *in* R^3, *the following properties hold.*

1. $\mathbf{a} \times \mathbf{a} = \mathbf{0}$

2. $\mathbf{a} \cdot (\mathbf{b} \times \mathbf{c}) = \mathbf{b} \cdot (\mathbf{c} \times \mathbf{a}) = \mathbf{c} \cdot (\mathbf{a} \times \mathbf{b})$

3. $(\mathbf{a} + \mathbf{b}) \times \mathbf{c} = (\mathbf{a} \times \mathbf{c}) + (\mathbf{b} \times \mathbf{c})$

4. $\mathbf{a} \times (\mathbf{b} + \mathbf{c}) = (\mathbf{a} \times \mathbf{b}) + (\mathbf{a} \times \mathbf{c})$

5. $\mathbf{a} \times \mathbf{b} = -(\mathbf{b} \times \mathbf{a})$

6. $\|\mathbf{a} \times \mathbf{b}\|^2 = \|\mathbf{a}\|^2\|\mathbf{b}\|^2 - (\mathbf{a} \cdot \mathbf{b})^2$

We verify statements (1) and (3); the others are left as exercises.

Proof. We begin with (1):

$$\mathbf{a} \times \mathbf{a} = \begin{bmatrix} a_1 \\ a_2 \\ a_3 \end{bmatrix} \times \begin{bmatrix} a_1 \\ a_2 \\ a_3 \end{bmatrix} = \begin{bmatrix} a_2a_3 - a_3a_2 \\ a_3a_1 - a_1a_3 \\ a_1a_2 - a_2a_1 \end{bmatrix} = \begin{bmatrix} 0 \\ 0 \\ 0 \end{bmatrix} = \mathbf{0}$$

To prove (3), let

$$\mathbf{a} = \begin{bmatrix} a_1 \\ a_2 \\ a_3 \end{bmatrix}, \quad \mathbf{b} = \begin{bmatrix} b_1 \\ b_2 \\ b_3 \end{bmatrix}, \quad \text{and} \quad \mathbf{c} = \begin{bmatrix} c_1 \\ c_2 \\ c_3 \end{bmatrix}$$

Then

$$\mathbf{a} + \mathbf{b} = \begin{bmatrix} a_1 + b_1 \\ a_2 + b_2 \\ a_3 + b_3 \end{bmatrix} \quad \text{while} \quad \mathbf{a} \times \mathbf{c} = \begin{bmatrix} a_2c_3 - a_3c_2 \\ a_3c_1 - a_1c_3 \\ a_1c_2 - a_2c_1 \end{bmatrix}$$

and

$$\mathbf{b} \times \mathbf{c} = \begin{bmatrix} b_2c_3 - b_3c_2 \\ b_3c_1 - b_1c_3 \\ b_1c_2 - b_2c_1 \end{bmatrix}$$

so

$$(\mathbf{a} \times \mathbf{c}) + (\mathbf{b} \times \mathbf{c}) = \begin{bmatrix} a_2c_3 - a_3c_2 + b_2c_3 - b_3c_2 \\ a_3c_1 - a_1c_3 + b_3c_1 - b_1c_3 \\ a_1c_2 - a_2c_1 + b_1c_2 - b_2c_1 \end{bmatrix}$$

$$= \begin{bmatrix} (a_2 + b_2)c_3 - (a_3 + b_3)c_2 \\ (a_3 + b_3)c_1 - (a_1 + b_1)c_3 \\ (a_1 + b_1)c_2 - (a_2 + b_2)c_1 \end{bmatrix} = (\mathbf{a} + \mathbf{b}) \times \mathbf{c}$$

REMARK The cross product works only in R^3. No useful generalization of the cross product to R^n, $n > 3$, has as yet been obtained. Moreover, it has been proved that no such generalization can exist.

SUMMARY

Terms to know:

Cross product of two vectors

EXERCISE SET 2.4

1. Find the following cross products.

a. $\begin{bmatrix} 1 \\ 2 \\ 0 \end{bmatrix} \times \begin{bmatrix} -3 \\ 1 \\ 0 \end{bmatrix}$ b. $\begin{bmatrix} -3 \\ 1 \\ -3 \end{bmatrix} \times \begin{bmatrix} 2 \\ 0 \\ 4 \end{bmatrix}$

c. $\begin{bmatrix} 1 \\ 5 \\ 2 \end{bmatrix} \times \begin{bmatrix} -1 \\ 3 \\ 0 \end{bmatrix}$ d. $\begin{bmatrix} -1 \\ 4 \\ 3 \end{bmatrix} \times \begin{bmatrix} 5 \\ 0 \\ 7 \end{bmatrix}$

e. $\begin{bmatrix} 1 \\ 2 \\ 3 \end{bmatrix} \times \begin{bmatrix} 1 \\ 2 \\ 3 \end{bmatrix}$

2. Find each value.

   a. $\mathbf{i} \times \mathbf{j}, \mathbf{j} \times \mathbf{k}, \mathbf{k} \times \mathbf{i}$

   b. $\mathbf{i} \times \mathbf{i}, \mathbf{j} \times \mathbf{j}, \mathbf{k} \times \mathbf{k}$

3. If $\mathbf{a}$ and $\mathbf{b}$ are parallel vectors in R^3, show that $\mathbf{a} \times \mathbf{b} = \mathbf{0}$. What can you say about $\mathbf{a}$ and $\mathbf{b}$ if $\mathbf{a} \times \mathbf{b} = \mathbf{0}$?

4. For vectors in R^3, verify that each of the following is true.

   a. $\mathbf{a} \times \mathbf{b} = -(\mathbf{b} \times \mathbf{a})$

   b. $\mathbf{a} \cdot (\mathbf{b} \times \mathbf{c}) = \mathbf{b} \cdot (\mathbf{c} \times \mathbf{a})$

   c. $\mathbf{a} \times (\mathbf{b} + \mathbf{c}) = (\mathbf{a} \times \mathbf{b}) + (\mathbf{a} \times \mathbf{c})$

5. a. Find $\|\mathbf{a} \times \mathbf{b}\|^2$.

   b. Find $\|\mathbf{a}\|^2\|\mathbf{b}\|^2 - (\mathbf{a} \cdot \mathbf{b})^2$.

   c. Do your answers to (a) and (b) agree?

6. Consider the parallelogram with the vertices shown in Figure 2.17.

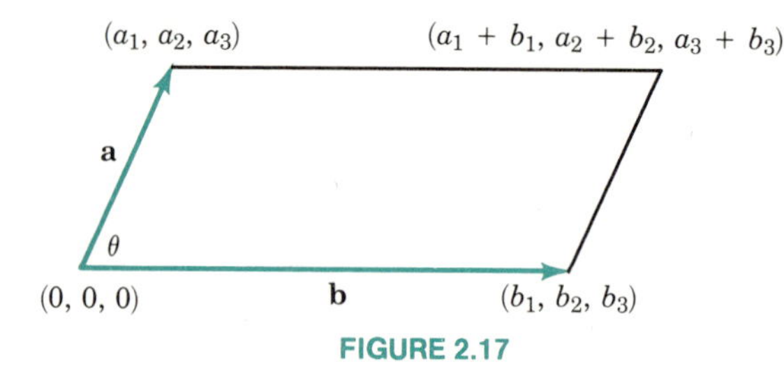

FIGURE 2.17

The sides may be considered to be vectors, as shown. The area of a parallelogram is

$$\|\mathbf{a}\|\,\|\mathbf{b}\|\sin\theta$$

Using Exercise 5(b) and the definition of $\mathbf{a} \cdot \mathbf{b}$, show that

$$\|\mathbf{a} \times \mathbf{b}\| = \|\mathbf{a}\|\,\|\mathbf{b}\|\sin\theta$$

7. Given two planes with normal vectors $\mathbf{N}_1$ and $\mathbf{N}_2$, the angle between the two planes is the angle between $\mathbf{N}_1$ and $\mathbf{N}_2$. Why? Find the cosine of the angle between two planes having equations

$$2x - y + z = -1 \quad \text{and} \quad x + 2y - z = 0$$

8. Show that the two planes having the equations

$$x + y + 2z = 5 \quad \text{and} \quad x - 7y + 3z = 1$$

are perpendicular.

9. Let

$$\mathbf{a} = \begin{bmatrix} 1 \\ 3 \\ 5 \end{bmatrix} \quad \text{and} \quad \mathbf{p} = \begin{bmatrix} -2 \\ 1 \\ 1 \end{bmatrix}$$

Find the point where the line through $\mathbf{a}$ in the direction $\mathbf{p}$ crosses the plane having the equation $2x + 3y - z = 1$.

10. In this problem, we develop a formula for the distance from a point to a plane. The diagram in R^3 shows a point $\mathbf{q}$ and a plane having the equation $\mathbf{N} \cdot \mathbf{x} = d$, where $d = \mathbf{N} \cdot \mathbf{p}$, and $\mathbf{p}$ is a point on the plane. The distance from the point $\mathbf{q}$ to the plane is the length of the line segment from $\mathbf{q}$ to the plane in the direction of the normal vector $\mathbf{N}$. The equation of the line through $\mathbf{q}$ in the direction of $\mathbf{N}$ is $\boldsymbol{\alpha}(t) = \mathbf{q} + t\mathbf{N}$.

    a. Find the number t_0 for which $\boldsymbol{\alpha}(t_0)$ is a point on the plane.

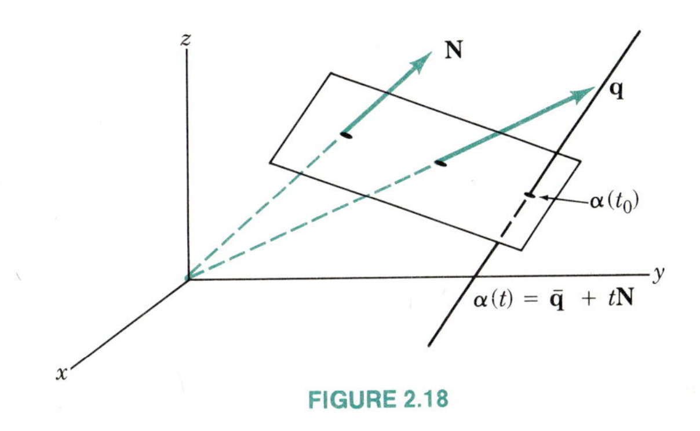

FIGURE 2.18

    b. Since the distance from $\mathbf{q}$ to the plane is $\|\mathbf{q} - \boldsymbol{\alpha}(t_0)\|$, prove that this distance is given by

$$\frac{|d - \mathbf{N} \cdot \mathbf{q}|}{\|\mathbf{N}\|}$$

    c. Does the formula in part (b) make sense in R^n for any n? Why? Explain your answer.

    d. Find the distance from the point

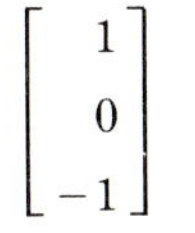

$$\begin{bmatrix} 1 \\ 0 \\ -1 \end{bmatrix}$$

to the plane having the equation $x + 2y + 3z = 1$.

CHAPTER THREE

VECTORS AND VECTOR SPACES

SECTION 3.1 VECTOR SPACES

The set of all $m \times n$ matrices and the set of all n-dimensional vectors, two apparently different sets, have similar mathematical structure. On each of these sets, the operations of addition and scalar multiplication have been defined. With respect to these operations, certain properties, such as the commutative property with respect to addition, are valid. These two mathematical structures are examples of a vector space. Vector spaces provide a body of concepts very useful in mathematics and have useful interpretations in the physical and social sciences. The historical name *vector space* suggests that we are dealing with geometric objects, which is not true. We develop algebraic ideas, with major interpretation in geometry.

We now give an indication how vectors are useful in information theory. The information and pictures that the Mariner space probe sent back to Earth when traveling past the planet Saturn were sent in a series of bursts. Each burst consisted of a string of 0s and 1s. If each string consisted of seven bits of information (0 or 1), one such burst could have been $(1, 0, 0, 1, 1, 0, 1)$. If $F = \{0, 1\}$ with the operation of addition defined by $0 + 0 = 1 + 1 = 0$, $1 + 0 = 0 + 1 = 1$ and multiplication by $1 \cdot 1 = 1$, $0 \cdot 0 = 0 \cdot 1 = 1 \cdot 0 = 0$, then F acts just like R. Hence, if we define F^7 to be the set of all 7-tuples with entries from F, then F^7 acts just like R^7, a vector space with which we are familiar. This vector space has been very useful not only in our space program but is also the basis for the new digital recording systems, which produce a very clear sound.

In this text we consider only vector spaces over R. However, the example just given from coding theory demonstrates that vector spaces over systems other than R are important.

We now give a precise definition of vector space.

Definition 3.1.1 A **vector space V over R** is a nonempty set V of elements called *vectors* with an operation of addition, denoted by $+$, satisfying the properties

A1. For all $\mathbf{v}_1, \mathbf{v}_2$ in V, $\mathbf{v}_1 + \mathbf{v}_2$ is a vector in V.

A2. For all $\mathbf{v}_1, \mathbf{v}_2$ in V, $\mathbf{v}_1 + \mathbf{v}_2 = \mathbf{v}_2 + \mathbf{v}_1$.

A3. For all $\mathbf{v}_1, \mathbf{v}_2, \mathbf{v}_3$ in V, $(\mathbf{v}_1 + \mathbf{v}_2) + \mathbf{v}_3 = \mathbf{v}_1 + (\mathbf{v}_2 + \mathbf{v}_3)$.

A4. There is a vector $\mathbf{0}$ in V such that $\mathbf{0} + \mathbf{v} = \mathbf{v} + \mathbf{0} = \mathbf{v}$ for all $\mathbf{v}$ in V.

A5. For each $\mathbf{v}$ in V, there is a vector $-\mathbf{v}$ in V such that $\mathbf{v} + (-\mathbf{v}) = (-\mathbf{v}) + \mathbf{v} = \mathbf{0}$.

and an operation of multiplication of a vector by a scalar (denoted by juxtaposition) satisfying the properties

M1. For all a in R and $\mathbf{v}$ in V, $a\mathbf{v}$ is a vector in V.

M2. For all a_1, a_2 in R and $\mathbf{v}$ in V, $(a_1a_2)\mathbf{v} = a_1(a_2\mathbf{v})$.

M3. For all a_1, a_2 in R and $\mathbf{v}$ in V, $(a_1 + a_2)\mathbf{v} = a_1\mathbf{v} + a_2\mathbf{v}$.

M4. For all $\mathbf{v}$ in V, $1\mathbf{v} = \mathbf{v}$.

M5. For all a in R and $\mathbf{v}_1, \mathbf{v}_2$ in V, $a(\mathbf{v}_1 + \mathbf{v}_2) = a\mathbf{v}_1 + a\mathbf{v}_2$.

Example 3.1.2 For each $n \geqq 1$, R^n is a vector space over R. ■

Example 3.1.3 Let $V = \{\mathbf{0}\}$; that is, V is a nonempty set containing only the zero vector. Define $\mathbf{0} + \mathbf{0} = \mathbf{0}$ and $r\mathbf{0} = \mathbf{0}$ for all r in R. This is a trivial example of a vector space V, but it is still a vector space with which you should be familiar. ■

Example 3.1.4 Let P_3 be the set of all polynomials of degree less than or equal to 3 over R (that is, the coefficients belong to R) with the addition and scalar multiplication as defined: If $\mathbf{p} = c_0 + c_1x + c_2x^2 + c_3x^3$ and $\mathbf{q} = k_0 + k_1x + k_2x^2 + k_3x^3$ are any polynomials in P_3, the sum of $\mathbf{p}$ and $\mathbf{q}$ is the polynomial

$$\mathbf{p} + \mathbf{q} = (c_0 + k_0) + (c_1 + k_1)x + (c_2 + k_2)x^2 + (c_3 + k_3)x^3$$

which is again in P_3; also, for any real number a, the scalar product of a and $\mathbf{p}$ is the polynomial

$$a\mathbf{p} = ac_0 + ac_1x + ac_2x^2 + ac_3x^3$$

which is again in P_3. Show that P_3 is a vector space.

Solution Properties A_1 and M_1 follow directly from our definitions of addition and scalar multiplication for P_3. Now we show that the remaining

Solution It is enough to show that at least one of A1 through A5 or one of M1 through M5 does not hold. We give several solutions, but any one is adequate.

First, notice that the operation of addition is not closed, since

$$(-3x^3 + 2x^2 + x + 4) + (3x^3 + 8x^2 - 12x + 7) = 10x^2 - 11x + 11$$

The sum of two third-degree polynomials is not necessarily a third-degree polynomial.

Second, the polynomial $\mathbf{0} = 0x^3 + 0x^2 + 0x + 0$ is the identity with respect to addition and it is not a third-degree polynomial.

Third, since there is no identity with respect to addition in the set of polynomials of fixed degree 3, there cannot be an additive inverse. ■

Example 3.1.8 The solution set of a system of an $m \times n$ nonhomogeneous linear equation with addition and scalar multiplication as defined on R^n is *not* a vector space.

Solution For vectors in R^n, the identity is the zero vector and the zero vector is not a solution of a nonhomogeneous system of linear equations. ■

In both examples that are not vector spaces, Examples 3.1.7 and 3.1.8, the defined set of vectors under consideration for a vector space did not contain the zero vector, since the zero polynomial has no degree. For any vector space V, $\mathbf{0}$ is in V by A4. We stress this fact by setting it off as a theorem.

Theorem 3.1.9 *If V is a vector space, then $\mathbf{0}$ belongs to V.*

This result, a restatement of A4 can be restated as follows.

Theorem 3.1.10 *If $\mathbf{0}$ does not belong to V, then V is not a vector space.*

Theorem 3.1.10 is useful to show that a set V is *not* a vector space.

Example 3.1.11 The set of all functions $f\colon R \to R$ is a vector space with vector addition $(f + g)(x) = f(x) + g(x)$ and scalar multiplication $(af)(x) = a \cdot f(x)$. The zero vector is the function $0\colon R \to R$ with $0(x) = 0$ for all real numbers x, whereas the negative of $f\colon R \to R$ is the function $(-f)\colon R \to R$, given by $(-f)(x) = -f(x)$ for all real numbers x. ■

Example 3.1.12 The set of all infinite sequences $(a_1, a_2, a_3, \ldots, a_n, \ldots)$ is a vector space with vector addition

$$(a_1, a_2, \ldots, a_n, \ldots) + (b_1, b_2, \ldots, b_n, \ldots)$$
$$= (a_1 + b_1, a_2 + b_2, \ldots, a_n + b_n, \ldots)$$

and scalar multiplication

$$a(b_1, b_2, \ldots, b_n, \ldots) = (ab_1, ab_2, \ldots, ab_n, \ldots)$$

The zero vector is the sequence $(0, 0, 0, \ldots, 0, \ldots)$, and the negative of the sequence $(a_1, a_2, \ldots, a_n, \ldots)$ is

$$-(a_1, a_2, \ldots, a_n, \ldots) = (-a_1, -a_2, \ldots, -a_n, \ldots)$$ ■

Example 3.1.13 The set P of all polynomials over R is a vector space with respect to the usual definition of addition and scalar multiplication. ■

Having seen a diverse collection of vector spaces, we now list some properties common to all vector spaces.

Theorem 3.1.14 *Let V be a vector space, $\mathbf{u} \in V$ and $a \in R$. Then*

1. $0\mathbf{u} = \mathbf{0}$.
2. $a\mathbf{0} = \mathbf{0}$.
3. *If* $a\mathbf{u} = \mathbf{0}$, *then* $a = 0$ *or* $\mathbf{u} = 0$.
4. $(-1)\mathbf{u} = -\mathbf{u}$.

Proof.

1. Let $0\mathbf{u} = \mathbf{v}$; then $\mathbf{v} + \mathbf{v} = 0\mathbf{u} + 0\mathbf{u} = (0 + 0)\mathbf{u} = 0\mathbf{u} = \mathbf{v}$. Thus $\mathbf{0} = \mathbf{v} - \mathbf{v} = (\mathbf{v} + \mathbf{v}) - \mathbf{v} = \mathbf{v} + (\mathbf{v} - \mathbf{v}) = \mathbf{v} + \mathbf{0} = \mathbf{v} = 0\mathbf{u}$.
3. If $a \neq 0$, then $(1/a)a\mathbf{u} = (1/a)\mathbf{0} = \mathbf{0}$. However, $(1/a)a\mathbf{u} = [(1/a)a]\mathbf{u} = 1\mathbf{u} = \mathbf{u}$, so $\mathbf{u} = \mathbf{0}$.

The proofs of (2) and (4) are left as exercises.

REMARK Part 3 of Theorem 3.1.14 is not intuitively obvious. While the product of two numbers a and b being 0 requires one of them to be zero, we have seen in Chapter 1 that the product of two nonzero matrices A and B can be zero with neither A nor B being 0.

SUMMARY

Terms to know:

Vector space over R

P_n

$R_{n \times n}$

Theorems to know:

If V is a vector space, then $\mathbf{0} \in V$.

Let V be a vector space, $\mathbf{u} \in V$ and $a \in R$. Then

1. $0\mathbf{u} = \mathbf{0}$.
2. $a\mathbf{0} = \mathbf{0}$.
3. If $a\mathbf{u} = \mathbf{0}$ then $a = 0$ or $\mathbf{u} = \mathbf{0}$.
4. $(-1)\mathbf{u} = -\mathbf{u}$.

EXERCISE SET 3.1

Determine whether the sets of vectors given in 1–13 are vector spaces.

1. $P_1 = \{p_0 + p_1x \mid p_0, p_1 \text{ in } R\}$ with the operations of Example 3.1.14.

2. $V = \left\{ \begin{bmatrix} x \\ y \\ z \end{bmatrix} \middle| \, x + 2y = 0 \right\}$ with the operations of R^3.

3. $V = \left\{ \begin{bmatrix} x \\ y \\ z \end{bmatrix} \middle| \, x + 2z = 0,\ x - y = 0 \right\}$ with the operations of R^3.

4. $U = \{\mathbf{p} \mid \mathbf{p} = (p_0, p_1, \ldots, p_n, \ldots),\ p_i \text{ in } R \text{ for all } i\}$ with $\mathbf{p} + \mathbf{q} = (p_0 + q_0, p_1 + q_1, \ldots, p_n + q_n, \ldots)$ and $a\mathbf{p} = (ap_0, ap_1, \ldots, ap_n, \ldots)$.

5. $F = \{f \mid f\colon R \to R \text{ such that } f(0) = 0\}$, where $(f + g)(x) = f(x) + g(x)$ and $(af)(x) = af(x)$.

6. $G = \left\{ \begin{bmatrix} a_{11} & a_{12} \\ a_{21} & a_{22} \end{bmatrix} \middle| \, a_{11}, a_{12}, a_{21}, a_{22} \text{ are in } R \right\}$ under the usual matrix addition and scalar multiplication.

7. $G = \left\{ A = \begin{bmatrix} a_{11} & a_{12} \\ a_{21} & a_{22} \end{bmatrix} \middle| \, a_{11}, a_{12}, a_{21}, a_{22} \text{ are in } R \text{ and } a_{11}a_{22} - a_{12}a_{21} = 0 \right\}$ with the usual matrix addition and scalar multiplication.

8. $U = \{\mathbf{p} = (p_0, p_1, \ldots, p_n, \ldots) \mid p_1 + p_2 + \cdots + p_n = 0\}$ with the operations as defined in Exercise 4.

9. $U = \left\{ \begin{bmatrix} x \\ y \end{bmatrix} \middle| \, x, y \text{ in } R \text{ and } x + y = 1 \right\}$ with the operations of R^2.

10. $U = \left\{ \begin{bmatrix} a_{11} & a_{12} & a_{13} \\ a_{21} & a_{22} & a_{23} \\ a_{31} & a_{32} & a_{33} \end{bmatrix} \middle| \, a_{11} + a_{22} + a_{33} = -1 \right\}$ with the usual operations on matrices.

11. $U = \left\{ \begin{bmatrix} x \\ y \end{bmatrix} \middle| \, x, y \text{ in } R \right\}$ and define

$$\begin{bmatrix} x_1 \\ y_1 \end{bmatrix} + \begin{bmatrix} x_2 \\ y_2 \end{bmatrix} = \begin{bmatrix} x_1 - x_2 \\ y_1 - y_2 \end{bmatrix} \quad \text{and} \quad a\begin{bmatrix} x_1 \\ y_1 \end{bmatrix} = \begin{bmatrix} ax_1 \\ ay_1 \end{bmatrix}$$

12. $U = \{f \mid f\colon R \to R \text{ and } f(1) = 1\}$ with the operations as defined in Exercise 5.

13. $V = \left\{ \begin{bmatrix} x \\ y \\ z \end{bmatrix} \middle| \, x > 0 \right\}$ with the addition and scalar multiplication of R^3.

For those students familiar with calculus, show that the sets of vectors in 14–18 are vector spaces and those in 19–20 are not vector spaces.

14. $U = \{f \mid f\colon R \to R$ and f is continuous$\}$ with the operations as defined in Exercise 5.

15. $U = \{f \mid f\colon [0,1] \to R$ and $\int_0^1 f(x)\,dx$ exists$\}$ with the operations as defined in Exercise 5.

16. $U = \{f \mid f\colon R \to R$ and $f'(x)$ exists for all $x\}$ with the operations as defined in Exercise 5.

17. $U = \{f \mid f\colon R \to R$ and $f'(0) = 0\}$ with the operations as defined in Exercise 5.

18. $U = \{f \mid f\colon R \to R$ and $f'' + f' = 0\}$ with the operations as defined in Exercise 5.

19. $U = \{f \mid f\colon R \to R$ and $f'(0) = 1\}$ with the operations as defined in Exercise 5.

20. $U = \{f \mid f\colon [0,1] \to R$ and $\int_0^1 f(x)\,dx = 3\}$ with the operations as defined in Exercise 5.

21. In a vector space V show that $-\mathbf{v} = -1\mathbf{v}$ for all $\mathbf{v}$ in V.

    Hint: Show that $\mathbf{v} + (-1)\mathbf{v} = \mathbf{0}$.

22. In a vector space V show that $a\mathbf{0} = \mathbf{0}$ for all $\mathbf{v}$ in V.

23. Prove that if $\mathbf{u} + \mathbf{v} = \mathbf{u} + \mathbf{w}$, then $\mathbf{v} = \mathbf{w}$.

24. Use Exercise 23 to prove that if $\mathbf{u} + \mathbf{v} = \mathbf{u}$ for all $\mathbf{u} \in V$, then $\mathbf{v} = \mathbf{0}$.

25. Let $\mathbf{N}$ be a nonzero vector in R^n. Show that the set of all vectors $\mathbf{x}$ in R^n for which $\mathbf{N} \cdot \mathbf{x} = 0$ is a vector space.

26. Is it true that a line through the origin in R^n is a vector space? Why?

SECTION 3.2 SUBSPACES

The vector space R^n contains other vector spaces. For example, the solutions of an $m \times n$ linear homogeneous system, $A\mathbf{x} = \mathbf{0}$, is a vector space that is contained in R^n.

In view of this, we give a definition.

Definition 3.2.1 Let S be a nonempty subset of a vector space V. S is a **subspace** of V if and only if S is a vector space with respect to the operations in V.

The set of all vectors in R^3 that have the form

$$\begin{bmatrix} 0 \\ x_2 \\ x_3 \end{bmatrix}$$

is a subspace of R^3. To prove this, we need to show that this set satisfies the definition of a vector space. Fortunately, there is a quicker way, as the following theorem indicates.

Theorem 3.2.2 *Let S be a nonempty subset of a vector space V. S is a subspace of V if and only if S satisfies the following conditions:*

1. *Whenever $\mathbf{x} \in S$ and $a \in R$, then $a\mathbf{x} \in S$.*
2. *Whenever $\mathbf{x}, \mathbf{y} \in S$, then $\mathbf{x} + \mathbf{y} \in S$.*

In other words, S is a subspace of V whenever S is nonempty and is closed under vector addition and scalar multiplication.

Proof. We first show that if S satisfies (1) and (2), then S is a vector space. Many of the properties required to be shown to prove that S is a vector space stem from the fact S is a subset of V and the operations on S are those in V. By (2), note that if $\mathbf{x}$ and $\mathbf{y}$ are in S, then $\mathbf{x} + \mathbf{y} \in S$. Thus, for example, to verify that the property A2 holds, note that if $\mathbf{x}, \mathbf{y} \in S$, then $\mathbf{x}, \mathbf{y} \in V$. But for $\mathbf{x}, \mathbf{y} \in V$, $\mathbf{x} + \mathbf{y} = \mathbf{y} + \mathbf{x}$. Since $\mathbf{x} + \mathbf{y}, \mathbf{y} + \mathbf{x} \in S$, and $\mathbf{x} + \mathbf{y} = \mathbf{y} + \mathbf{x}$, A2 holds in S. In the same manner, A3 can be verified.

Turning to scalar multiplication, note first that for $r \in R$ and $\mathbf{x} \in S$, $r\mathbf{x} \in S$ by (1). Thus, for example, to verify M2, if $a_1, a_2 \in R$ and $\mathbf{x} \in S$, then by (1), $(a_1a_2)\mathbf{x}$ and $a_2\mathbf{x}$ both lie in S. Applying (1) again to $a_2\mathbf{x}$, $a_1(a_2\mathbf{x}) \in S$. But by M2 for V, $(a_1a_2)\mathbf{x} = a_1(a_2\mathbf{x})$. Since both lie in S, M2 holds for S. In a similar manner, M3, M4, and M5 can be verified to hold in S. Thus we need only verify the presence in S of the zero vector and that the inverse of each vector in S lies in S. Since S is closed under scalar multiplication by (1), $(-1)\mathbf{x}$ lies in S. But $(-1)\mathbf{x} = -\mathbf{x}$, and so $-\mathbf{x} \in S$. Thus the inverse of any vector in S is also in S. The closure of S with respect to addition and the assumption that S is nonempty imply, by (2), that if $\mathbf{x} \in S$, then $\mathbf{x} + (-\mathbf{x}) = \mathbf{0}$ is in S.

Conversely, if S is a vector space, then S is nonempty and closed under vector addition and scalar multiplication, making S a subspace. Thus properties (1) and (2) are satisfied by the vectors in S.

To show that an arbitrary nonempty set of vectors is a vector space under addition and scalar multiplication we must verify that properties A1 through A5 as well as properties M1 through M5 are satisfied by the vectors in S. However, to verify that an arbitrary nonempty subset of a vector space is a subspace, we need only show that this set is closed under addition and scalar multiplication (i.e., satisfies properties (1) and (2) of Theorem 3.2.2). To show this is usually much less time consuming than is verifying that the ten properties of a vector space are satisfied.

Example 3.2.3 The set S of all vectors in R^3 that have the form

$$\begin{bmatrix} 0 \\ x_2 \\ x_3 \end{bmatrix}$$

is a subspace of R^3 since:

First, S is *nonempty* since $\begin{bmatrix} 0 \\ 0 \\ 0 \end{bmatrix}$ is in S.

S is closed under addition. Let

$$\begin{bmatrix} 0 \\ x_2 \\ x_3 \end{bmatrix} \quad \text{and} \quad \begin{bmatrix} 0 \\ y_2 \\ y_3 \end{bmatrix}$$

be arbitrary vectors in S. Then

$$\begin{bmatrix} 0 \\ x_2 \\ x_3 \end{bmatrix} + \begin{bmatrix} 0 \\ y_2 \\ y_3 \end{bmatrix} = \begin{bmatrix} 0 + 0 \\ x_2 + y_2 \\ x_3 + y_3 \end{bmatrix} = \begin{bmatrix} 0 \\ x_2 + y_2 \\ x_3 + y_3 \end{bmatrix}$$

is a vector in S since it has the form that is required to be in S.

S is closed under scalar multiplication. Let

$$\begin{bmatrix} 0 \\ x_2 \\ x_3 \end{bmatrix}$$

be an arbitrary vector in S and a be any real number. Then

$$a \begin{bmatrix} 0 \\ x_2 \\ x_3 \end{bmatrix} = \begin{bmatrix} 0 \\ ax_2 \\ ax_3 \end{bmatrix}$$

is a vector in S since it has the required form to be in S. ■

Example 3.2.4 Show that

$$S = \left\{ \begin{bmatrix} x_1 \\ x_2 \\ x_3 \end{bmatrix} \middle| \, x_1 + x_2 = 0 \right\}$$

is a subspace of R^3.

Solution First, S is nonempty since it contains

$$\begin{bmatrix} 0 \\ 0 \\ 0 \end{bmatrix}$$

Let

$$\begin{bmatrix} x_1 \\ x_2 \\ x_3 \end{bmatrix} \quad \text{and} \quad \begin{bmatrix} y_1 \\ y_2 \\ y_3 \end{bmatrix}$$

be in S. Then $x_1 + x_2 = 0$ and $y_1 + y_2 = 0$. Their sum,

$$\begin{bmatrix} x_1 \\ x_2 \\ x_3 \end{bmatrix} + \begin{bmatrix} y_1 \\ y_2 \\ y_3 \end{bmatrix} = \begin{bmatrix} x_1 + y_1 \\ x_2 + y_2 \\ x_3 + y_3 \end{bmatrix}$$

is a vector in S since $(x_1 + y_1) + (x_2 + y_2) = (x_1 + x_2) + (y_1 + y_2) = 0 + 0 = 0$.

If

$$\begin{bmatrix} x_1 \\ x_2 \\ x_3 \end{bmatrix}$$

is an arbitrary vector in S, then $x_1 + x_2 = 0$ and

$$a\begin{bmatrix} x_1 \\ x_2 \\ x_3 \end{bmatrix} = \begin{bmatrix} ax_1 \\ ax_2 \\ ax_3 \end{bmatrix}$$

is in S since $(ax_1) + (ax_2) = a(x_1 + x_2) = a \cdot 0 = 0$. ■

Example 3.2.5 If V is any vector space, then V is a subspace of itself. Furthermore, the set $S = \{\mathbf{0}\}$ is a subspace of V. This shows every vector space does contain at least one vector subspace. ■

Example 3.2.6 Let S be the set of all vectors of the form

$$a\begin{bmatrix} 1 \\ 0 \\ 1 \\ 0 \\ 1 \end{bmatrix} + b\begin{bmatrix} -2 \\ -1 \\ 0 \\ 1 \\ 2 \end{bmatrix}$$

where a and b are arbitrary real numbers. Show that S is a subspace of R^5.

Solution First we note that S is nonempty, since S contains the zero vector of R^5 obtained by letting $a = b = 0$. Let

$$a_1\begin{bmatrix} 1 \\ 0 \\ 1 \\ 0 \\ 1 \end{bmatrix} + b_1\begin{bmatrix} -2 \\ -1 \\ 0 \\ 1 \\ 2 \end{bmatrix} \quad \text{and} \quad a_2\begin{bmatrix} 1 \\ 0 \\ 1 \\ 0 \\ 1 \end{bmatrix} + b_2\begin{bmatrix} -2 \\ -1 \\ 0 \\ 1 \\ 2 \end{bmatrix}$$

be arbitrary vectors in S. Their sum,

$$\left(a_1\begin{bmatrix}1\\0\\1\\0\\1\end{bmatrix} + b_1\begin{bmatrix}-2\\-1\\0\\1\\2\end{bmatrix}\right) + \left(a_2\begin{bmatrix}1\\0\\1\\0\\1\end{bmatrix} + b_2\begin{bmatrix}-2\\-1\\0\\1\\2\end{bmatrix}\right)$$

$$= (a_1 + a_2)\begin{bmatrix}1\\0\\1\\0\\1\end{bmatrix} + (b_1 + b_2)\begin{bmatrix}-2\\-1\\0\\1\\2\end{bmatrix}$$

is a vector belonging to S. Furthermore,

$$\alpha\left(a\begin{bmatrix}1\\0\\1\\0\\1\end{bmatrix} + b\begin{bmatrix}-2\\-1\\0\\1\\2\end{bmatrix}\right) = (\alpha a)\begin{bmatrix}1\\0\\1\\0\\1\end{bmatrix} + (\alpha b)\begin{bmatrix}-2\\-1\\0\\1\\2\end{bmatrix}$$

is a vector belonging to S. Thus S is a subspace of R^5. ■

Example 3.2.7 The set S of polynomials ax^3 for an arbitrary number a is a subspace of all polynomials of degree less than or equal to 3 since

$$0x^3 = 0 \quad \text{is in } S$$

$$ax^3 + bx^3 = (a + b)x^3$$

and

$$\alpha(ax^3) = (\alpha a)x^3$$ ■

Example 3.2.8 The set S of 2×2 matrices

$$\begin{bmatrix} a & b \\ c & d \end{bmatrix}$$

having the property $a + d = 0$ (sum of diagonal elements is 0) is a subspace of $M_{2\times 2}$ because

$$\begin{bmatrix} 0 & 0 \\ 0 & 0 \end{bmatrix}$$

is in S and if

$$\begin{bmatrix} a & b \\ c & d \end{bmatrix} \quad \text{and} \quad \begin{bmatrix} x & y \\ z & w \end{bmatrix}$$

are in S ($a + d = 0$ and $x + w = 0$), then

$$\begin{bmatrix} a & b \\ c & d \end{bmatrix} + \begin{bmatrix} x & y \\ z & w \end{bmatrix} = \begin{bmatrix} a + x & b + y \\ c + z & d + w \end{bmatrix}$$

is in S, since $(a + x) + (d + w) = (a + d) + (x + w) = 0 + 0 = 0$. Also

$$\alpha \begin{bmatrix} a & b \\ c & d \end{bmatrix} = \begin{bmatrix} \alpha a & \alpha b \\ \alpha c & \alpha d \end{bmatrix}$$

is in S since $(\alpha a) + (\alpha d) = \alpha(a + d) = \alpha(0) = 0$. ■

Not all subsets of vector spaces are subspaces, as the following examples show.

Example 3.2.9 Let

$$S = \left\{ \begin{bmatrix} x \\ y \\ z \end{bmatrix} \middle| x + y = 1 \right\}$$

Then S is not a subspace of R^3 because if

$$\begin{bmatrix} x \\ y \\ z \end{bmatrix}$$

is in S, then

$$0 \begin{bmatrix} x \\ y \\ z \end{bmatrix} = \begin{bmatrix} 0 \\ 0 \\ 0 \end{bmatrix}$$

must belong to S. The sum of the first and second entries of

$$\begin{bmatrix} 0 \\ 0 \\ 0 \end{bmatrix}$$

is not 1 (the requirement needed to be placed in S) and so

$$0 \begin{bmatrix} x \\ y \\ z \end{bmatrix}$$

is not in S. ■

Example 3.2.10 The set of all polynomials of degree 3 is not a subspace of the set of polynomials of degree less than or equal to 3 since $2x^3$ and $-2x^3$ are polynomials of degree 3 and $(2x^3) + (-2x^3) = 0$ is not a polynomial of degree 3. (Why is this example different from Example 3.2.7?) ■

The fact that the zero vector, $\mathbf{0}$, always belongs to a vector space is very useful in showing that a certain subset S of a vector space V is not a subspace of V by showing that $\mathbf{0}$ does not belong to S.

Example 3.2.11 The set S of all

$$\begin{bmatrix} x \\ y \\ z \end{bmatrix}$$

such that $x - y + z = 1$ is not a subspace of R^3, since the zero vector,

$$\begin{bmatrix} 0 \\ 0 \\ 0 \end{bmatrix}$$

is not in S. It does not satisfy the property $0 - 0 + 0 = 1$. ■

The set of solutions of an $m \times n$ homogeneous system of linear equations $A\mathbf{x} = \mathbf{0}$ is a subspace of R^n. Since the equation of a line through the origin in R^2 is $y = mx$, or $y - mx = 0$ (a homogeneous system), a line through the origin in R^2 is a subspace. In a like manner, since the equations of a plane through the origin is $ax + by + cz = 0$ (a homogeneous system), a plane is a subspace of R^3.

SUMMARY

Terms to know:

Subspace

Theorems to know:

S is a subspace of the vector space V if and only if

1. $\mathbf{v}$ is in S and r is a real number, then $r\mathbf{v}$ is in S.
2. $\mathbf{v}_1$ and $\mathbf{v}_2$ are in S, then $\mathbf{v}_1 + \mathbf{v}_2$ is in S.
3. S is nonempty.

EXERCISE SET 3.2

1. Which of the following subsets of R^2 are subspaces of R^2?

a. $\left\{\begin{bmatrix} x \\ y \end{bmatrix} \middle| x = y\right\}$ b. $\left\{\begin{bmatrix} x \\ y \end{bmatrix} \middle| y = 2x\right\}$

c. $\left\{\begin{bmatrix} x \\ y \end{bmatrix} \middle| x + y = 0\right\}$ d. $\left\{\begin{bmatrix} x \\ y \end{bmatrix} \middle| y = 0\right\}$

e. $\left\{\begin{bmatrix} x \\ y \end{bmatrix} \middle| x + y = 1\right\}$ f. $\left\{\begin{bmatrix} x \\ y \end{bmatrix} \middle| \alpha x + \beta y = 0 \text{ for fixed } \alpha, \beta \in R\right\}$

2. Determine whether the following are subspaces of R^3.

a. $\left\{ \begin{bmatrix} x \\ y \\ z \end{bmatrix} \middle| z = 0 \right\}$ b. $\left\{ \begin{bmatrix} x \\ y \\ z \end{bmatrix} \middle| x + y + z = 1 \right\}$

c. $\left\{ \begin{bmatrix} x \\ y \\ z \end{bmatrix} \middle| x + y + z = 0 \right\}$ d. $\left\{ \begin{bmatrix} x \\ y \\ z \end{bmatrix} \middle| 3x - y = 0 \right\}$

e. $\left\{ \begin{bmatrix} x \\ y \\ z \end{bmatrix} \middle| x - y + z = 1 \right\}$ f. $\left\{ \begin{bmatrix} x \\ y \\ z \end{bmatrix} \middle| x \geqq y \geqq z \right\}$

3. Which of the following subsets of R^4 are subspaces, where

$$X = \begin{bmatrix} x_1 \\ x_2 \\ x_3 \\ x_4 \end{bmatrix}$$

a. all X, where $x_2 = 2x_1$, $x_3 = x_1 + x_2$

b. all X, where $x_2 = x_3 = 0$

c. all X, where $3x_1 - 2x_2 = 0$

4. Determine whether the following are subspaces of the vector space of all 2×2 matrices.

a. $\left\{ \begin{bmatrix} a & b \\ c & d \end{bmatrix} \middle| b = 0,\ d = a + c \right\}$

b. $\left\{ \begin{bmatrix} a & b \\ c & d \end{bmatrix} \middle| a = d = 0 \right\}$

c. $\left\{ \begin{bmatrix} a & b \\ c & d \end{bmatrix} \middle| a = c = 0,\ b = d = 1 \right\}$

d. $\left\{ \begin{bmatrix} a & b \\ c & d \end{bmatrix} \middle| a + b = c + d = a + c = b + d = 0 \right\}$

5. Are the following subsets subspaces of P?

a. $\{p \in P \mid p = p_0 + p_2x^2 + \cdots + p_kx^k\}$

b. $\{p \in P \mid p_0 + p_1 + p_4 = 0\}$

6. Which of the following are subspaces of the space of all functions?

a. all functions $f(x)$ with $2f(0) = f(1)$

b. all functions $f(x)$ with $0 + f(1) = f(0) + 1$

c. all functions $f(x)$ with $f(x) = f(x - 1)$

d. all positive functions

e. all even functions (i.e., $f(-x) = f(x)$ for all x)

f. all odd functions (i.e., $f(-x) = -f(x)$ for all x)

7. a. Let $\mathbf{x}$ be a specific vector in a vector space V. Show that $\{a\mathbf{x} \mid a \in R\}$ is a subspace of V.

   b. Let $\mathbf{x}$ and $\mathbf{y}$ be two specific vectors in a vector space V. Show that $\{a\mathbf{x} + b\mathbf{y}\colon a, b \in R\}$ is a subspace of V.

   c. Let

$$V = R^2 \quad \text{and} \quad x = \begin{bmatrix} 2 \\ -3 \end{bmatrix}$$

   Describe

$$\left\{ a\begin{bmatrix} 2 \\ -3 \end{bmatrix} \middle| \, a \in R \right\}$$

   geometrically.

   d. Let $V = R^3$,

$$\mathbf{x} = \begin{bmatrix} 2 \\ -3 \\ 1 \end{bmatrix} \quad \text{and} \quad \mathbf{y} = \begin{bmatrix} 1 \\ 0 \\ 1 \end{bmatrix}$$

   Describe

$$\left\{ a\begin{bmatrix} 2 \\ -3 \\ 1 \end{bmatrix} + b\begin{bmatrix} 1 \\ 0 \\ 1 \end{bmatrix} \middle| \, a, b \in R \right\}$$

   geometrically.

8. Show that if S and T are two subspaces of the same vector space V, then $S \cap T$ is also a subspace but $S \cup T$ is not.

9. Let $\mathbf{N}$ be a nonzero vector in R^n. Prove that the set of vectors $\mathbf{x}$ in R^n for which $\mathbf{N} \cdot \mathbf{x} = 0$ is a subspace of R^n.

SECTION 3.3 SPANNING SETS

A vector space may be regarded as the set of all combinations of sums and scalar multiples of vectors.

Definition 3.3.1 A vector of the form $a_1\mathbf{x}_1 + a_2\mathbf{x}_2 + \cdots + a_t\mathbf{x}_t$ is called a **linear combination** of the vectors $\mathbf{x}_1, \mathbf{x}_2, \ldots, \mathbf{x}_t$.

Example 3.3.2 The vector

$$\begin{bmatrix} 7 \\ 10 \\ 16 \end{bmatrix}$$

in R^3 is a linear combination of the vectors

$$\begin{bmatrix} 2 \\ 0 \\ 2 \end{bmatrix}, \quad \begin{bmatrix} 1 \\ 2 \\ 0 \end{bmatrix}, \quad \text{and} \quad \begin{bmatrix} 0 \\ 1 \\ 3 \end{bmatrix}$$

since there exist scalars $a_1 = 2$, $a_2 = 3$, and $a_3 = 4$ such that

$$\begin{bmatrix} 7 \\ 10 \\ 16 \end{bmatrix} = 2\begin{bmatrix} 2 \\ 0 \\ 2 \end{bmatrix} + 3\begin{bmatrix} 1 \\ 2 \\ 0 \end{bmatrix} + 4\begin{bmatrix} 0 \\ 1 \\ 3 \end{bmatrix}$$ ■

Example 3.3.3 The vector $-2 + 3x + x^3$ belonging to P_3 is a linear combination of the vectors 1, $1 + x$, $1 + x + x^2$, and $1 + x + x^2 + x^3$, since

$$-2 + 3x + x^3 = -5(1) + 3(1 + x) + (-1)(1 + x + x^2) + 1(1 + x + x^2 + x^3)$$ ■

Example 3.3.4 Show that the vector

$$\begin{bmatrix} 3 \\ 6 \end{bmatrix}$$

is a linear combination of the vectors

$$\begin{bmatrix} 2 \\ 4 \end{bmatrix} \quad \text{and} \quad \begin{bmatrix} 1 \\ 5 \end{bmatrix}$$

Solution We need to determine scalars a_1 and a_2 such that

$$a_1\begin{bmatrix} 2 \\ 4 \end{bmatrix} + a_2\begin{bmatrix} 1 \\ 5 \end{bmatrix} = \begin{bmatrix} 3 \\ 6 \end{bmatrix}$$

This equation can be rewritten as

$$\begin{aligned} 2a_1 + a_2 &= 3 \\ 4a_1 + 5a_2 &= 6 \end{aligned}$$

or as

$$\begin{bmatrix} 2 & 1 \\ 4 & 5 \end{bmatrix}\begin{bmatrix} a_1 \\ a_2 \end{bmatrix} = \begin{bmatrix} 3 \\ 6 \end{bmatrix}$$

(The solution of this system exists and is $a_1 = \frac{3}{2}$ and $a_2 = 0$.) ■

Example 3.3.5 The vector

$$\begin{bmatrix} 7 & 6 \\ 9 & 4 \end{bmatrix}$$

in the set of all 2×2 matrices is a linear combination of

$$\begin{bmatrix} 1 & 0 \\ 0 & 0 \end{bmatrix}, \quad \begin{bmatrix} 0 & -2 \\ 0 & 0 \end{bmatrix}, \quad \begin{bmatrix} 1 & 0 \\ 1 & 0 \end{bmatrix}, \quad \text{and} \quad \begin{bmatrix} 0 & 0 \\ -1 & -1 \end{bmatrix}$$

since for scalars $a_1 = 2$, $a_2 = -3$, $a_3 = 5$, and $a_4 = -4$, we have

$$\begin{bmatrix} 7 & 6 \\ 9 & 4 \end{bmatrix} = 2\begin{bmatrix} 1 & 0 \\ 0 & 0 \end{bmatrix} + (-3)\begin{bmatrix} 0 & -2 \\ 0 & 0 \end{bmatrix} + 5\begin{bmatrix} 1 & 0 \\ 1 & 0 \end{bmatrix} + (-4)\begin{bmatrix} 0 & 0 \\ -1 & -1 \end{bmatrix}$$ ■

Example 3.3.6 The infinite sequence $\mathbf{v} = (1, 2, 3, 4, 5, 6, \ldots)$ is a linear combination of $\mathbf{v}_1 = (1, 1, 1, \ldots)$, $\mathbf{v}_2 = (0, 1, 1, 1, \ldots)$, $\mathbf{v}_3 = (0, 0, 1, 1, 1, \ldots)$, $\mathbf{v}_4 = (0, 0, 0, 1, 1, 1, \ldots)$, $\mathbf{v}_5 = (0, 0, 0, 0, 1, 1, \ldots), \ldots$, since

$$\mathbf{v} = (1)\mathbf{v}_1 + 1(\mathbf{v}_2) + (1)\mathbf{v}_3 + 1(\mathbf{v}_4) + (1)\mathbf{v}_5 + \cdots$$ ■

The following example shows how linear combinations of vectors can be useful in describing a vector space.

Example 3.3.7 The vector space

$$a\begin{bmatrix} 3 \\ -2 \\ 1 \\ 0 \end{bmatrix} + b\begin{bmatrix} -1 \\ 0 \\ 0 \\ 1 \end{bmatrix}, \qquad a, b \in R$$

of all solutions of the system

$$\begin{aligned} 3x_1 + 5x_2 + x_3 + 3x_4 &= 0 \\ 2x_2 + 4x_3 &= 0 \end{aligned}$$

is the set of all linear combinations of

$$\begin{bmatrix} 3 \\ -2 \\ 1 \\ 0 \end{bmatrix} \quad \text{and} \quad \begin{bmatrix} -1 \\ 0 \\ 0 \\ 1 \end{bmatrix}$$

In this situation, we say that the set of vectors

$$\left\{ \begin{bmatrix} 3 \\ -2 \\ 1 \\ 0 \end{bmatrix}, \begin{bmatrix} -1 \\ 0 \\ 0 \\ 1 \end{bmatrix} \right\}$$

spans the vector space. ■

The following definition makes this idea precise.

Definition 3.3.8 Let $\mathbf{x}_1, \mathbf{x}_2, \ldots, \mathbf{x}_t$ be elements of a vector space V. The set **spanned** by $\{\mathbf{x}_1, \mathbf{x}_2, \ldots, \mathbf{x}_t\}$ is defined to be the set of all linear combinations

$$a_1\mathbf{x}_1 + a_2\mathbf{x}_2 + \cdots + a_t\mathbf{x}_t$$

for scalars any $a_1, a_2, \ldots, a_t$, under the addition and scalar multiplication operations of the vector space V.

The set spanned by $\{\mathbf{x}_1, \ldots, \mathbf{x}_t\}$ is denoted by $\langle \mathbf{x}_1, \mathbf{x}_2, \ldots, \mathbf{x}_t \rangle$.

Example 3.3.9 Let

$$x_1 = \begin{bmatrix} 2 \\ 5 \\ 0 \end{bmatrix} \quad \text{and} \quad x_2 = \begin{bmatrix} -1 \\ 0 \\ 3 \end{bmatrix}$$

Find $\langle x_1, x_2 \rangle$.

Solution By the definition,

$$\langle x_1, x_2 \rangle = \left\{ a\begin{bmatrix} 2 \\ 5 \\ 0 \end{bmatrix} + b\begin{bmatrix} -1 \\ 0 \\ 3 \end{bmatrix} \middle| \, a, b \in R \right\}$$

$$= \left\{ \begin{bmatrix} 2a - b \\ 5a \\ 3b \end{bmatrix} \middle| \, a, b \in R \right\}$$

■

Theorem 3.3.10 shows that the space spanned by a set of vectors is always a subspace.

Theorem 3.3.10 *Let $S = \{\mathbf{x}_1, \ldots, \mathbf{x}_t\}$ be a nonempty subset of a space V. Then the set $\langle S \rangle$ spanned by S is a subspace of V.*

Proof. Let $a_1\mathbf{x}_1 + a_2\mathbf{x}_2 + \cdots + a_t\mathbf{x}_t$ and $b_1\mathbf{x}_1 + b_2\mathbf{x}_2 + \cdots + b_t\mathbf{x}_t$ be two arbitrary vectors in $\langle S \rangle$. Then the sum

$$(a_1\mathbf{x}_1 + a_2\mathbf{x}_2 + \cdots + a_t\mathbf{x}_t) + (b_1\mathbf{x}_1 + b_2\mathbf{x}_2 + \cdots + b_t\mathbf{x}_t)$$

$$= (a_1 + b_1)\mathbf{x}_1 + (a_2 + b_2)\mathbf{x}_2 + \cdots + (a_t + b_t)\mathbf{x}_t$$

is a linear combination of $x_1, x_2, \ldots, x_t$ and is thus $\langle S \rangle$. Therefore, $\langle S \rangle$ is closed under addition.

Furthermore, the scalar multiple

$$\alpha(a_1\mathbf{x}_1 + a_2\mathbf{x}_2 + \cdots + a_t\mathbf{x}_t) = (\alpha a_1)\mathbf{x}_1 + (\alpha a_2)\mathbf{x}_2 + \cdots + (\alpha a_t)\mathbf{x}_t$$

is a linear combination of $\mathbf{x}_1, \mathbf{x}_2, \ldots, \mathbf{x}_t$ and is in $\langle S \rangle$. Thus $\langle S \rangle$ is closed under scalar multiplication.

This theorem shows that the set $\langle \mathbf{x}_1, \mathbf{x}_2, \ldots, \mathbf{x}_t \rangle$ spanned by $S = \{x_1, x_2, \ldots, x_t\}$ is a vector space. The set

$$S = \{x_1, x_2, \ldots, x_t\}$$

is called a **set of generators**, or **spanning set** for $\langle S \rangle$ since it generates the subspace $\langle S \rangle = \langle \mathbf{x}_1, \mathbf{x}_2, \ldots, \mathbf{x}_t \rangle$ under the two vector space operations of V.

R^2 is spanned by

$$\left\{ \begin{bmatrix} 1 \\ 0 \end{bmatrix}, \begin{bmatrix} 0 \\ 1 \end{bmatrix} \right\}$$

since an arbitrary vector

$$\begin{bmatrix} x \\ y \end{bmatrix}$$

in R^2 can be written as the linear combination

$$\begin{bmatrix} x \\ y \end{bmatrix} = x \begin{bmatrix} 1 \\ 0 \end{bmatrix} + y \begin{bmatrix} 0 \\ 1 \end{bmatrix}$$

The next two examples show that other sets of vectors span R^2.

Example 3.3.11 Show that R^2 is spanned by the set

$$\left\{ \begin{bmatrix} 1 \\ 0 \end{bmatrix}, \begin{bmatrix} 1 \\ 1 \end{bmatrix} \right\}$$

Solution We need to verify that any arbitrary vector

$$\begin{bmatrix} x \\ y \end{bmatrix}$$

can be written as a linear combination of

$$\begin{bmatrix} 1 \\ 0 \end{bmatrix} \quad \text{and} \quad \begin{bmatrix} 1 \\ 1 \end{bmatrix}$$

We need to show there exist scalars a_1 and a_2 such that

$$a_1 \begin{bmatrix} 1 \\ 0 \end{bmatrix} + a_2 \begin{bmatrix} 1 \\ 1 \end{bmatrix} = \begin{bmatrix} x \\ y \end{bmatrix}$$

Rewriting this equation yields the system

$$a_1 + a_2 = x$$

$$a_2 = y$$

and our problem reduces to determining whether or not the system has a solution. There is a solution for any choice of x and y. To see this, choose $a_2 = y$ and $a_1 = x - y$; then

$$(x-y)\begin{bmatrix}1\\0\end{bmatrix} + y\begin{bmatrix}1\\1\end{bmatrix} = \begin{bmatrix}x-y\\0\end{bmatrix} + \begin{bmatrix}y\\y\end{bmatrix} = \begin{bmatrix}x\\y\end{bmatrix}$$ ■

Example 3.3.12 Verify that R^2 is also spanned by

$$\left\{\begin{bmatrix}2\\0\end{bmatrix}, \begin{bmatrix}0\\1\end{bmatrix}, \begin{bmatrix}3\\4\end{bmatrix}\right\}$$

Solution This set of vectors will span R^2 provided every vector can be written as a linear combination of the given vectors. There must exist scalars a_1, a_2, and a_3 such that

$$a_1\begin{bmatrix}2\\0\end{bmatrix} + a_2\begin{bmatrix}0\\1\end{bmatrix} + a_3\begin{bmatrix}3\\4\end{bmatrix} = \begin{bmatrix}x\\y\end{bmatrix}$$

Equivalently, the system

$$\begin{aligned} 2a_1 \qquad + 3a_3 &= x \\ a_2 + 4a_3 &= y \end{aligned}$$

must have a solution for any choice of x and y in order for the given vectors to be a spanning set for R^2. This system does have a solution; namely, choose a_3 arbitrarily and then

$$a_1 = -\frac{3}{2}a_3 + \frac{x}{2}$$

$$a_2 = -4a_3 + y$$

For example, suppose

$$\begin{bmatrix}x\\y\end{bmatrix} = \begin{bmatrix}4\\3\end{bmatrix}$$

that is, $x = 4$ and $y = 3$. Then, choosing $a_3 = 2$ yields

$$a_1 = -\frac{3}{2}(2) + \frac{4}{2} = -1$$

$$a_2 = -4(2) + 3 = -5$$

Thus

$$\begin{bmatrix}4\\3\end{bmatrix} = -1\begin{bmatrix}2\\0\end{bmatrix} - 5\begin{bmatrix}0\\1\end{bmatrix} + 2\begin{bmatrix}3\\4\end{bmatrix}$$ ■

The preceding examples show that the set of vectors that spans R^2 is not unique. In fact, two spanning sets had two elements and the other had three elements. This is true for other vector spaces, as the next few examples demonstrate.

Example 3.3.13 $P_3 = \{p_0 + p_1x + p_2x^2 + p_3x^3 \mid p_0, p_1, p_2, p_3 \in R\}$ is spanned by $\{\mathbf{v}_1 = 1, \mathbf{v}_2 = x, \mathbf{v}_3 = x^2, \mathbf{v}_4 = x^3\}$, since every vector in P_3 is a linear combination of 1, x, x^2, and x^3.

P_3 is also spanned by $\{\mathbf{v}_1 = 1, \mathbf{v}_2 = 1 + x, \mathbf{v}_3 = 1 + x + x^2, \mathbf{v}_4 = 1 + x + x^2 + x^3\}$. To see this, let $p = p_0 + p_1x + p_2x^2 + p_3x^3$ be any polynomial of degree less than or equal 3. We must find scalars a_1, a_2, a_3, and a_4 for which

$$\begin{aligned} p &= a_1(1) + a_2(1 + x) + a_3(1 + x + x^2) + a_4(1 + x + x^2 + x^3) \\ &= (a_1 + a_2 + a_3 + a_4) + (a_2 + a_3 + a_4)x + (a_3 + a_4)x^2 + a_4x^3 \end{aligned}$$

By equality of polynomials

$$\begin{aligned} p_0 &= a_1 + a_2 + a_3 + a_4 \\ p_1 &= \phantom{a_1 + {}} a_2 + a_3 + a_4 \\ p_2 &= \phantom{a_1 + a_2 + {}} a_3 + a_4 \\ p_3 &= \phantom{a_1 + a_2 + a_3 + {}} a_4 \end{aligned}$$

Backsolving, we obtain

$$a_4 = p_3, a_3 = p_2 - p_3, a_2 = p_1 - p_2, \quad \text{and} \quad a_1 = p_0 - p_1 \qquad \blacksquare$$

Example 3.3.14 The set of matrices

$$\left\{\begin{bmatrix} 1 & 0 \\ 0 & 0 \end{bmatrix}, \begin{bmatrix} 0 & 1 \\ 0 & 0 \end{bmatrix}, \begin{bmatrix} 0 & 0 \\ 1 & 0 \end{bmatrix}, \begin{bmatrix} 0 & 0 \\ 0 & 1 \end{bmatrix}\right\}$$

spans the space of all 2×2 matrices, since for arbitrary

$$\begin{bmatrix} a & b \\ c & d \end{bmatrix}$$

we have

$$\begin{bmatrix} a & b \\ c & d \end{bmatrix} = a\begin{bmatrix} 1 & 0 \\ 0 & 0 \end{bmatrix} + b\begin{bmatrix} 0 & 1 \\ 0 & 0 \end{bmatrix} + c\begin{bmatrix} 0 & 0 \\ 1 & 0 \end{bmatrix} + d\begin{bmatrix} 0 & 0 \\ 0 & 1 \end{bmatrix}$$

The set of matrices

$$\left\{\begin{bmatrix} 1 & 0 \\ 0 & 1 \end{bmatrix}, \begin{bmatrix} 1 & 0 \\ 0 & -1 \end{bmatrix}, \begin{bmatrix} 1 & 1 \\ 0 & 0 \end{bmatrix}, \begin{bmatrix} 1 & 0 \\ 1 & 0 \end{bmatrix}\right\}$$

also spans this space, since for arbitrary

$$\begin{bmatrix} a & b \\ c & d \end{bmatrix}$$

we can find scalars a_1, a_2, a_3, and a_4 such that

$$\begin{bmatrix} a & b \\ c & d \end{bmatrix} = a_1\begin{bmatrix} 1 & 0 \\ 0 & 1 \end{bmatrix} + a_2\begin{bmatrix} 1 & 0 \\ 0 & -1 \end{bmatrix} + a_3\begin{bmatrix} 1 & 1 \\ 0 & 0 \end{bmatrix} + a_4\begin{bmatrix} 1 & 0 \\ 1 & 0 \end{bmatrix}$$

Solving this last equation is equivalent to solving the system

$$\begin{aligned} a_1 + a_2 + a_3 + a_4 &= a \\ a_3 \qquad &= b \\ a_4 &= c \\ a_1 - a_2 \qquad\qquad &= d \end{aligned}$$

The solutions exist, and they are

$$\begin{aligned} a_1 &= \frac{1}{2}(a - b - c + d) \\ a_2 &= \frac{1}{2}(a - b - c - d) \\ a_3 &= b \\ a_4 &= c \end{aligned}$$

■

One problem that arises concerning spanning sets is whether a given vector space is spanned by some given set of vectors in that space. The solution to this problem, as we have seen, reduces to determining whether a system of linear equations has a solution. Another more difficult problem is to find a set of vectors that spans V for a given a vector space V.

Example 3.3.15 Find a set of vectors that spans the solution space of

$$\begin{aligned} x_1 \qquad - x_3 \qquad + x_5 &= 0 \\ 2x_1 + x_2 - 4x_3 \qquad + 4x_5 &= 0 \\ x_1 \qquad + 8x_3 + x_4 - 2x_5 &= 0 \end{aligned}$$

Solution The solution of this system is

$$\begin{aligned} x_1 &= x_3 - x_5 \\ x_2 &= 2x_3 - 2x_5 \\ x_4 &= -9x_3 + 3x_5 \end{aligned}$$

written as

$$\begin{bmatrix} x_1 \\ x_2 \\ x_3 \\ x_4 \\ x_5 \end{bmatrix} = \begin{bmatrix} x_3 - x_5 \\ 2x_3 - 2x_5 \\ x_3 \\ -9x_3 + 3x_5 \\ x_5 \end{bmatrix} = \begin{bmatrix} x_3 \\ 2x_3 \\ x_3 \\ -9x_3 \\ 0 \end{bmatrix} + \begin{bmatrix} -x_5 \\ -2x_5 \\ 0 \\ 3x_5 \\ x_5 \end{bmatrix} = x_3 \begin{bmatrix} 1 \\ 2 \\ 1 \\ -9 \\ 0 \end{bmatrix} + x_5 \begin{bmatrix} -1 \\ -2 \\ 0 \\ 3 \\ 1 \end{bmatrix}$$

Since every solution is a linear combination of

$$\begin{bmatrix} 1 \\ 2 \\ 1 \\ -9 \\ 0 \end{bmatrix} \quad \text{and} \quad \begin{bmatrix} -1 \\ -2 \\ 0 \\ 3 \\ 1 \end{bmatrix}$$

then the solution space is spanned by the set

$$\left\{ \begin{bmatrix} 1 \\ 2 \\ 1 \\ -9 \\ 0 \end{bmatrix}, \begin{bmatrix} -1 \\ -2 \\ 0 \\ 3 \\ 1 \end{bmatrix} \right\}$$

■

In the next section we determine the smallest number of vectors possible in a spanning set of a vector space V. However, in contrast to this, we can now determine that $S = \langle \mathbf{x}_1, \mathbf{x}_2, \ldots, \mathbf{x}_n \rangle$ is the smallest vector space containing $\{\mathbf{x}_1, \mathbf{x}_2, \ldots, \mathbf{x}_n\}$.

Theorem 3.3.16 *If V is a vector space containing the set $\{\mathbf{x}_1, \mathbf{x}_2, \ldots, \mathbf{x}_n\}$, then V contains $\langle \mathbf{x}_1, \mathbf{x}_2, \ldots, \mathbf{x}_n \rangle$.*

Proof. Let $\mathbf{x} = a_1\mathbf{x}_1 + a_2\mathbf{x}_2 + \cdots + a_n\mathbf{x}_n \in \langle \mathbf{x}_1, \mathbf{x}_2, \ldots, \mathbf{x}_n \rangle$. Since V contains $\{\mathbf{x}_1, \mathbf{x}_2, \ldots, \mathbf{x}_n\}$, then V contains $a_1\mathbf{x}_1, a_2\mathbf{x}_2, \ldots, a_n\mathbf{x}_n$ and also $a_1\mathbf{x}_1 + a_2\mathbf{x}_2 + \cdots + a_n\mathbf{x}_n$. Thus V contains $\mathbf{x}$, and since $\mathbf{x} \in \langle \mathbf{x}_1, \mathbf{x}_2, \ldots, \mathbf{x}_n \rangle$ was arbitrary, $\langle \mathbf{x}_1, \mathbf{x}_2, \ldots, \mathbf{x}_n \rangle \subset V$.

SUMMARY

Terms to know:

Linear combination

Spanning set

$\langle \mathbf{x}_1, \mathbf{x}_2, \ldots, \mathbf{x}_n \rangle$

$\langle S \rangle$

Theorems to know:

If V is a vector space and $\mathbf{x}_1, \mathbf{x}_2, \ldots, \mathbf{x}_n \in V$, then $\langle \mathbf{x}_1, \mathbf{x}_2, \ldots, \mathbf{x}_n \rangle$ is a subspace of V.

Methods to know:

Determine whether or not $\mathbf{v} \in \langle \mathbf{x}_1, \mathbf{x}_2, \ldots, \mathbf{x}_n \rangle$.

EXERCISE SET 3.3

In Exercises 1–6 *determine whether* $\{\mathbf{x}_1, \mathbf{x}_2, \ldots, \mathbf{x}_t\}$ *spans* S.

1. $\mathbf{x}_1 = \begin{bmatrix} 1 \\ 1 \end{bmatrix}$, $\mathbf{x}_2 = \begin{bmatrix} 1 \\ -1 \end{bmatrix}$, and $S = R^2$

2. $\mathbf{x}_1 = \begin{bmatrix} 1 \\ 1 \\ 1 \end{bmatrix}$, $\mathbf{x}_2 = \begin{bmatrix} 1 \\ -1 \\ 1 \end{bmatrix}$, $\mathbf{x}_3 = \begin{bmatrix} -1 \\ 1 \\ 1 \end{bmatrix}$ and $S = R^3$

3. $\mathbf{x}_1 = \begin{bmatrix} 1 \\ 1 \\ 0 \end{bmatrix}$, $\mathbf{x}_2 = \begin{bmatrix} 1 \\ 0 \\ 0 \end{bmatrix}$, and $S = \left\{ \begin{bmatrix} x_1 \\ x_2 \\ 0 \end{bmatrix} \middle| \, x_1, x_2 \in R \right\}$

4. $\mathbf{x}_1 = \begin{bmatrix} 0 \\ 1 \\ 1 \\ 1 \end{bmatrix}$, $\mathbf{x}_2 = \begin{bmatrix} 0 \\ 1 \\ 1 \\ 0 \end{bmatrix}$, $\mathbf{x}_3 = \begin{bmatrix} 0 \\ 1 \\ 0 \\ 0 \end{bmatrix}$, $\mathbf{x}_4 = \begin{bmatrix} 1 \\ 0 \\ 0 \\ 0 \end{bmatrix}$ and $S = R^4$

5. $\mathbf{x}_1 = 1 + x + x^2$, $\mathbf{x}_2 = x + x^2$, $\mathbf{x}_3 = x^2$ and $S = P_3$

6. $\mathbf{x}_1 = \begin{bmatrix} 0 & 1 \\ 1 & 0 \end{bmatrix}$, $\mathbf{x}_2 = \begin{bmatrix} 1 & 0 \\ 1 & 1 \end{bmatrix}$, $\mathbf{x}_3 = \begin{bmatrix} 1 & 1 \\ 0 & 1 \end{bmatrix}$, $\mathbf{x}_4 = \begin{bmatrix} 1 & 1 \\ 1 & 0 \end{bmatrix}$ and $S = R_{2 \times 2}$

7. Let

$$\mathbf{x}_1 = \begin{bmatrix} 3 \\ 1 \end{bmatrix} \quad \text{and} \quad \mathbf{x}_2 = \begin{bmatrix} 1 \\ 3 \end{bmatrix}$$

a. Find $\langle \mathbf{x}_1, \mathbf{x}_2 \rangle$

b. Represent $\begin{bmatrix} 7 \\ -5 \end{bmatrix}$ as a linear combination of $\mathbf{x}_1$ and $\mathbf{x}_2$.

8. a. Show that $\begin{bmatrix} 3a + b \\ -a + 5b \\ 2a - 2b \end{bmatrix}$ is a linear combination of $\begin{bmatrix} 3 \\ -1 \\ 2 \end{bmatrix}$ and $\begin{bmatrix} 1 \\ 5 \\ -2 \end{bmatrix}$.

b. Express the vector

$$\begin{bmatrix} a - 2b + 3c \\ -\frac{1}{2}a + \frac{1}{3}b + \frac{1}{4}c \\ -3a + 2b - c \end{bmatrix}$$

as a linear combination of three nonzero vectors.

9. Let

$$\mathbf{x}_1 = \begin{bmatrix} 2 \\ -1 \end{bmatrix} \quad \text{and} \quad \mathbf{x}_2 = \begin{bmatrix} -6 \\ 3 \end{bmatrix}$$

Show that $\langle \mathbf{x}_1, \mathbf{x}_2 \rangle = \langle \mathbf{x}_1 \rangle$. In general, show that if $\mathbf{x}_2$ is a nonzero scalar multiple of $\mathbf{x}_1$, then $\langle \mathbf{x}_1, \mathbf{x}_2 \rangle = \langle \mathbf{x}_1 \rangle$.

10. Let

$$\mathbf{x}_1 = \begin{bmatrix} 2 \\ 0 \end{bmatrix}, \quad \mathbf{x}_2 = \begin{bmatrix} 0 \\ 3 \end{bmatrix}, \quad \text{and} \quad \mathbf{x}_3 = \begin{bmatrix} 6 \\ -6 \end{bmatrix}$$

Show that $\langle \mathbf{x}_1, \mathbf{x}_2, \mathbf{x}_3 \rangle = \langle \mathbf{x}_1, \mathbf{x}_2 \rangle$. In general, show that if $\mathbf{x}_3$ is a linear combination of $\mathbf{x}_1$ and $\mathbf{x}_2$, then $\langle \mathbf{x}_1, \mathbf{x}_2, \mathbf{x}_3 \rangle = \langle \mathbf{x}_1, \mathbf{x}_2 \rangle$.

11. Show that $\{1, 1 + x, 1 + x^2\}$ spans P_2.

12. Show that $\langle 1, 1 + x, 1 - x \rangle = \langle 1, 1 - x \rangle$.

13. Show that

$$\left\langle \begin{bmatrix} 1 & 1 \\ 0 & 1 \end{bmatrix}, \begin{bmatrix} 0 & 1 \\ 0 & 1 \end{bmatrix}, \begin{bmatrix} 1 & 2 \\ 0 & 1 \end{bmatrix} \right\rangle = \left\langle \begin{bmatrix} 1 & 1 \\ 0 & 0 \end{bmatrix}, \begin{bmatrix} 0 & 1 \\ 0 & 1 \end{bmatrix} \right\rangle$$

14. Are the following spanning sets for R^3? If not, what subspace of R^3 do they span?

a. $$\left\langle \begin{bmatrix} 1 \\ 5 \\ 4 \end{bmatrix}, \begin{bmatrix} 2 \\ 1 \\ 1 \end{bmatrix}, \begin{bmatrix} 3 \\ 4 \\ 2 \end{bmatrix} \right\rangle$$

b. $$\left\langle \begin{bmatrix} 3 \\ 2 \\ 1 \end{bmatrix}, \begin{bmatrix} 1 \\ 2 \\ 3 \end{bmatrix}, \begin{bmatrix} 12 \\ 16 \\ 16 \end{bmatrix} \right\rangle$$

c. $$\left\langle \begin{bmatrix} 1 \\ 2 \\ 0 \end{bmatrix}, \begin{bmatrix} 1 \\ 0 \\ 0 \end{bmatrix}, \begin{bmatrix} 0 \\ 2 \\ 0 \end{bmatrix} \right\rangle$$

15. Find vectors that span the solution space of each system.

a. $$\begin{aligned} x_1 - x_2 - x_3 + x_4 &= 0 \\ x_1 + 2x_2 - 3x_3 - 2x_4 &= 0 \end{aligned}$$

b. $$\begin{aligned} x_1 + x_2 - x_3 + x_4 + x_5 &= 0 \\ x_3 + x_4 + x_5 &= 0 \\ 2x_1 + 2x_2 - x_3 + x_5 &= 0 \\ 2x_3 - 4x_4 - 2x_5 &= 0 \\ x_1 - x_2 - 2x_3 - 3x_4 + x_5 &= 0 \end{aligned}$$

SECTION 3.4 LINEAR INDEPENDENCE OF VECTORS

Let $\mathbf{x}_1, \ldots, \mathbf{x}_n$ be vectors in a vector space V. If $\mathbf{v}$ is a vector in the subspace spanned by $\{\mathbf{x}_1, \mathbf{x}_2, \ldots, \mathbf{x}_n\}$, there are scalar coefficients $a_1, \ldots, a_n$ for which

$$\mathbf{v} = a_1\mathbf{x}_1 + \cdots + a_n\mathbf{x}_n$$

In this section, we consider whether the scalar coefficients $a_1, a_2, \ldots, a_n$ are unique. In other words, if

$$\mathbf{v} = a_1\mathbf{x}_1 + \cdots + a_n\mathbf{x}_n$$

and

$$\mathbf{v} = b_1\mathbf{x}_1 + \cdots + b_n\mathbf{x}_n$$

must

$$a_1 = b_1, \ldots, a_n = b_n$$

If $\mathbf{x}$ can be expressed in both ways, then

$$\begin{aligned} \mathbf{0} = \mathbf{x} - \mathbf{x} &= (a_1\mathbf{x}_1 + \cdots + a_n\mathbf{x}_n) - (b_1\mathbf{x}_1 + \cdots + b_n\mathbf{x}_n) \\ &= (a_1 - b_1)\mathbf{x}_1 + \cdots + (a_n - b_n)\mathbf{x}_n \end{aligned}$$

Note that $\mathbf{0} = 0\mathbf{x}_1 + \cdots + 0\mathbf{x}_n$. Thus we reduce the original question to that of whether the zero vector, $\mathbf{0}$, can be uniquely expressed as a linear combination of $\mathbf{x}_1, \ldots, \mathbf{x}_n$.

Definition 3.4.1 The **set of vectors** $\{\mathbf{x}_1, \ldots, \mathbf{x}_n\}$ is **linearly independent** if whenever

$$\mathbf{0} = a_1\mathbf{x}_1 + \cdots + a_n\mathbf{x}_n$$

then $a_1 = a_2 = \cdots = a_n = 0$ are the only constants for which the equation is true.

If $\{\mathbf{x}_1, \ldots, \mathbf{x}_n\}$ is not a linearly independent set, the set $\{\mathbf{x}_1, \ldots, \mathbf{x}_n\}$ is said to be **linearly dependent.**

If each vector is uniquely expressible as a linear combination of $\mathbf{x}_1, \ldots, \mathbf{x}_n$, then if $\mathbf{v} = a_1\mathbf{x}_1 + \cdots + a_1\mathbf{x}_1$ and $\mathbf{v} = b_1\mathbf{x}_1 + \cdots + b_n\mathbf{x}_n$, so that $\mathbf{v} = a_1\mathbf{x}_1 + \cdots + a_n\mathbf{x}_n = b_1\mathbf{x}_1 + \cdots + b_n\mathbf{x}_n$, then by uniqueness $a_1 = b_1, \ldots, a_n = b_n$. With this condition, $\mathbf{0} = 0\mathbf{x}_1 + \cdots + 0\mathbf{x}_n = a_1\mathbf{x}_1 + \cdots + a_n\mathbf{x}_n$ implies $a_1 = a_2 = \cdots = a_n = 0$, or, in other words, $\{\mathbf{x}_1, \ldots, \mathbf{x}_n\}$ is linearly independent.

Conversely, if $\{\mathbf{x}_1, \ldots, \mathbf{x}_n\}$ is linearly independent and $\mathbf{v} = a_1\mathbf{x}_1 + \cdots + a_n\mathbf{x}_n$ and $\mathbf{v} = b_1\mathbf{x}_1 + \cdots + b_n\mathbf{x}_n$, we have

$$\begin{aligned}\mathbf{0} &= \mathbf{v} - \mathbf{v} \\ &= (a_1\mathbf{x}_1 + \cdots + a_n\mathbf{x}_n) - (b_1\mathbf{x}_1 + \cdots + b_n\mathbf{x}_n) \\ &= (a_1 - b_1)\mathbf{x}_1 + \cdots + (a_n - b_n)\mathbf{x}_n\end{aligned}$$

so

$$a_1 - b_1 = 0,\ a_2 - b_2 = 0, \ldots, a_n - b_n = 0$$

or $a_1 = b_1$, $a_2 = b_2, \ldots, a_n = b_n$. Thus, each linear combination of $\mathbf{x}_1, \ldots, \mathbf{x}_n$ is unique.

Theorem 3.4.2 *The set $\{\mathbf{x}_1, \ldots, \mathbf{x}_n\}$ is linearly independent if and only if each $\mathbf{v} \in \langle \mathbf{x}_1, \ldots, \mathbf{x}_n \rangle$ can be uniquely expressed as*

$$\mathbf{v} = v_1\mathbf{x}_1 + v_2\mathbf{x}_2 + \cdots + v_n\mathbf{x}_n$$

The value of Theorem 3.4.2 is that it is much easier to determine whether $\{\mathbf{x}_1, \ldots, \mathbf{x}_n\}$ is linearly independent than to check whether each linear combination $a_1\mathbf{x}_1 + \cdots + a_n\mathbf{x}_n$ is unique.

One method used to determine whether a set $\{\mathbf{x}_1, \ldots, \mathbf{x}_n\}$ of vectors is linearly independent is to solve a system of equations, as shown in the following examples.

Example 3.4.3 Determine whether or not the set

$$\left\{ \begin{bmatrix} 1 \\ 1 \\ 1 \end{bmatrix}, \begin{bmatrix} 1 \\ -1 \\ 1 \end{bmatrix}, \begin{bmatrix} -1 \\ 1 \\ 1 \end{bmatrix} \right\}$$

is linearly independent.

Solution Suppose

$$a\begin{bmatrix} 1 \\ 1 \\ 1 \end{bmatrix} + b\begin{bmatrix} 1 \\ -1 \\ 1 \end{bmatrix} + c\begin{bmatrix} -1 \\ 1 \\ 1 \end{bmatrix} = \begin{bmatrix} 0 \\ 0 \\ 0 \end{bmatrix} \qquad \text{for some } a, b, c \in R$$

Then

$$\begin{bmatrix} a + b - c \\ a - b + c \\ a + b + c \end{bmatrix} = \begin{bmatrix} 0 \\ 0 \\ 0 \end{bmatrix}$$

or

$$\begin{aligned} a + b - c &= 0 \\ a - b + c &= 0 \\ a + b + c &= 0 \end{aligned}$$

Thus we are reduced to solving the above system of equations, obtaining $a = b = c = 0$ as the only solution. The set is linearly independent. ■

Example 3.4.4 Determine whether the set

$$\left\{\begin{bmatrix}1\\0\\1\end{bmatrix}, \begin{bmatrix}1\\1\\0\end{bmatrix}, \begin{bmatrix}5\\2\\3\end{bmatrix}\right\}$$

is linearly independent.

Solution Suppose

$$a\begin{bmatrix}1\\0\\1\end{bmatrix} + b\begin{bmatrix}1\\1\\0\end{bmatrix} + c\begin{bmatrix}5\\2\\3\end{bmatrix} = \begin{bmatrix}0\\0\\0\end{bmatrix} \qquad \text{for some } a, b, c \in R$$

Then

$$\begin{bmatrix}a + b + 5c\\ b + 2c\\ a + \quad 3c\end{bmatrix} = \begin{bmatrix}0\\0\\0\end{bmatrix}$$

and so

$$\begin{aligned} a + b + 5c &= 0\\ b + 2c &= 0\\ a + \quad 3c &= 0 \end{aligned}$$

This system has solutions $a = -3c$, $b = -2c$, where c is arbitrarily chosen. Thus if $c = 1$, then $b = -2$, $a = -3$, and

$$-3\begin{bmatrix}1\\0\\1\end{bmatrix} - 2\begin{bmatrix}1\\1\\0\end{bmatrix} + 1\begin{bmatrix}5\\2\\3\end{bmatrix} = \begin{bmatrix}0\\0\\0\end{bmatrix}$$

The set is linearly dependent. ■

Example 3.4.5 a. If

$$\mathbf{x}_1 = \begin{bmatrix}2\\4\end{bmatrix}, \qquad \mathbf{x}_2 = \begin{bmatrix}3\\6\end{bmatrix}, \qquad \mathbf{x}_3 = \begin{bmatrix}1\\5\end{bmatrix}$$

show that the set $\{\mathbf{x}_1, \mathbf{x}_2, \mathbf{x}_3\}$ is linearly dependent.

b. Express each of $\mathbf{x}_1$ and $\mathbf{x}_2$ as a linear combination of the other two.

c. Show that $\mathbf{x}_3$ cannot be expressed as a linear combination of the other two.

d. Show that $\{\mathbf{x}_1, \mathbf{x}_3\}$ is a linearly independent set.

e. Show that $\{\mathbf{x}_2, \mathbf{x}_3\}$ is a linearly independent set.

f. Show that $\{\mathbf{x}_1, \mathbf{x}_2\}$ is a linearly dependent set.

Solution a. Suppose

$$a\begin{bmatrix}2\\4\end{bmatrix} + b\begin{bmatrix}3\\6\end{bmatrix} + c\begin{bmatrix}1\\5\end{bmatrix} = \begin{bmatrix}0\\0\end{bmatrix}$$

Then the system

$$\begin{aligned} 2a + 3b + c &= 0 \\ 4a + 6b + 5c &= 0 \end{aligned}$$

has solutions $c = 0$ and $a = -\frac{3}{2}b$, where b is arbitrarily chosen. For $c = 0$, if $b = 2$, then $a = -3$ and

$$-3\begin{bmatrix}2\\4\end{bmatrix} + 2\begin{bmatrix}3\\6\end{bmatrix} + 0\begin{bmatrix}1\\5\end{bmatrix} = \begin{bmatrix}0\\0\end{bmatrix}$$

The set is linearly dependent.

b. Let $a\mathbf{x}_2 + b\mathbf{x}_3 = \mathbf{x}_1$. Then

$$a\begin{bmatrix}3\\6\end{bmatrix} + b\begin{bmatrix}1\\5\end{bmatrix} = \begin{bmatrix}2\\4\end{bmatrix}$$

or

$$\begin{aligned} 3a + b &= 2 \\ 6a + 5b &= 4 \end{aligned}$$

The solution of this system is $a = \frac{2}{3}$ and $b = 0$. Thus $\mathbf{x}_1$ is a linear combination of $\mathbf{x}_2$ and $\mathbf{x}_3$.

Now let $c\mathbf{x}_1 + d\mathbf{x}_3 = \mathbf{x}_2$. Then

$$c\begin{bmatrix}2\\4\end{bmatrix} + d\begin{bmatrix}1\\5\end{bmatrix} = \begin{bmatrix}3\\6\end{bmatrix}$$

or

$$\begin{aligned} 2c + d &= 3 \\ 4c + 5d &= 6 \end{aligned}$$

The solution of this system is $c = \frac{3}{2}$ and $d = 0$. Thus $\mathbf{x}_2$ is a linear combination of $\mathbf{x}_1$ and $\mathbf{x}_3$.

c. Let $a\mathbf{x}_1 + b\mathbf{x}_2 = \mathbf{x}_3$. Then

$$a\begin{bmatrix}2\\4\end{bmatrix} + b\begin{bmatrix}3\\6\end{bmatrix} = \begin{bmatrix}1\\5\end{bmatrix}$$

or

$$2a + 3b = 1$$
$$4a + 6b = 5$$

This system has no solution and so $\mathbf{x}_3$ cannot be written as a linear combination of $\mathbf{x}_1$ and $\mathbf{x}_2$.

d. Let $a\mathbf{x}_1 + b\mathbf{x}_3 = \mathbf{0}$. Then

$$a\begin{bmatrix} 2 \\ 4 \end{bmatrix} + b\begin{bmatrix} 1 \\ 5 \end{bmatrix} = \begin{bmatrix} 0 \\ 0 \end{bmatrix}$$

or

$$2a + b = 0$$
$$4a + 5b = 0$$

The only solution of this system is $a = b = 0$, and so the set $\{\mathbf{x}_1, \mathbf{x}_3\}$ is linearly independent.

e. Let $a\mathbf{x}_2 + b\mathbf{x}_3 = \mathbf{0}$. Then

$$a\begin{bmatrix} 3 \\ 6 \end{bmatrix} + b\begin{bmatrix} 1 \\ 5 \end{bmatrix} = \begin{bmatrix} 0 \\ 0 \end{bmatrix}$$

or

$$3a + b = 0$$
$$6a + 5b = 0$$

The only solution of this system is $a = b = 0$, and so the set $\{\mathbf{x}_2, \mathbf{x}_3\}$ is linearly independent.

f. Let $a\mathbf{x}_1 + b\mathbf{x}_2 = \mathbf{0}$. Then

$$a\begin{bmatrix} 2 \\ 4 \end{bmatrix} + b\begin{bmatrix} 3 \\ 6 \end{bmatrix} = \begin{bmatrix} 0 \\ 0 \end{bmatrix}$$

or

$$2a + 3b = 0$$
$$4a + 6b = 0$$

There are infinitely many solutions to this system, since b is arbitrary and $a = -\frac{3}{2}b$. For example, when $b = 2$, $a = -3$ and

$$-3\begin{bmatrix} 2 \\ 4 \end{bmatrix} + 2\begin{bmatrix} 3 \\ 6 \end{bmatrix} = \begin{bmatrix} 0 \\ 0 \end{bmatrix}$$

Thus the set $\{\mathbf{x}_1, \mathbf{x}_2\}$ is linearly dependent. ■

REMARK If two vectors are proportional, then they form a linearly dependent set. For example

$$\begin{bmatrix} 2 \\ 4 \end{bmatrix} \quad \text{and} \quad \begin{bmatrix} 3 \\ 6 \end{bmatrix}$$

are proportional, since

$$\begin{bmatrix} 2 \\ 4 \end{bmatrix} = \frac{2}{3}\begin{bmatrix} 3 \\ 6 \end{bmatrix}$$

Furthermore,

$$1\begin{bmatrix} 2 \\ 4 \end{bmatrix} - \frac{2}{3}\begin{bmatrix} 3 \\ 6 \end{bmatrix} = \begin{bmatrix} 0 \\ 0 \end{bmatrix}$$

Hence linear dependence may be thought of as a generalization of the concept of proportionality.

Example 3.4.6 Show that the set $\{1, 1 + x, 1 + x + x^2\}$ is linearly independent in P_2.

Solution Suppose $a(1) + b(1 + x) + c(1 + x + x^2) = \mathbf{0}$, then $0 + 0x + 0x^2 = (a + b + c)1 + (b + c)x + cx^2$. Hence we solve the system of equations

$$\begin{aligned} a + b + c &= 0 \\ b + c &= 0 \\ c &= 0 \end{aligned}$$

obtaining $a = b = c = 0$ as the only solution. ■

Example 3.4.7 Show that the set

$$\left\{\begin{bmatrix} 1 & 0 \\ 0 & 1 \end{bmatrix}, \begin{bmatrix} 0 & 1 \\ 1 & 0 \end{bmatrix}, \begin{bmatrix} 0 & 1 \\ 1 & 1 \end{bmatrix}\right\}$$

is independent in $R_{2\times 2}$.

Solution Let

$$a\begin{bmatrix} 1 & 0 \\ 0 & 1 \end{bmatrix} + b\begin{bmatrix} 0 & 1 \\ 1 & 0 \end{bmatrix} + c\begin{bmatrix} 0 & 1 \\ 1 & 1 \end{bmatrix} = \begin{bmatrix} 0 & 0 \\ 0 & 0 \end{bmatrix}$$

Then

$$\begin{bmatrix} a & b + c \\ b + c & a + c \end{bmatrix} = \begin{bmatrix} 0 & 0 \\ 0 & 0 \end{bmatrix}$$

We solve the system of equations

$$\begin{aligned} a &= 0 \\ b + c &= 0 \\ a + c &= 0 \end{aligned}$$

obtaining $a = b = c = 0$ as the only solution. The set is linearly independent. ■

Example 3.4.8 Determine whether the set of vectors

$$\left\{ \begin{bmatrix} 1 \\ 0 \\ 1 \end{bmatrix}, \begin{bmatrix} 1 \\ 1 \\ 0 \end{bmatrix}, \begin{bmatrix} 0 \\ 1 \\ 1 \end{bmatrix}, \begin{bmatrix} -2 \\ 1 \\ 3 \end{bmatrix} \right\}$$

is linearly independent or dependent.

Solution Let

$$\mathbf{x}_1 = \begin{bmatrix} 1 \\ 0 \\ 1 \end{bmatrix}, \quad \mathbf{x}_2 = \begin{bmatrix} 1 \\ 1 \\ 0 \end{bmatrix}, \quad \mathbf{x}_3 = \begin{bmatrix} 0 \\ 1 \\ 1 \end{bmatrix}, \text{ and } \mathbf{x}_4 = \begin{bmatrix} -2 \\ 1 \\ 3 \end{bmatrix}$$

To determine whether the set $\{\mathbf{x}_1, \mathbf{x}_2, \mathbf{x}_3, \mathbf{x}_4\}$ is linearly independent, we examine the equation

$$a_1 \begin{bmatrix} 1 \\ 0 \\ 1 \end{bmatrix} + a_2 \begin{bmatrix} 1 \\ 1 \\ 0 \end{bmatrix} + a_3 \begin{bmatrix} 0 \\ 1 \\ 1 \end{bmatrix} + a_4 \begin{bmatrix} -2 \\ 1 \\ 3 \end{bmatrix} = \begin{bmatrix} 0 \\ 0 \\ 0 \end{bmatrix}$$

In order to solve this equation for a_1, a_2, a_3, and a_4 we must solve the system of linear equations

$$\begin{aligned} a_1 + a_2 \qquad - 2a_4 &= 0 \\ a_2 + a_3 + \ a_4 &= 0 \\ a_1 + \qquad a_3 + 3a_4 &= 0 \end{aligned}$$

In matrix form, this system becomes

$$\begin{bmatrix} 1 & 1 & 0 & -2 \\ 0 & 1 & 1 & 1 \\ 1 & 0 & 1 & 3 \end{bmatrix} \begin{bmatrix} a_1 \\ a_2 \\ a_3 \\ a_4 \end{bmatrix} = \begin{bmatrix} 0 \\ 0 \\ 0 \end{bmatrix}$$

Note that the jth column of the matrix is the jth vector, $\mathbf{x}_j$. We can show that this system has a nonzero solution. Since we have more unknowns than equations, the system has nonzero solutions, which means the set of vectors is linearly dependent. ■

Note that the jth column of the coefficient matrix is the jth vector $\mathbf{v}_j$. This always happens. In general the problem of finding whether or not the vectors $\mathbf{v}_1, \ldots, \mathbf{v}_t$ in R^n are linearly independent is the same problem as finding whether or not the n by t homogeneous system $A\mathbf{x} = \mathbf{0}$ has nontrivial solutions when $A = [\mathbf{v}_1, \ldots, \mathbf{v}_t]$.

The following theorem generalizes the result of Example 3.4.8.

Theorem 3.4.9 *Every set of more than n vectors in R^n is linearly dependent.*

Proof. Suppose $\mathbf{x}_1, \mathbf{x}_2, \ldots, \mathbf{x}_m$ are vectors in R^n and $m > n$. Let A be the $n \times m$ matrix whose jth column is the vector $\mathbf{x}_j$. Then the vectors will be dependent if we can show $A\mathbf{x} = \mathbf{0}$ has nontrivial solutions, but this system has more unknowns than equations. Hence, $A\mathbf{x} = \mathbf{0}$ must have nontrivial solutions.

SUMMARY

Terms to know:

Linearly independent set of vectors

Linearly dependent set of vectors

Theorems to know:

The set $\{\mathbf{x}_1, \ldots, \mathbf{x}_n\}$ is a linearly independent set of vectors if and only if each $\mathbf{v} \in \langle \mathbf{x}_1, \ldots, \mathbf{x}_n \rangle$ can be uniquely expressed as

$$\mathbf{v} = a_1\mathbf{x}_1 + \cdots + a_n\mathbf{x}_n$$

Every set of more than n vectors in R^n is linearly dependent.

Methods to know:

Determine whether a set of vectors is linearly independent.

EXERCISE SET 3.4

Determine if the following sets are linearly independent.

1. $\left\{\begin{bmatrix}1\\0\end{bmatrix}, \begin{bmatrix}2\\1\end{bmatrix}\right\}$

2. $\left\{\begin{bmatrix}1\\0\\1\end{bmatrix}, \begin{bmatrix}2\\1\\4\end{bmatrix}, \begin{bmatrix}3\\2\\0\end{bmatrix}\right\}$

3. $\left\{\begin{bmatrix}1\\1\end{bmatrix}, \begin{bmatrix}1\\0\end{bmatrix}, \begin{bmatrix}2\\1\end{bmatrix}\right\}$

4. $\left\{\begin{bmatrix}1\\1\\1\end{bmatrix}, \begin{bmatrix}0\\0\\0\end{bmatrix}, \begin{bmatrix}1\\2\\-1\end{bmatrix}\right\}$

SECTION 3.5 BASIS OF A VECTOR SPACE

As we noted in Section 2.5, there is nothing unique about the number of generators of a vector space. For example, the set

$$\left\{\begin{bmatrix}1\\0\end{bmatrix},\begin{bmatrix}1\\1\end{bmatrix}\right\}$$

generated R^2 as well as the set

$$\left\{\begin{bmatrix}1\\2\end{bmatrix},\begin{bmatrix}-1\\1\end{bmatrix},\begin{bmatrix}2\\3\end{bmatrix}\right\}$$

The important distinction between these two sets is that the first set is linearly independent, whereas the other one is not. If a space V is generated by a linearly independent set of vectors, B, then B is called a basis for V. This idea is made precise in the following definition.

Definition 3.5.1 Let V be a vector space. The set $\{\mathbf{x}_1, \ldots, \mathbf{x}_k\}$ of vectors is a **basis** of V if

1. $\{\mathbf{x}_1, \ldots, \mathbf{x}_k\}$ is a linearly independent set.
2. $\langle \mathbf{x}_1, \ldots, \mathbf{x}_k \rangle = V$.

Each linearly independent subset $\{\mathbf{x}_1, \mathbf{x}_2, \ldots, \mathbf{x}_k\}$ of a vector space V that spans V is a basis for V. Furthermore, since any basis for the subspace V must be a linearly independent set, it will never contain the zero vector.

Example 3.5.2 The set

$$\left\{\begin{bmatrix}1\\0\end{bmatrix},\begin{bmatrix}0\\1\end{bmatrix}\right\}$$

is a basis for R^2.

The set

$$\left\{\begin{bmatrix}1\\0\\0\end{bmatrix},\begin{bmatrix}0\\1\\0\end{bmatrix},\begin{bmatrix}0\\0\\1\end{bmatrix}\right\}$$

is a basis for R^3.

In general, the set of n vectors

$$\left\{\begin{bmatrix}1\\0\\0\\\vdots\\0\end{bmatrix}, \begin{bmatrix}0\\1\\0\\\vdots\\0\end{bmatrix}, \ldots, \begin{bmatrix}0\\0\\0\\\vdots\\1\end{bmatrix}\right\}$$

is a basis for R^n. We call each of these bases the **standard basis** for the vector space which they span. ■

Example 3.5.3 a. The set $\{1, x, x^2, \ldots, x^n\}$ is the standard basis for P_n.

b. The set

$$\left\{\begin{bmatrix}1 & 0\\0 & 0\\0 & 0\end{bmatrix}, \begin{bmatrix}0 & 1\\0 & 0\\0 & 0\end{bmatrix}, \begin{bmatrix}0 & 0\\1 & 0\\0 & 0\end{bmatrix}, \begin{bmatrix}0 & 0\\0 & 1\\0 & 0\end{bmatrix}, \begin{bmatrix}0 & 0\\0 & 0\\1 & 0\end{bmatrix}, \begin{bmatrix}0 & 0\\0 & 0\\0 & 1\end{bmatrix}\right\}$$

is the standard basis for $M_{3\times 2}$. ■

Example 3.5.4 Determine whether

$$\left\{\begin{bmatrix}1\\1\\1\end{bmatrix}, \begin{bmatrix}1\\0\\1\end{bmatrix}, \begin{bmatrix}1\\1\\0\end{bmatrix}\right\}$$

is a basis for R^3.

Solution First we determine if the set is linearly independent. Suppose

$$a\begin{bmatrix}1\\1\\1\end{bmatrix} + b\begin{bmatrix}1\\0\\1\end{bmatrix} + c\begin{bmatrix}1\\1\\0\end{bmatrix} = \begin{bmatrix}0\\0\\0\end{bmatrix}$$

Then we are reduced to solving the system

$$\begin{aligned} a + b + c &= 0 \\ a + \phantom{b +{}} c &= 0 \\ a + b \phantom{{}+ c} &= 0 \end{aligned}$$

whose only solution is $a = b = c = 0$, and so the set is linearly independent.

Next we determine whether

$$\left\langle \begin{bmatrix}1\\1\\1\end{bmatrix}, \begin{bmatrix}1\\0\\1\end{bmatrix}, \begin{bmatrix}1\\1\\0\end{bmatrix} \right\rangle = R^3$$

Let

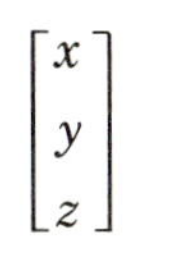

be an arbitrary vector. If there exist scalars a, b, and c such that

$$a\begin{bmatrix}1\\1\\1\end{bmatrix} + b\begin{bmatrix}1\\0\\1\end{bmatrix} + c\begin{bmatrix}1\\1\\0\end{bmatrix} = \begin{bmatrix}x\\y\\z\end{bmatrix}$$

then we know the set spans R^3. The corresponding system of equations

$$\begin{aligned} a + b + c &= x \\ a + \phantom{b +{}} c &= y \\ a + b \phantom{{}+ c} &= z \end{aligned}$$

has the solution $c = x - z$, $b = x - y$ and $a = -x + y + z$. Thus

$$(-x + y + z)\begin{bmatrix}1\\1\\1\end{bmatrix} + (x - y)\begin{bmatrix}1\\0\\1\end{bmatrix} + (x - z)\begin{bmatrix}1\\1\\0\end{bmatrix} = \begin{bmatrix}x\\y\\z\end{bmatrix}$$

and so every vector

$$\begin{bmatrix}x\\y\\z\end{bmatrix}$$

in R^3 can be written uniquely as a linear combination of

$$\begin{bmatrix}1\\1\\1\end{bmatrix}, \quad \begin{bmatrix}1\\0\\1\end{bmatrix}, \quad \text{and} \quad \begin{bmatrix}1\\1\\0\end{bmatrix}$$

Thus the set is a basis of R^3. ■

Example 3.5.5 Determine if

$$\left\{\begin{bmatrix}1\\1\\0\end{bmatrix}, \begin{bmatrix}-2\\1\\-1\end{bmatrix}\right\}$$

is a basis for R^3.

Solution First we decide whether or not the set is linearly independent. Suppose

$$a\begin{bmatrix}1\\1\\0\end{bmatrix} + b\begin{bmatrix}-2\\1\\-1\end{bmatrix} = \begin{bmatrix}0\\0\\0\end{bmatrix}$$

This equation is equivalent to solving the system

$$\begin{aligned} a - 2b &= 0 \\ a + b &= 0 \\ - b &= 0 \end{aligned}$$

The last equation says $b = 0$ and so $a + b = 0$ yields $a = 0$. Thus the vectors are linearly independent.

Next, we answer the question, Does

$$\left\langle\begin{bmatrix}1\\1\\0\end{bmatrix}, \begin{bmatrix}-2\\1\\-1\end{bmatrix}\right\rangle = R^3$$

Let

$$\begin{bmatrix}x\\y\\z\end{bmatrix}$$

be an arbitrary vector. From

$$a\begin{bmatrix}1\\1\\0\end{bmatrix} + b\begin{bmatrix}-2\\1\\-1\end{bmatrix} = \begin{bmatrix}x\\y\\z\end{bmatrix}$$

we obtain the system

$$\begin{aligned} a - 2b &= x \\ a + b &= y \\ - b &= z \end{aligned}$$

This system reduces to

$$\begin{aligned} a - 2b &= x \\ 3b &= y - x \\ 0 &= z + \tfrac{1}{3}(y - x) \end{aligned}$$

This system has a solution precisely when

$$z + \tfrac{1}{3}(y - x) = 0$$

Choosing

$$\begin{bmatrix} x \\ y \\ z \end{bmatrix} = \begin{bmatrix} 1 \\ 2 \\ 0 \end{bmatrix}$$

$$z + \tfrac{1}{3}(y - x) = 0 + \tfrac{1}{3}(2 - 1) \neq 0$$

Thus

$$\left\{ \begin{bmatrix} 1 \\ 1 \\ 0 \end{bmatrix}, \begin{bmatrix} -2 \\ 1 \\ -1 \end{bmatrix} \right\}$$

does not span R^3 and, consequently, this set is not a basis for R^3. ■

Example 3.5.6 Does the set

$$\left\{ \begin{bmatrix} 1 \\ 1 \\ 0 \end{bmatrix}, \begin{bmatrix} 0 \\ 1 \\ 1 \end{bmatrix}, \begin{bmatrix} 1 \\ 0 \\ 1 \end{bmatrix}, \begin{bmatrix} -2 \\ -1 \\ 1 \end{bmatrix} \right\}$$

form a basis for R^3?

Solution Let us reverse the order of the three preceding examples and determine first whether or not this set spans R^3. Let

$$\begin{bmatrix} x \\ y \\ z \end{bmatrix}$$

be an arbitrary vector. From the equation

$$a \begin{bmatrix} 1 \\ 1 \\ 0 \end{bmatrix} + b \begin{bmatrix} 0 \\ 1 \\ 1 \end{bmatrix} + c \begin{bmatrix} 1 \\ 0 \\ 1 \end{bmatrix} + d \begin{bmatrix} -2 \\ -1 \\ 1 \end{bmatrix} = \begin{bmatrix} x \\ y \\ z \end{bmatrix}$$

we have the following system:

$$\begin{aligned} a + \quad\quad c - 2d &= x \\ a + b \quad\quad - \ d &= y \\ b + c + \ d &= z \end{aligned}$$

whose solution is

$$\begin{aligned} a &= \tfrac{1}{2}(x + y - z) + 2d \\ b &= -\tfrac{1}{2}(x - y - z) - d \\ c &= \tfrac{1}{2}(x - y + z) \quad \text{where } d \text{ is arbitrary} \end{aligned}$$

Thus choosing a value for d yields values for a, b, and c by the above equations, and so the given set of vectors spans R^3.

Is the given set linearly independent? Let

$$a\begin{bmatrix}1\\1\\0\end{bmatrix} + b\begin{bmatrix}0\\1\\1\end{bmatrix} + c\begin{bmatrix}1\\0\\1\end{bmatrix} + d\begin{bmatrix}-2\\-1\\1\end{bmatrix} = \begin{bmatrix}0\\0\\0\end{bmatrix}$$

The corresponding system of equations is

$$\begin{aligned} a \quad\quad + c - 2d &= 0 \\ a + b \quad\quad - \ d &= 0 \\ b + c + \ d &= 0 \end{aligned}$$

with solution

$$\begin{aligned} a &= 2d \\ b &= -d \\ c &= 0 \\ d &\quad \text{arbitrary} \end{aligned}$$

Since any nonzero choice for d yields nonzero values for a and b, the set is not linearly independent. For example, with $d = 1$ and $a = 2$, $b = -1$ and $c = 0$,

$$2\begin{bmatrix}1\\1\\0\end{bmatrix} - 1\begin{bmatrix}0\\1\\1\end{bmatrix} + 0\begin{bmatrix}1\\0\\1\end{bmatrix} + \begin{bmatrix}-2\\-1\\1\end{bmatrix} = \begin{bmatrix}0\\0\\0\end{bmatrix}$$

Therefore, the set in question does not form a basis for R^3. ■

By Theorem 3.4.9, we know that this set is dependent and so is not a basis of R^3. This shows that we could have saved a lot of effort by applying our theorems. However, we chose this longer way to do this example to demonstrate the general method of doing this type of problem.

Example 3.5.7 Show that $\{\mathbf{v}_1 = 1 + x, \mathbf{v}_2 = x + x^2, \mathbf{v}_3 = x^2 + x^3, \mathbf{v}_4 = 1 + x^3\}$ is not a basis for P_3.

Solution We first check whether $\{\mathbf{v}_1, \mathbf{v}_2, \mathbf{v}_3, \mathbf{v}_4\}$ spans P_3. Let $p_0 + p_1x + p_2x^2 + p_3x^3$ be an arbitrary polynomial of degree less than or equal to 3. We must find real numbers a, b, c, and d satisfying

$$a(1 + x) + b(x + x^2) + c(x^2 + x^3) + d(1 + x^3)$$

$$= p_0 + p_1x + p_2x^2 + p_3x^3$$

or

$$(a + d)1 + (a + b)x + (b + c)x^2 + (c + d)x^3$$

$$= p_0 + p_1x + p_2x^2 + p_3x^3$$

Hence, we must solve the system of equations

$$\begin{aligned} a \qquad\qquad + d &= p_0 \\ a + b \qquad\qquad &= p_1 \\ b + c \qquad &= p_2 \\ c + d &= p_3 \end{aligned}$$

This system is inconsistent, and so $\{\mathbf{v}_1, \mathbf{v}_2, \mathbf{v}_3, \mathbf{v}_4\}$ is not a basis of P_3. ■

Example 3.5.8 Show that

$$\left\{\mathbf{v}_1 = \begin{bmatrix} 1 & 1 \\ 0 & 0 \end{bmatrix}, \mathbf{v}_2 = \begin{bmatrix} 0 & 1 \\ 0 & 1 \end{bmatrix}, \mathbf{v}_3 = \begin{bmatrix} 0 & 0 \\ 1 & 1 \end{bmatrix}, \mathbf{v}_4 = \begin{bmatrix} 1 & 0 \\ 1 & 1 \end{bmatrix}\right\}$$

is a basis for $R_{2\times 2}$.

Solution We first check that $\{\mathbf{v}_1, \mathbf{v}_2, \mathbf{v}_3, \mathbf{v}_4\}$ spans $R_{2\times 2}$. To this end, let

$$\begin{bmatrix} a & b \\ c & d \end{bmatrix}$$

be an arbitrary 2×2 matrix. We must find $x_1, x_2, x_3, x_4 \in R$ satisfying

$$x_1\mathbf{v}_1 + x_2\mathbf{v}_2 + x_3\mathbf{v}_3 + x_4\mathbf{v}_4 = \begin{bmatrix} a & b \\ c & d \end{bmatrix}$$

Thus we must solve the system of equations

$$\begin{aligned} x_1 \qquad\qquad\quad + x_4 &= a \\ x_1 + x_2 \qquad\qquad\quad &= b \\ x_3 + x_4 &= c \\ x_2 + x_3 + x_4 &= d \end{aligned}$$

The solution to this system is

$$\begin{aligned} x_1 &= \quad\qquad b + \ \ c - d \\ x_2 &= \qquad\qquad -c + d \\ x_3 &= -a + b + 2c - d \\ x_4 &= \quad a - b - \ \ c + d \end{aligned}$$

So $R_{2\times 2} = \langle \mathbf{v}_1, \mathbf{v}_2, \mathbf{v}_3, \mathbf{v}_4 \rangle$.

Next we must check the linear independence of $\{\mathbf{v}_1, \mathbf{v}_2, \mathbf{v}_3, \mathbf{v}_4\}$. Suppose

$$x_1\mathbf{v}_1 + x_2\mathbf{v}_2 + x_3\mathbf{v}_3 + x_4\mathbf{v}_4 = \begin{bmatrix} 0 & 0 \\ 0 & 0 \end{bmatrix}$$

Thus we must solve the homogeneous system of equations

$$\begin{aligned} x_1 \qquad\qquad\quad + x_4 &= 0 \\ x_1 + x_2 \qquad\qquad\quad &= 0 \\ x_3 + x_4 &= 0 \\ x_2 + x_3 + x_4 &= 0 \end{aligned}$$

The only solution to this system of equations is $x_1 = x_2 = x_3 = x_4 = 0$, so the set $\{\mathbf{v}_1, \mathbf{v}_2, \mathbf{v}_3, \mathbf{v}_4\}$ is linearly independent and is a basis for $R_{2\times 2}$. ■

In Example 3.5.9, we show how to enlarge a linearly independent set $\{\mathbf{x}_1, \ldots, \mathbf{x}_k\} \subseteq V$ to a basis of V.

Example 3.5.9 Verify that the linearly independent set

$$\left\{ \begin{bmatrix} 3 \\ -1 \\ 2 \end{bmatrix}, \begin{bmatrix} 1 \\ 5 \\ -2 \end{bmatrix} \right\}$$

does not span R^3. Then enlarge the set to a basis for R^3.

Solution In order for

$$\left\{\begin{bmatrix} 3 \\ -1 \\ 2 \end{bmatrix}, \begin{bmatrix} 1 \\ 5 \\ -2 \end{bmatrix}\right\}$$

to be a basis for R^3, every vector in R^3 must be written as a linear combination of the two vectors. For example, choose

$$\begin{bmatrix} 1 \\ 0 \\ 0 \end{bmatrix}$$

and see if a and b exist such that

$$a\begin{bmatrix} 3 \\ -1 \\ 2 \end{bmatrix} + b\begin{bmatrix} 1 \\ 5 \\ -2 \end{bmatrix} = \begin{bmatrix} 1 \\ 0 \\ 0 \end{bmatrix}$$

or

$$\begin{aligned} 3a + b &= 1 \\ -a + 5b &= 0 \\ 2a - 2b &= 0 \end{aligned}$$

From the last equation, $2a - 2b = 0$, we have $a = b$; from the second, we have $a = -5b$. The only way this can happen is for $a = b = 0$, but this is not so according to the first equation. Thus

$$\begin{bmatrix} 3 \\ -1 \\ 2 \end{bmatrix} \text{ and } \begin{bmatrix} 1 \\ 5 \\ -2 \end{bmatrix}$$

do not form a basis for R^3.

To enlarge the set to a basis for R^3, a vector

$$\begin{bmatrix} x_1 \\ y_1 \\ z_1 \end{bmatrix}$$

must be chosen so that it is not a linear combination of the given two vectors, that is,

$$\begin{bmatrix} x_1 \\ y_1 \\ z_1 \end{bmatrix} \neq a\begin{bmatrix} 3 \\ -1 \\ 2 \end{bmatrix} + b\begin{bmatrix} 1 \\ 5 \\ -2 \end{bmatrix} = \begin{bmatrix} 3a + b \\ -a + 5b \\ 2a - 2b \end{bmatrix}$$

For example, we can choose

$$\begin{bmatrix} 1 \\ 0 \\ 0 \end{bmatrix}$$

for the third vector because

$$\begin{bmatrix} 1 \\ 0 \\ 0 \end{bmatrix} \neq \begin{bmatrix} 3a + b \\ -a + 5b \\ 2a - 2b \end{bmatrix}$$

which we verified earlier. Thus

$$\left\{ \begin{bmatrix} 3 \\ -1 \\ 2 \end{bmatrix}, \begin{bmatrix} 1 \\ 5 \\ 2 \end{bmatrix}, \begin{bmatrix} 1 \\ 0 \\ 0 \end{bmatrix} \right\}$$

is linearly independent. It turns out that these three vectors span R^3. (See Exercise 8). ■

REMARK In this example we chose

$$\begin{bmatrix} 1 \\ 0 \\ 0 \end{bmatrix}$$

arbitrarily. We could have chosen any vector $\mathbf{x}$ as long as

$$\mathbf{x} \notin \left\langle \begin{bmatrix} 3 \\ -1 \\ 2 \end{bmatrix}, \begin{bmatrix} 1 \\ 5 \\ -2 \end{bmatrix} \right\rangle$$

REMARK The set $\{\mathbf{0}\}$ is the only vector space that does not possess a basis.

We state two theorems. The first says that a set of n linearly independent vectors in R^n is a basis for R^n. The second says every nonzero vector space V has a basis.

Theorem 3.5.10 *Let $\{\mathbf{x}_1, \mathbf{x}_2, \ldots, \mathbf{x}_n\}$ be a linearly independent set of vectors in R^n. Then $\{\mathbf{x}_1, \mathbf{x}_2, \ldots, \mathbf{x}_n\}$ is a basis for R^n.*

The proof is left as an exercise.

Theorem 3.5.11 *Every nonzero vector space spanned by a finite set has a basis.* (This theorem is tedious to prove, and so it is only stated.)

Since a subspace S of a vector space V is itself a vector space, a basis for a subspace S is a linearly independent set of vectors from S that spans S. In the last two examples, we find a basis of a subspace of a vector space.

Example 3.5.12 Find a basis for the set of solutions to the following system of equations:

$$x_1 \quad + x_3 = 0$$
$$x_2 + x_3 = 0$$

Solution For a vector

$$\mathbf{x} = \begin{bmatrix} x_1 \\ x_2 \\ x_3 \end{bmatrix}$$

to be a solution to this system of equations, we must have $x_1 = -x_3$ and $x_2 = -x_3$. Hence the solution space of this system is spanned by the vector

$$\mathbf{x} = \begin{bmatrix} x_1 \\ x_2 \\ x_3 \end{bmatrix} = x_3 \begin{bmatrix} -1 \\ -1 \\ 1 \end{bmatrix}$$

Thus the solution space of this system has a basis where the set consists of the one vector

$$\mathbf{x} = \begin{bmatrix} -1 \\ -1 \\ 1 \end{bmatrix}$$ ■

Example 3.5.13 Find a basis for the subspace

$$S = \left\{ \begin{bmatrix} a & b \\ b & a \end{bmatrix} \middle| \, a, b \text{ are real numbers} \right\}$$

of $R_{2\times 2}$.

Solution Since

$$\begin{bmatrix} a & b \\ b & a \end{bmatrix} = a \begin{bmatrix} 1 & 0 \\ 0 & 1 \end{bmatrix} + b \begin{bmatrix} 0 & 1 \\ 1 & 0 \end{bmatrix}$$

S is spanned by the two matrices

$$\begin{bmatrix} 1 & 0 \\ 0 & 1 \end{bmatrix} \text{ and } \begin{bmatrix} 0 & 1 \\ 1 & 0 \end{bmatrix}$$

Since these two matrices are linearly independent, they form a basis for S. ■

SUMMARY

Terms to know:

Basis of a vector space

Standard basis for R^n, P_n, $R_{m\times n}$

Theorems to know:

If $\{\mathbf{x}_1, \ldots, \mathbf{x}_n\}$ is a linearly independent set of vectors in R^n, then $\{\mathbf{x}_1, \ldots, \mathbf{x}_n\}$ is a basis for R^n.

Every nonzero vector space spanned by a finite set has a basis.

Methods to know:

Determine whether the set $\{\mathbf{x}_1, \ldots, \mathbf{x}_n\}$ is a basis of V.

EXERCISE SET 3.5

Determine if the following sets are a basis for the indicated vector space.

1. $\left\{\begin{bmatrix}1\\1\\1\end{bmatrix}, \begin{bmatrix}-1\\1\\1\end{bmatrix}, \begin{bmatrix}1\\-1\\1\end{bmatrix}\right\}$ for R^3

2. $\left\{\begin{bmatrix}1\\0\\1\end{bmatrix}, \begin{bmatrix}1\\1\\0\end{bmatrix}, \begin{bmatrix}0\\1\\1\end{bmatrix}\right\}$ for R^3

3. $\left\{\begin{bmatrix}-1\\1\end{bmatrix}, \begin{bmatrix}-1\\0\end{bmatrix}\right\}$ for R^2

4. $\left\{\begin{bmatrix}1\\1\end{bmatrix}, \begin{bmatrix}0\\1\end{bmatrix}, \begin{bmatrix}-1\\0\end{bmatrix}\right\}$ for R^2

5. $\left\{\begin{bmatrix}1\\2\\1\end{bmatrix}, \begin{bmatrix}1\\1\\1\end{bmatrix}, \begin{bmatrix}0\\1\\1\end{bmatrix}, \begin{bmatrix}1\\1\\0\end{bmatrix}\right\}$ for R^3

6. $\left\{\begin{bmatrix}1\\1\\1\\1\end{bmatrix}, \begin{bmatrix}1\\1\\1\\0\end{bmatrix}, \begin{bmatrix}1\\1\\0\\0\end{bmatrix}, \begin{bmatrix}1\\0\\0\\0\end{bmatrix}\right\}$ for R^4

7. $\left\{\begin{bmatrix}2\\0\\0\end{bmatrix}, \begin{bmatrix}0\\0\\2\end{bmatrix}, \begin{bmatrix}0\\0\\0\end{bmatrix}\right\}$ for R^3

8. Verify that

$$\left\{\begin{bmatrix}3\\-1\\2\end{bmatrix}, \begin{bmatrix}1\\5\\2\end{bmatrix}, \begin{bmatrix}1\\0\\0\end{bmatrix}\right\}$$

spans R^3.

9. a. Verify that

$$\left\{\begin{bmatrix}3\\6\\7\end{bmatrix}, \begin{bmatrix}1\\-3\\5\end{bmatrix}\right\}$$

is not a basis for R^3. Enlarge

$$\left\{\begin{bmatrix}3\\6\\7\end{bmatrix}, \begin{bmatrix}1\\-3\\5\end{bmatrix}\right\}$$

to a basis for R^3.

b. Expand $\{1, 2 - x, x^2 + 1\}$ to a basis of P_3 and then express the polynomial $2 - 3x + x^2 + 4x^3$ in terms of the basis found.

10. Let

$$S = \left\{\begin{bmatrix}a\\b\end{bmatrix} \middle| \begin{bmatrix}a\\b\end{bmatrix} \in R^2 \quad \text{and} \quad b = 3a\right\}$$

Find a basis for S.

11. Let

$$S = \left\{\begin{bmatrix}a\\b\\c\end{bmatrix} \middle| \begin{bmatrix}a\\b\\c\end{bmatrix} \in R^3 \quad \text{and} \quad 2a + b + 3c = 0\right\}$$

Find a basis for S.

12. Let

$$S = \left\{\begin{bmatrix}a\\b\\c\\d\end{bmatrix} \middle| \begin{bmatrix}a\\b\\c\\d\end{bmatrix} \in R^4 \quad \text{and} \quad a + b + c + d = 0\right\}$$

Find a basis for S.

13. Prove: If $\{\mathbf{x}_1, \mathbf{x}_2, \ldots, \mathbf{x}_t\}$ is a basis for S, a subspace of R^n, then any vector $\mathbf{x} \in S$ can be expressed uniquely as a linear combination of the $\mathbf{x}_i$'s.

14. Show that

$$\left\{\begin{bmatrix}1 & 0\\0 & 1\end{bmatrix}, \begin{bmatrix}1 & 0\\1 & 1\end{bmatrix}, \begin{bmatrix}0 & 1\\1 & 1\end{bmatrix}, \begin{bmatrix}0 & 1\\1 & 0\end{bmatrix}\right\}$$

is a basis for $R_{2\times 2}$.

15. a. Show that

$$\left\{\begin{bmatrix}1 & 0\\0 & 1\end{bmatrix}, \begin{bmatrix}1 & 1\\1 & 1\end{bmatrix}, \begin{bmatrix}0 & 1\\0 & 1\end{bmatrix}\right\}$$

is not a basis for $R_{2\times 2}$.

b. Is this set of vectors linearly independent? If so, extend this set to a basis of $R_{2\times 2}$.

16. Show that

$$\left\{\begin{bmatrix}\frac{1}{2} & 0\\0 & 1\end{bmatrix}, \begin{bmatrix}0 & 1\\1 & 0\end{bmatrix}, \begin{bmatrix}0 & \frac{1}{3}\\0 & \frac{1}{3}\end{bmatrix}, \begin{bmatrix}0 & 1\\0 & 0\end{bmatrix}\right\}$$

is a basis for $R_{2\times 2}$.

17. Is $\{1 + x^3, x^2 - 1, 1 + x, x^3\}$ a basis for P_3?

18. Show that $\{1, x, x^2, x^2 + x\}$ is not a basis for P_2. Why is this set not a basis?

19. Is $\{1 - x^3, x - x^2, x + x^2, x^3 + 1\}$ a basis for P_3?

20. Find a basis for $\{p_0 + p_1x + p_2x^2 + p_3x^3 \mid p_0 + p_1 + p_2 + p_3 = 0\}$.

21. Find a basis for $\{p_0 + p_1x + p_2x^2 + p_3x^3 \mid p_1 = 3p_0\}$.

22. Find a basis for

$$\left\{\begin{bmatrix} a & b \\ c & d \end{bmatrix} \middle| \, 2a + 3b + 3c = 0\right\}$$

23. Find a basis for the set of solutions of the given systems

a.
$$\begin{aligned} 2x_1 + 3x_2 + 4x_3 &= 0 \\ x_1 + x_2 + x_3 &= 0 \\ 4x_1 + 5x_2 + 6x_3 &= 0 \end{aligned}$$

b.
$$\begin{aligned} x_1 + x_2 - x_3 + x_4 \qquad &= 0 \\ x_3 + x_4 + x_5 &= 0 \\ 2x_1 + 2x_2 - x_3 + \qquad x_5 &= 0 \\ x_1 + x_2 - 2x_3 - \qquad x_5 &= 0 \\ 2x_3 - 4x_4 + 2x_5 &= 0 \\ x_1 - x_2 - 2x_3 - 3x_4 + x_5 &= 0 \end{aligned}$$

c.
$$\begin{aligned} 2x_1 + x_2 + \qquad 6x_4 + x_5 + x_6 + 2x_7 &= 0 \\ 4x_1 + x_2 + \qquad 6x_4 + 2x_5 + 2x_6 + x_7 &= 0 \end{aligned}$$

24. Prove that if $S = \{\mathbf{x}_1, \mathbf{x}_2, \ldots, \mathbf{x}_n\}$ is a linearly independent set of vectors in R^n, then S is a basis for R^n.

25. A basis $\{\mathbf{x}_1, \mathbf{x}_2, \ldots, \mathbf{x}_n\}$ for a subspace V of R^n is an **orthonormal** basis if $\mathbf{x}_i \cdot \mathbf{x}_j = 0$ for $i \neq j$ and $\mathbf{x}_i \cdot \mathbf{x}_i = 1$. If $\{\mathbf{x}_1, \mathbf{x}_2, \ldots, \mathbf{x}_n\}$ is an orthonormal basis for V and $\mathbf{x}$ is any vector in V, show that

$$\mathbf{x} = \beta_1\mathbf{x}_1 + \beta_2\mathbf{x}_2 + \cdots + \beta_n\mathbf{x}_n$$

where $\beta_i = \mathbf{x}_i \cdot \mathbf{x}_i$.

SECTION 3.6 BASIS AND DIMENSION

We use the results of Section 3.5 to simplify greatly the amount of work needed to find a basis of a vector space. Naturally, we reduce everything to solving a system of equations.

First, recall the following definition:

A *basis* $\{\mathbf{v}_1, \ldots, \mathbf{v}_k\}$ of a space S is a linearly independent set that spans S.

The main result of this section rests on the following lemma.

Lemma 3.6.1

Let $\{\mathbf{v}_1, \mathbf{v}_2, \ldots, \mathbf{v}_k\}$ be a basis of S and $\{\mathbf{w}_1, \mathbf{w}_2, \ldots, \mathbf{w}_n\}$ be a subset of S with $n > k$. Then the set $\{\mathbf{w}_1, \mathbf{w}_2, \ldots, \mathbf{w}_n\}$ is linearly dependent.

Proof. Since $\{\mathbf{v}_1, \mathbf{v}_2, \ldots, \mathbf{v}_k\}$ is a basis, each $\mathbf{w}_i$ can be uniquely written as

$$\mathbf{w}_i = a_{i1}\mathbf{v}_1 + a_{i2}\mathbf{v}_2 + \cdots + a_{ik}\mathbf{v}_k$$

To show that $\{\mathbf{w}_1, \mathbf{w}_2, \ldots, \mathbf{w}_n\}$ is linearly dependent, we must show

$$c_1\mathbf{w}_1 + c_2\mathbf{w}_2 + \cdots + c_n\mathbf{w}_n = \mathbf{0}$$

has a nontrivial solution. Now

$$\begin{aligned}
c_1\mathbf{w}_1 &= c_1a_{11}\mathbf{v}_1 + c_1a_{12}\mathbf{v}_2 + \cdots + c_1a_{1k}\mathbf{v}_k \\
c_2\mathbf{w}_2 &= c_2a_{21}\mathbf{v}_1 + c_2a_{22}\mathbf{v}_2 + \cdots + c_2a_{2k}\mathbf{v}_k \\
&\ \ \vdots \\
c_n\mathbf{w}_n &= c_na_{n1}\mathbf{v}_1 + c_na_{n2}\mathbf{v}_2 + \cdots + c_na_{nk}\mathbf{v}_k
\end{aligned}$$

Adding both columns we obtain

$$\begin{aligned}
\mathbf{0} &= c_1\mathbf{w}_1 + c_2\mathbf{w}_2 + \cdots + c_n\mathbf{w}_n \\
&= (c_1a_{11} + c_2a_{21} + \cdots + c_na_{n1})\mathbf{v}_1 \\
&\quad + (c_1a_{12} + c_2a_{22} + \cdots + c_na_{n2})\mathbf{v}_2 \\
&\quad + \cdots + (c_1a_{1k} + c_2a_{2k} + \cdots + c_na_{nk})\mathbf{v}_k
\end{aligned}$$

Since $\{\mathbf{v}_1, \mathbf{v}_2, \ldots, \mathbf{v}_k\}$ is linearly independent, we have

$$\begin{aligned}
c_1a_{11} + c_2a_{21} + \cdots + c_na_{n1} &= 0 \\
c_1a_{12} + c_2a_{22} + \cdots + c_na_{n2} &= 0 \\
&\ \ \vdots \\
c_1a_{1k} + c_2a_{2k} + \cdots + c_na_{nk} &= 0
\end{aligned}$$

a homogeneous system of k equations in n unknowns with $k < n$. Thus the system has nonzero solutions, and so $\{\mathbf{w}_1, \mathbf{w}_2, \ldots, \mathbf{w}_n\}$ is linearly dependent.

This result gives us a multitude of ways to find a basis of a subspace of a vector space. First, note the following theorem.

Theorem 3.6.2 *If $\{\mathbf{v}_1, \mathbf{v}_2, \ldots, \mathbf{v}_m\}$ and $\{\mathbf{u}_1, \mathbf{u}_2, \ldots, \mathbf{u}_n\}$ are bases for a subspace S, then $m = n$.*

Proof. Since $\{\mathbf{v}_1, \mathbf{v}_2, \ldots, \mathbf{v}_m\}$ is a basis and $\{\mathbf{u}_1, \mathbf{u}_2, \ldots, \mathbf{u}_n\}$ is a set of linearly independent vectors, we cannot have $n > m$ by Lemma 3.6.1. Thus $m \geqq n$.

Since $\{\mathbf{u}_1, \mathbf{u}_2, \ldots, \mathbf{u}_n\}$ is a basis and $\{\mathbf{v}_1, \mathbf{v}_2, \ldots, \mathbf{v}_m\}$ is a set of linearly independent vectors, we cannot have $m > n$. Thus $n \geqq m$, also. Since $m \geqq n$ and $n \geqq m$ then $m = n$.

Definition 3.6.3 The number of vectors in a basis of a subspace S is called the **dimension** of S and is denoted by $\dim(S)$. If $\dim(S) = n$ for some positive integer n, then S is said to be **finite dimensional**. The vector space $\{\mathbf{0}\}$ is defined to have dimension 0.

Example 3.6.4 a. The dimension of R^2 is 2 since

$$\left\{\begin{bmatrix}1\\0\end{bmatrix}, \begin{bmatrix}0\\1\end{bmatrix}\right\}$$

is a basis.

b. The dimension of R^3 is 3 since

$$\left\{\begin{bmatrix}1\\0\\0\end{bmatrix}, \begin{bmatrix}0\\1\\0\end{bmatrix}, \begin{bmatrix}0\\0\\1\end{bmatrix}\right\}$$

is a basis.

c. The dimension of R^n is n since $\{\mathbf{e}_1, \mathbf{e}_2, \ldots, \mathbf{e}_n\}$ is a basis, where

$$\mathbf{e}_i = \begin{bmatrix}0\\0\\\vdots\\1\\0\\0\\\vdots\\0\end{bmatrix} \rightarrow i\text{th row}$$

■

Now we see how to find a basis for a space S.

Theorem 3.6.5 *Let $S = \langle \mathbf{v}_1, \mathbf{v}_2, \ldots, \mathbf{v}_k \rangle$. Then there is a subset T of $\{\mathbf{v}_1, \mathbf{v}_2, \ldots, \mathbf{v}_k\}$ that is a basis of S.*

Proof. If $\{\mathbf{v}_1, \mathbf{v}_2, \ldots, \mathbf{v}_k\}$ is linearly independent, then $\{\mathbf{v}_1, \mathbf{v}_2, \ldots, \mathbf{v}_k\}$ is a basis, since it spans S. If the set $\{\mathbf{v}_1, \mathbf{v}_2, \ldots, \mathbf{v}_k\}$ is not linearly independent (i.e., it is linearly dependent), renumber the vectors so that $\mathbf{v}_k \in \langle \mathbf{v}_1, \mathbf{v}_2, \ldots, \mathbf{v}_{k-1} \rangle$ (or $\mathbf{v}_k$ is a linear combination of $\{\mathbf{v}_1, \ldots, \mathbf{v}_{k-1}\}$). Then $S = \langle \mathbf{v}_1, \mathbf{v}_2, \ldots, \mathbf{v}_{k-1} \rangle$. If $\{\mathbf{v}_1, \mathbf{v}_2, \ldots, \mathbf{v}_{k-1}\}$ is linearly independent, we are done. If not, $\mathbf{v}_{k-1} \in \langle \mathbf{v}_1, \mathbf{v}_2, \ldots, \mathbf{v}_{k-2} \rangle$ (after

is a linear combination of the other two. The set

$$\left\{\begin{bmatrix} 0 \\ 0 \\ -8 \end{bmatrix}, \begin{bmatrix} 1 \\ 2 \\ -9 \end{bmatrix}\right\}$$

is linearly independent and thus is a basis for S and $\dim(S) = 2$. ■

The procedure for determining a basis for S in Example 3.6.7 is long. It was done this way in order to illustrate the proof of Theorem 3.6.6. We illustrate an easier method to find basis for S in Example 3.6.8.

Example 3.6.8 Find a basis for

$$S = \left\langle \begin{bmatrix} 1 \\ 2 \\ -1 \end{bmatrix}, \begin{bmatrix} 2 \\ 4 \\ 6 \end{bmatrix}, \begin{bmatrix} 0 \\ 0 \\ -8 \end{bmatrix}, \begin{bmatrix} 1 \\ 2 \\ -9 \end{bmatrix} \right\rangle$$

Solution First, we place these vectors as the rows of a matrix

$$A = \begin{bmatrix} 1 & 2 & -1 \\ 2 & 4 & 6 \\ 0 & 0 & -8 \\ 1 & 2 & -9 \end{bmatrix}$$

Then (-2)R1 to R2, (-1)R1 to R4, (1)R2 to R3, and (1)R2 to R4 applied in order to A yield

$$B = \begin{bmatrix} 1 & 2 & -1 \\ 0 & 0 & 8 \\ 0 & 0 & 0 \\ 0 & 0 & 0 \end{bmatrix}$$

The set

$$\left\{\begin{bmatrix} 1 \\ 2 \\ -1 \end{bmatrix}, \begin{bmatrix} 0 \\ 0 \\ 8 \end{bmatrix}\right\}$$

is linearly independent, since if

$$a\begin{bmatrix} 1 \\ 2 \\ -1 \end{bmatrix} + b\begin{bmatrix} 0 \\ 0 \\ 8 \end{bmatrix} = \begin{bmatrix} a \\ 2a \\ -a + 8b \end{bmatrix} = \begin{bmatrix} 0 \\ 0 \\ 0 \end{bmatrix}$$

then $a = 0$, and $0 = -a + 8b$, or $b = 0$. The elementary row operations that we used to reduce A to row echelon form involved only linear combinations of the rows of A. Thus the nonzero row vectors of the row echelon form span the same subspace of R^3 as the original rows of A, namely, S.

Thus $\dim(S) = 2$ and a basis for S is

$$\left\langle \begin{bmatrix} 1 \\ 2 \\ -1 \end{bmatrix}, \begin{bmatrix} 0 \\ 0 \\ 8 \end{bmatrix} \right\rangle$$

■

We formalize the method used in Example 3.6.8 in the following definition.

Definition 3.6.9 The **row space** of the matrix A is the subspace of R^n spanned by the rows of A. The row space of A is denoted by $\mathrm{RS}(A)$.

The row space of a matrix is useful in finding a basis for $S = \langle \mathbf{v}_1, \mathbf{v}_2, \ldots, \mathbf{v}_n \rangle \subseteq R^n$ by performing the following steps:

1. Form a matrix A whose first row is $\mathbf{v}_1$, second row is $\mathbf{v}_2, \ldots$.
2. Use elementary row operations to reduce A to its row echelon form, B.
3. The nonzero rows of B form a basis for S.

We illustrate this procedure in Examples 3.6.10 and 3.6.11.

Example 3.6.10 Let

$$S = \left\langle \begin{bmatrix} 1 \\ 1 \\ 0 \\ 2 \end{bmatrix}, \begin{bmatrix} 0 \\ 1 \\ 2 \\ 2 \end{bmatrix}, \begin{bmatrix} 1 \\ 2 \\ 2 \\ 4 \end{bmatrix} \right\rangle$$

Form the matrix A and apply row reduction operations on A.

Solution The generating vectors of S become rows of A. Thus

$$A = \begin{bmatrix} 1 & 1 & 0 & 2 \\ 0 & 1 & 2 & 2 \\ 1 & 2 & 2 & 4 \end{bmatrix}$$

Now (-1)R1 to R3 gives

$$\begin{bmatrix} 1 & 1 & 0 & 2 \\ 0 & 1 & 2 & 2 \\ 0 & 1 & 2 & 2 \end{bmatrix}$$

and (-1)R2 to R3 results in

$$\begin{bmatrix} 1 & 1 & 0 & 2 \\ 0 & 1 & 2 & 2 \\ 0 & 0 & 0 & 0 \end{bmatrix}$$

A basis for S is

$$\left\{ \begin{bmatrix} 1 \\ 1 \\ 0 \\ 2 \end{bmatrix}, \begin{bmatrix} 0 \\ 1 \\ 2 \\ 2 \end{bmatrix} \right\}$$

■

Example 3.6.11 Let

$$S = \left\langle \begin{bmatrix} 0 \\ 0 \\ 1 \\ -2 \\ 3 \end{bmatrix}, \begin{bmatrix} 0 \\ 0 \\ -2 \\ 1 \\ 0 \end{bmatrix}, \begin{bmatrix} 0 \\ 0 \\ 1 \\ 0 \\ 3 \end{bmatrix} \right\rangle$$

Form the matrix A whose row space is S and obtain its row echelon form.

Solution The generating vectors of S become row vectors of A, resulting in

$$A = \begin{bmatrix} 0 & 0 & 1 & -2 & 3 \\ 0 & 0 & -2 & 1 & 0 \\ 0 & 0 & 1 & 0 & 3 \end{bmatrix}$$

Thus (2)R1 to R2, (-1)R1 to R3, and $(\frac{2}{3})$R2 to R3, applied in order to A, results in its row echelon form,

$$B = \begin{bmatrix} 0 & 0 & 1 & -2 & 3 \\ 0 & 0 & 0 & -3 & 6 \\ 0 & 0 & 0 & 0 & 4 \end{bmatrix}$$

A basis for S is

$$\left\{ \begin{bmatrix} 0 \\ 0 \\ 1 \\ -2 \\ 3 \end{bmatrix}, \begin{bmatrix} 0 \\ 0 \\ 0 \\ -3 \\ 6 \end{bmatrix}, \begin{bmatrix} 0 \\ 0 \\ 0 \\ 0 \\ 4 \end{bmatrix} \right\}$$

■

The reason for reducing the matrix A to row echelon form B is that the nonzero rows of B form a basis for $S = \mathrm{RS}(A)$. The verification of this is left as an exercise. In general, the reason this process yields a basis for S is shown by the following theorems.

Theorem 3.6.12 *Let A be an $m \times n$ matrix and let B be the result of performing a row operation on A. Then $RS(A) = RS(B)$.*

We sketch a proof of Theorem 3.6.12. Let B be obtained from A by adding (2)R4 to R6. Let the rows of A be denoted by $\mathbf{A}_1, \ldots, \mathbf{A}_n$ and the rows of B be denoted by $\mathbf{B}_1, \ldots, \mathbf{B}_n$. To show that $\mathrm{RS}(A) = \mathrm{RS}(B)$, we must show that any vector written as a linear combination of the rows of A can be written as a linear combination of the rows of B, and vice-versa. To this end we suppose

$$\mathbf{v} = a_1\mathbf{A}_1 + \cdots + a_n\mathbf{A}_n$$

Then since row 6 of B is (2)R4 of A added to row 6 of A,

$$\mathbf{B}_6 = 2\mathbf{A}_4 + \mathbf{A}_6$$

Therefore, since for k different from 6, $\mathbf{A}_k = \mathbf{B}_k$, we have

$$\begin{aligned}
\mathbf{v} &= a_1\mathbf{A}_1 + \cdots + a_n\mathbf{A}_n \\
&= a_1\mathbf{B}_1 + \cdots + a_6\mathbf{A}_6 + \cdots + a_n\mathbf{B}_n \\
&= a_1\mathbf{B}_1 + \cdots + a_6(\mathbf{B}_6 - 2\mathbf{A}_4) + \cdots + a_n\mathbf{B}_n \\
&= a_1\mathbf{B}_1 + \cdots + a_6(\mathbf{B}_6 - 2\mathbf{B}_4) + \cdots + a_n\mathbf{B}_n \\
&= a_1\mathbf{B}_1 + \cdots + (a_4 - 2a_6)\mathbf{B}_4 + \cdots + a_6\mathbf{B}_6 + \cdots + a_n\mathbf{B}_n
\end{aligned}$$

This shows that $\mathbf{v}$ is in $\mathrm{RS}(B)$.

To show that $\mathrm{RS}(B)$ is a subspace of $\mathrm{RS}(A)$, we merely reverse the steps in the above process, noting that $\mathbf{B}_6 = 2\mathbf{A}_4 + \mathbf{A}_6$. To modify the sketch of the proof to a full-blown proof, merely change the 2 to k, the 4 to i, and the 6 to j, and you are done for the case of adding a multiple k of one row to another. To cover the case of multiplying row i of A by k, change the 2 to k and the 4 to i and erase A_6.

Theorem 3.6.13 *After reducing A to row echelon form B, the nonzero rows of B are linearly independent.*

The proof of this theorem consists of repeated applications of Theorem 3.6.12 to $\mathrm{RS}(A)$. A complete proof requires use of mathematical induction.

Example 3.6.14

Let

$$S = \left\langle \begin{bmatrix} 1 \\ 1 \\ 1 \end{bmatrix}, \begin{bmatrix} -1 \\ 1 \\ -1 \end{bmatrix}, \begin{bmatrix} 1 \\ 1 \\ 3 \end{bmatrix}, \begin{bmatrix} 0 \\ 2 \\ 1 \end{bmatrix} \right\rangle$$

Find a basis for S.

Solution Let

$$A = \begin{bmatrix} 1 & 1 & 1 \\ -1 & 1 & -1 \\ 1 & 1 & 3 \\ 0 & 2 & 1 \end{bmatrix}$$

That is, put the vectors in the spanning set for S into the rows of A. The following sequence of row operations changes A to B: (1)R1 to R2, (-1)R1 to R3, (-1)R2 to R4, and $(-\frac{1}{2})$R3 to R4.

$$B = \begin{bmatrix} 1 & 1 & 1 \\ 0 & 2 & 0 \\ 0 & 0 & 2 \\ 0 & 0 & 0 \end{bmatrix}$$

Thus a basis for S is

$$\left\{ \begin{bmatrix} 1 \\ 1 \\ 1 \end{bmatrix}, \begin{bmatrix} 0 \\ 2 \\ 0 \end{bmatrix}, \begin{bmatrix} 0 \\ 0 \\ 2 \end{bmatrix} \right\}$$

Another way to find a basis is to build up a basis from a linearly independent set or to extend a linearly independent set to a basis.

Theorem 3.6.15 *Let $\{\mathbf{v}_1, \mathbf{v}_2, \ldots, \mathbf{v}_k\}$ be a linearly independent set in R^n. Then $\mathbf{v}_{k+1}, \mathbf{v}_{k+2}, \ldots, \mathbf{v}_n$ can be found, making $\{\mathbf{v}_1, \mathbf{v}_2, \ldots, \mathbf{v}_k, \mathbf{v}_{k+1}, \ldots, \mathbf{v}_n\}$ a basis for R^n.*

Proof. If $k = n$, we are done, for if $\mathbf{v} \in R^n$, then $\{\mathbf{v}_1, \ldots, \mathbf{v}_n, \mathbf{v}\}$ is linearly dependent (since $n + 1 > n$). Thus $R^n = \langle \mathbf{v}_1, \mathbf{v}_2, \ldots, \mathbf{v}_n \rangle$.

If $k < n$, choose $\mathbf{v}_{k+1}$ in R^n such that $\mathbf{v}_{k+1}$ is not in $\langle \mathbf{v}_1, \mathbf{v}_2, \ldots, \mathbf{v}_k \rangle$. We can do this because if $\langle \mathbf{v}_1, \ldots, \mathbf{v}_k \rangle = R^n$, then $\{\mathbf{v}_1, \mathbf{v}_2, \ldots, \mathbf{v}_k\}$ is a basis of R^n, so $k = n$. Then $\{\mathbf{v}_1, \mathbf{v}_2, \ldots, \mathbf{v}_k, \mathbf{v}_{k+1}\}$ is linearly independent. If $k + 1 = n$, we are done, since $\{\mathbf{v}_1, \mathbf{v}_2, \ldots, \mathbf{v}_k, \mathbf{v}_{k+1}\}$ spans R^n. If $k + 1 < n$, choose $\mathbf{v}_{k+2}$ in R^n such that $\mathbf{v}_{k+2}$ is not in $\langle \mathbf{v}_1, \ldots, \mathbf{v}_k, \mathbf{v}_{k+1} \rangle$. After $n - k$ steps, $\{\mathbf{v}_1, \mathbf{v}_2, \ldots, \mathbf{v}_k, \mathbf{v}_{k+1}, \ldots, \mathbf{v}_n\}$ is a basis for R^n.

Corollary 3.6.16
A set $B = \{\mathbf{v}_1, \ldots, \mathbf{v}_n\}$ is a basis of V if and only if whenever $C = \{\mathbf{v}_1, \ldots, \mathbf{v}_n, \mathbf{v}_{n+1}, \ldots, \mathbf{v}_k\}$ and C is linearly independent, then $k = n$ and $B = C$.

Example 3.6.17 Extend

$$\left\{ \begin{bmatrix} 1 \\ 1 \\ 1 \end{bmatrix}, \begin{bmatrix} -1 \\ 1 \\ 2 \end{bmatrix} \right\}$$

to a basis of R^3.

Solution The two given vectors are linearly independent, since one is not a multiple of the other.

Now, if

$$\mathbf{v} \in \left\langle \begin{bmatrix} 1 \\ 1 \\ 1 \end{bmatrix}, \begin{bmatrix} -1 \\ 1 \\ 2 \end{bmatrix} \right\rangle$$

there exist constants a and b such that

$$\mathbf{v} = a\begin{bmatrix} 1 \\ 1 \\ 1 \end{bmatrix} + b\begin{bmatrix} -1 \\ 1 \\ 2 \end{bmatrix} = \begin{bmatrix} a - b \\ a + b \\ a + 2b \end{bmatrix}$$

To find a vector not in

$$\left\langle \begin{bmatrix} 1 \\ 1 \\ 1 \end{bmatrix}, \begin{bmatrix} -1 \\ 1 \\ 2 \end{bmatrix} \right\rangle$$

we choose a, b, and c so that the

$$\begin{bmatrix} a - b \\ a + b \\ c \end{bmatrix} \quad \text{and} \quad \begin{bmatrix} a - b \\ a + b \\ a + 2b \end{bmatrix}$$

are not equal. For example, if $a = 1$ and $b = 1$, then $a - b = 0$, $a + b = 2$, and $a + 2b = 3$. Thus

$$\begin{bmatrix} 0 \\ 2 \\ 3 \end{bmatrix} \quad \text{would belong to} \quad \left\langle \begin{bmatrix} 1 \\ 1 \\ 1 \end{bmatrix}, \begin{bmatrix} -1 \\ 1 \\ 2 \end{bmatrix} \right\rangle$$

but

$$\begin{bmatrix} 0 \\ 2 \\ c \end{bmatrix}$$

would not for any $c \neq 3$, since if the first and second entries are 0 and 2, respectively, then 3 has to be the third entry in order for the vector to belong to

$$\left\langle \begin{bmatrix} 1 \\ 1 \\ 1 \end{bmatrix}, \begin{bmatrix} -1 \\ 1 \\ 2 \end{bmatrix} \right\rangle$$

Thus choosing $c = 2$, the set

$$\left\{\begin{bmatrix}1\\1\\1\end{bmatrix}, \begin{bmatrix}-1\\1\\2\end{bmatrix}, \begin{bmatrix}0\\2\\2\end{bmatrix}\right\}$$

is linearly independent and, since $\dim(R^3) = 3$, this set spans R^3 and so is a basis for R^3. ■

SUMMARY

Terms to know:

Dimension of a vector space

Finite dimensional subspace

Row space of a matrix

Theorems to know:

If $\{\mathbf{v}_1, \ldots, \mathbf{v}_m\}$ and $\{\mathbf{u}_1, \ldots, \mathbf{u}_n\}$ are bases of a vector space V, then $m = n$.

If $S = \langle \mathbf{v}_1, \ldots, \mathbf{v}_n \rangle$, there is a subset of $\{\mathbf{v}_1, \ldots, \mathbf{v}_n\}$ that is a basis of V.

If $\{\mathbf{v}_1, \ldots, \mathbf{v}_k\}$ is a linearly independent set in R^n, then $\mathbf{v}_{k+1}, \ldots, \mathbf{v}_n$ can be found, making $\{\mathbf{v}_1, \ldots, \mathbf{v}_n\}$ a basis of R^n.

Methods to know:

Extend the linearly independent set $\{\mathbf{v}_1, \ldots, \mathbf{v}_k\}$ to a basis of a vector space.

Find a linearly independent subset of $\{\mathbf{v}_1, \ldots, \mathbf{v}_k\}$ that is a basis of $S = \langle \mathbf{v}_1, \ldots, \mathbf{v}_k \rangle$.

EXERCISE SET 3.6

1. The set

$$S_1 = \left\{\begin{bmatrix}1\\0\\1\end{bmatrix}, \begin{bmatrix}-1\\1\\0\end{bmatrix}, \begin{bmatrix}0\\-1\\1\end{bmatrix}\right\}$$

is a basis of R^3. By Lemma 3.6.1, we know the set

$$S_2 = \left\{\begin{bmatrix}1\\0\\1\end{bmatrix}, \begin{bmatrix}2\\1\\4\end{bmatrix}, \begin{bmatrix}-2\\1\\3\end{bmatrix}, \begin{bmatrix}4\\0\\-1\end{bmatrix}\right\}$$

is linearly dependent.

a. Express each vector in S_2 as a linear combination of the vectors in S_1. This gives us the a_{ij} used in the proof of Lemma 3.6.1. Denote the matrix with a_{ij} in the ith row and jth column by A.

b. Write the entries of the matrix A and then find A^T.

c. Solve $A^T\mathbf{k} = \mathbf{0}$ for $\mathbf{k}$.

d. Show that

$$k_1\begin{bmatrix}1\\0\\1\end{bmatrix} + k_2\begin{bmatrix}2\\1\\4\end{bmatrix} + k_3\begin{bmatrix}-2\\1\\3\end{bmatrix} + k_4\begin{bmatrix}4\\0\\-1\end{bmatrix} = \begin{bmatrix}0\\0\\0\end{bmatrix}$$

2. Find the dimension of the row space of each of the following matrices.

a. $\begin{bmatrix}1 & 0 & 0\\-1 & 2 & 1\\1 & 0 & 3\end{bmatrix}$ b. $\begin{bmatrix}1 & 2 & 1\\-1 & 3 & 1\\1 & 7 & 3\end{bmatrix}$ c. $\begin{bmatrix}1 & 2 & 0 & 3\\4 & 0 & 1 & 5\\0 & 1 & -1 & 1\\1 & -1 & 1 & 0\end{bmatrix}$

d. $\begin{bmatrix}1 & 2\\2 & 1\end{bmatrix}$ e. $\begin{bmatrix}1 & 0 & 1 & 0\\-1 & 1 & 1 & 1\\1 & -1 & 1 & 0\end{bmatrix}$

3. a. Show that

$$\left\{\begin{bmatrix}1\\0\\-2\\0\end{bmatrix}, \begin{bmatrix}0\\1\\2\\2\end{bmatrix}\right\}$$

is a basis for

$$S = \left\langle \begin{bmatrix}1\\1\\0\\2\end{bmatrix}, \begin{bmatrix}0\\1\\2\\2\end{bmatrix}, \begin{bmatrix}1\\2\\2\\4\end{bmatrix} \right\rangle$$

b. Show that

$$\left\{\begin{bmatrix}1\\1\\0\\2\end{bmatrix}, \begin{bmatrix}0\\1\\2\\2\end{bmatrix}\right\}$$

is a basis for

$$S = \left\langle \begin{bmatrix}1\\1\\0\\2\end{bmatrix}, \begin{bmatrix}0\\1\\2\\2\end{bmatrix}, \begin{bmatrix}1\\2\\2\\4\end{bmatrix} \right\rangle$$

c. Show that

$$\left\{\begin{bmatrix}0\\0\\1\\0\\0\end{bmatrix},\begin{bmatrix}0\\0\\0\\1\\0\end{bmatrix},\begin{bmatrix}0\\0\\0\\0\\1\end{bmatrix}\right\}$$

is a basis for

$$S = \left\langle\begin{bmatrix}0\\0\\-1\\2\\3\end{bmatrix},\begin{bmatrix}0\\0\\-2\\1\\2\end{bmatrix},\begin{bmatrix}0\\0\\1\\0\\3\end{bmatrix}\right\rangle$$

d. Show that

$$\left\{\begin{bmatrix}0\\0\\1\\-2\\3\end{bmatrix},\begin{bmatrix}0\\0\\0\\-3\\6\end{bmatrix},\begin{bmatrix}0\\0\\0\\2\\0\end{bmatrix}\right\}$$

is a basis for

$$S = \left\langle\begin{bmatrix}0\\0\\-1\\2\\3\end{bmatrix},\begin{bmatrix}0\\0\\-2\\1\\2\end{bmatrix},\begin{bmatrix}0\\0\\1\\0\\3\end{bmatrix}\right\rangle$$

4. Find a basis for the row space of each of the following matrices.

a. $\begin{bmatrix}1 & 3\\0 & 1\end{bmatrix}$ b. $\begin{bmatrix}1 & 3 & -1\\1 & 0 & 2\end{bmatrix}$ c. $\begin{bmatrix}1 & 0 & -1 & 2\\0 & 1 & 3 & 4\\2 & 1 & -1 & -1\end{bmatrix}$

d. $\begin{bmatrix}-1 & 1 & -1 & 1\\1 & 2 & -2 & 3\\2 & 4 & 0 & 2\end{bmatrix}$ e. $\begin{bmatrix}1 & -1 & 1\\2 & -1 & 1\\1 & 3 & 2\\0 & -3 & 1\end{bmatrix}$

5. Find a basis for the space S spanned by each set.

a. $\left\{\begin{bmatrix}1\\2\\3\end{bmatrix},\begin{bmatrix}1\\-1\\4\end{bmatrix}\right\}$ b. $\left\{\begin{bmatrix}1\\-1\\1\end{bmatrix},\begin{bmatrix}2\\-1\\1\end{bmatrix},\begin{bmatrix}0\\-3\\1\end{bmatrix}\right\}$

c. $\left\{\begin{bmatrix}-1\\1\\1\\0\end{bmatrix}, \begin{bmatrix}1\\-1\\0\\0\end{bmatrix}, \begin{bmatrix}1\\-1\\-1\\1\end{bmatrix}, \begin{bmatrix}4\\3\\2\\1\end{bmatrix}\right\}$ d. $\left\{\begin{bmatrix}1\\2\end{bmatrix}, \begin{bmatrix}2\\1\end{bmatrix}, \begin{bmatrix}1\\1\end{bmatrix}\right\}$

e. $\left\{\begin{bmatrix}1\\-1\\1\\0\end{bmatrix}, \begin{bmatrix}-1\\1\\0\\1\end{bmatrix}, \begin{bmatrix}1\\0\\0\\1\end{bmatrix}, \begin{bmatrix}0\\1\\1\\0\end{bmatrix}\right\}$

6. Extend the following linearly independent sets to a basis of R^4.

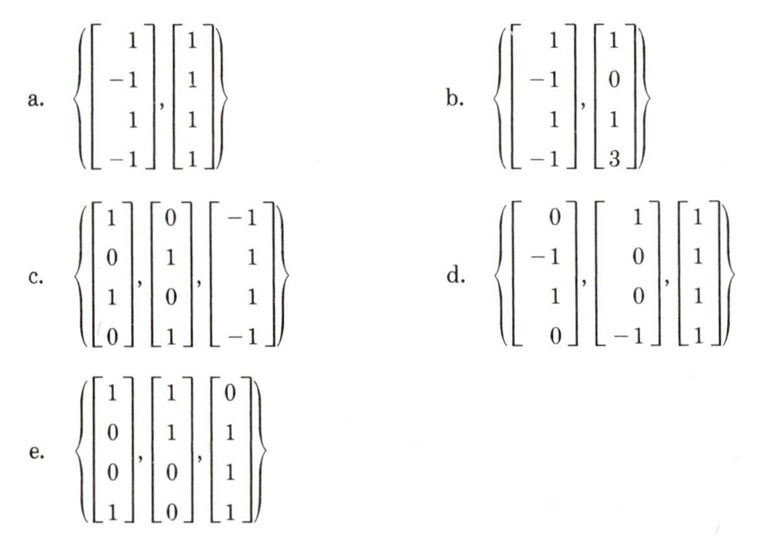

a. $\left\{\begin{bmatrix}1\\-1\\1\\-1\end{bmatrix}, \begin{bmatrix}1\\1\\1\\1\end{bmatrix}\right\}$ b. $\left\{\begin{bmatrix}1\\-1\\1\\-1\end{bmatrix}, \begin{bmatrix}1\\0\\1\\3\end{bmatrix}\right\}$

c. $\left\{\begin{bmatrix}1\\0\\1\\0\end{bmatrix}, \begin{bmatrix}0\\1\\0\\1\end{bmatrix}, \begin{bmatrix}-1\\1\\1\\-1\end{bmatrix}\right\}$ d. $\left\{\begin{bmatrix}0\\-1\\1\\0\end{bmatrix}, \begin{bmatrix}1\\0\\0\\-1\end{bmatrix}, \begin{bmatrix}1\\1\\1\\1\end{bmatrix}\right\}$

e. $\left\{\begin{bmatrix}1\\0\\0\\1\end{bmatrix}, \begin{bmatrix}1\\1\\0\\0\end{bmatrix}, \begin{bmatrix}0\\1\\1\\1\end{bmatrix}\right\}$

7. a. Show that if U and V are finite-dimensional vector spaces and if $U \subseteq V$ (i.e., U is a subspace of V), then $\dim(U) \leq \dim(V)$.
   b. Find all types of subspaces of R^3 and describe them geometrically.

   Hint: Any subspace of R^3 has dimension 0, 1, 2, or 3.

8. Explain why the space of all polynomials has no finite basis.

9. Let V be a vector space of dimension n. Prove each of the following.
   a. Any subset of V with more than n vectors is linearly dependent.
   b. Any subset of V with fewer than n vectors is not a basis.
   c. Any set with n vectors that spans V is a basis.
   d. Any linearly independent set of n vectors is a basis.

SECTION 3.7 INFINITE DIMENSIONAL VECTOR SPACES (OPTIONAL)

In the previous sections, we considered finite subsets $\{\mathbf{v}_1, \ldots, \mathbf{v}_k\}$ of a vector space V. Our definitions of spanning sets and linearly independent sets required finite sets of vectors. The purpose of this section is to extend these definitions to infinite sets and to demonstrate the existence of infinite-dimensional vector spaces.

Notice first that only a finite number of vectors can be added together. (To add an infinite number of vectors would require the concept of convergence and a discussion of the sum of an infinite series, which is not a vector space operation.) However, this restriction is not fatal, as shown by the following definitions, which extend the concepts of spanning set, linearly independent set, and basis.

Definition 3.7.1 The set $\{\mathbf{v}_1, \mathbf{v}_2, \ldots, \mathbf{v}_n, \ldots\}$ of vectors in V **spans** V if for each $\mathbf{v} \in V$, there is a finite subset $\{\mathbf{v}_1, \mathbf{v}_2, \ldots, \mathbf{v}_k\}$ of $\{\mathbf{v}_1, \mathbf{v}_2, \ldots, \mathbf{v}_n, \ldots\}$ for which

$$\mathbf{v} = a_1\mathbf{v}_1 + a_2\mathbf{v}_2 + \cdots + a_k\mathbf{v}_k$$

Example 3.7.2 The set $\{1, x, x^2, x^3, \ldots, x^n, \ldots\}$ spans the vector space P of all polynomials with real coefficients, since for each $\mathbf{p} = p_0 + p_1x + p_2x^2 + \cdots + p_kx^k \in P$,

$$\mathbf{p} = p_0(1) + p_1(x) + p_2(x^2) + \cdots + p_k(x^k)$$

Thus $\mathbf{v} = \mathbf{p}, \mathbf{v}_1 = 1, \mathbf{v}_2 = x, \mathbf{v}_3 = x^2, \ldots,$ in the notation of the definition. ■

Example 3.7.3 The vector space of all infinite sequences having real entries with only a finite number of entries nonzero is spanned by the set $\{\mathbf{v}_1, \mathbf{v}_2, \ldots, \mathbf{v}_n, \ldots\}$ with $\mathbf{v}_1 = (1, 0, 0, \ldots)$, $\mathbf{v}_2 = (0, 1, 0, \ldots), \ldots$, since

$$(a_1, a_2, \ldots, a_k, 0, 0, \ldots) = a_1\mathbf{v}_1 + a_2\mathbf{v}_2 + \cdots + a_k\mathbf{v}_k$$ ■

The concept of linear independence of a set may be extended to infinite sets as follows.

Definition 3.7.4 The subset $\{\mathbf{v}_1, \mathbf{v}_2, \ldots, \mathbf{v}_n, \ldots\}$ of a vector space V is **linearly independent** if each finite subset $\{\mathbf{v}_1, \mathbf{v}_2, \ldots, \mathbf{v}_k\}$ is linearly independent.

The sets of Examples 3.7.2 and 3.7.3 are linearly independent, as shown in Examples 3.7.5 and 3.7.6.

Example 3.7.5 The set $\{1, x, x^2, \ldots, x^n, \ldots\}$ of P is linearly independent. To see this, suppose for some k,

$$\mathbf{0} = a_1(1) + a_2(x) + a_3(x^3) + \cdots + a_k(x^k)$$

Then $0 = a_1 = a_2 = \cdots = a_k$ by equality of polynomials. ■

Example 3.7.6 The set $\{\mathbf{v}_1, \mathbf{v}_2, \ldots, \mathbf{v}_n, \ldots\}$ of Example 3.7.3 is linearly independent, since if

$$(0,0,\ldots) = a_1\mathbf{v}_1 + \cdots + a_k\mathbf{v}_k \text{ for some } k$$

we have

$$(0,0,\ldots) = (a_1, a_2, \ldots, a_k, 0, \ldots)$$

which requires

$$a_1 = a_2 = \cdots = a_k = 0$$

As before, these definitions lead to the concept of a basis.

Definition 3.7.7 A subset B of a vector space V is a **basis** if and only if

1. B spans V.
2. B is linearly independent.

According to this definition, the set in Example 3.7.2 is a basis for P, whereas the set of Example 3.7.3 is a basis for $V = \{(a_1, a_2, \ldots) \mid a_i \in R$ for all i and for some n, $a_k = 0$ for $k \geqq n\}$.

Using Zorn's lemma and other axioms of naive set theory, which will not be discussed in this text, the following theorem remains true in vector spaces having an infinite basis.

Theorem 3.7.8 *Let B be an infinite subset of a vector space V. The following statements are equivalent.*

1. B is a basis for V.

2. B spans V and no proper subset of B spans V.

3. B is linearly independent and if $B \subseteq C \subseteq V$ and C is linearly independent, then $B = C$.

Finally, we state one theorem, which is the basis of linear algebra. However, the proof—or even a discussion of this theorem for infinite dimensional vector spaces—is not considered in this text.

Theorem 3.7.9 *Every nonzero vector space has a basis.*

SUMMARY

Terms to know:

Spanning set
Linearly independent set
Basis

Theorems to know:

Every nonzero vector space has a basis.

CHAPTER FOUR

LINEAR TRANSFORMATIONS AND MATRICES

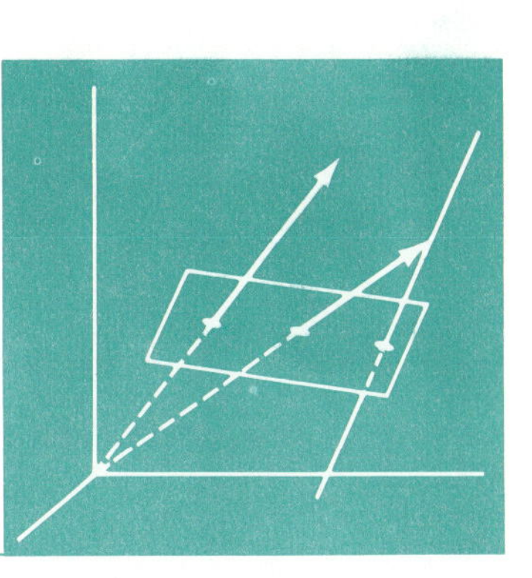

SECTION 4.1 LINEAR TRANSFORMATIONS

Much of mathematics is concerned with algebraic structures and their mappings—especially those mappings that preserve structure. In linear algebra, we are concerned with mappings that preserve addition and scalar multiplication. Such mappings (or functions) are called *linear transformations.*

Definition 4.1.1 A function f from a vector space U to a vector space V, denoted by $f: U \to V$, is called a **linear transformation** from U to V if and only if

1. For arbitrary vectors $\mathbf{u}_1$ and $\mathbf{u}_2$ in U,

$$f(\mathbf{u}_1 + \mathbf{u}_2) = f(\mathbf{u}_1) + f(\mathbf{u}_2)$$

2. For an arbitrary scalar a and arbitrary vector $\mathbf{u}$ in U,

$$f(a\mathbf{u}) = af(\mathbf{u})$$

The two requirements for a linear transformation can be replaced by a single condition:

$$f(a\mathbf{u}_1 + b\mathbf{u}_2) = af(\mathbf{u}_1) + bf(\mathbf{u}_2)$$

where $\mathbf{u}_1$ and $\mathbf{u}_2$ are arbitrary vectors in U and a and b are arbitrary real numbers. We say that f *preserves linear combinations.* (We leave the verification of this as an exercise.)

Example 4.1.2 Show that the mapping $f: R^3 \to R^2$ defined by

$$f\left(\begin{bmatrix} x_1 \\ x_2 \\ x_3 \end{bmatrix}\right) = \begin{bmatrix} x_1 + x_2 \\ x_2 - x_3 \end{bmatrix}$$

is a linear transformation.

Solution We verify that (1) and (2) of the definition are satisfied. First, we verify (1). Let

$$\mathbf{x} = \begin{bmatrix} x_1 \\ x_2 \\ x_3 \end{bmatrix} \quad \text{and} \quad \mathbf{y} = \begin{bmatrix} y_1 \\ y_2 \\ y_3 \end{bmatrix}$$

be in R^3. Then

$$\mathbf{x} + \mathbf{y} = \begin{bmatrix} x_1 + y_1 \\ x_2 + y_2 \\ x_3 + y_3 \end{bmatrix}$$

and

$$f(\mathbf{x} + \mathbf{y}) = f\left(\begin{bmatrix} x_1 + y_1 \\ x_2 + y_2 \\ x_3 + y_3 \end{bmatrix}\right) = \begin{bmatrix} (x_1 + y_1) + (x_2 + y_2) \\ (x_2 + y_2) - (x_3 + y_3) \end{bmatrix}$$

$$= \begin{bmatrix} x_1 + y_1 + x_2 + y_2 \\ x_2 + y_2 - x_3 - y_3 \end{bmatrix}$$

$$= \begin{bmatrix} (x_1 + x_2) + (y_1 + y_2) \\ (x_2 - x_3) + (y_2 - x_3) \end{bmatrix} = \begin{bmatrix} x_1 + x_2 \\ x_2 - x_3 \end{bmatrix} + \begin{bmatrix} y_1 + y_2 \\ y_2 - y_3 \end{bmatrix}$$

$$= f\left(\begin{bmatrix} x_1 \\ x_2 \\ x_3 \end{bmatrix}\right) + f\left(\begin{bmatrix} y_1 \\ y_2 \\ y_3 \end{bmatrix}\right) = f(\mathbf{x}) + f(\mathbf{y})$$

To verify (2), let a be any real number and

$$\mathbf{x} = \begin{bmatrix} x_1 \\ x_2 \\ x_3 \end{bmatrix}$$

Then

$$a\begin{bmatrix} x_1 \\ x_2 \\ x_3 \end{bmatrix} = \begin{bmatrix} ax_1 \\ ax_2 \\ ax_3 \end{bmatrix} \quad \text{and} \quad f(a\mathbf{x}) = f\left(\begin{bmatrix} ax_1 \\ ax_2 \\ ax_3 \end{bmatrix}\right) = \begin{bmatrix} ax_1 + ax_2 \\ ax_2 - ax_3 \end{bmatrix}$$

$$= \begin{bmatrix} a(x_1 + x_2) \\ a(x_2 - x_3) \end{bmatrix} = a\begin{bmatrix} x_1 + x_2 \\ x_2 - x_3 \end{bmatrix} = af(\mathbf{x})$$ ■

Example 4.1.3 Let the mapping $f: P_3 \to P_2$ be defined by

$$f(\mathbf{p}) = f(p_0 + p_1x + p_2x^2 + p_3x^3) = (p_0 - p_1) + (p_2 + p_3)x^2$$

Verify that f is a linear transformation.

Solution Again, we verify that (1) and (2) hold. To verify (1), we note that

$$\mathbf{p} + \mathbf{q} = (p_0 + q_0) + (p_1 + q_1)x + (p_2 + q_2)x^2 + (p_3 + q_3)x^3$$

so

$$\begin{aligned}
f(\mathbf{p} + \mathbf{q}) &= [(p_0 + q_0) - (p_1 + q_1)] + [(p_2 + q_2) + (p_3 + q_3)]x^2 \\
&= [p_0 + q_0 - p_1 - q_1] + [p_2 + q_2 + p_3 + q_3]x^2 \\
&= [(p_0 - p_1) + (q_0 - q_1)] + [(p_2 + p_3) + (q_2 + q_3)]x^2 \\
&= [(p_0 - p_1) + (p_2 + p_3)x^2] + [(q_0 - q_1) + (q_2 + q_3)x^2] \\
&= f(\mathbf{p}) + f(\mathbf{q})
\end{aligned}$$

To verify (2), observe that since $a\mathbf{p} = ap_0 + ap_1x + ap_2x^2 + ap_3x^3$, then

$$\begin{aligned}
f(a\mathbf{p}) &= (ap_0 - ap_1) + (ap_2 + ap_3)x^2 \\
&= a(p_0 - p_1) + a(p_2 + p_3)x^2 \\
&= a[(p_0 - p_1) + (p_2 + p_3)x^2] = af(\mathbf{p})
\end{aligned}$$ ■

Example 4.1.4 Show that the function $f(x) = 5x + 3$ is not a linear transformation of R into R.

Solution Property (1) is not satisfied, since

$$\begin{aligned}
f(x_1 + x_2) = 5(x_1 + x_2) + 3 = (5x_1 + 3) + (5x_2 + 0) &\neq f(x_1) + f(x_2) \\
&= (5x_1 + 3) + (5x_2 + 3)
\end{aligned}$$

so that f is not linear. Property (2) is not satisfied either. ■

Example 4.1.5 Show that the function $f(x) = 7x$ is a linear transformation of R into R.

Solution We have

$$f(x_1 + x_2) = 7(x_1 + x_2) = 7x_1 + 7x_2 = f(x_1) + f(x_2)$$

$$f(ax_1) = 7(ax_1) = a(7x_1) = af(x_1)$$ ■

The following theorem is often useful to show that a mapping $f: U \to V$ is *not* a linear transformation.

Theorem 4.1.6 *If $f: U \to V$ is a linear transformation, then $f(\mathbf{0}) = \mathbf{0}$.*

Proof. $f(\mathbf{0}) = f(0\mathbf{u}) = 0f(\mathbf{u}) = \mathbf{0}$.

Hence, if $f(\mathbf{0}) \neq \mathbf{0}$, f *cannot* be a linear transformation. In Example 4.1.4, since $\mathbf{0} = 0$ and $f(\mathbf{0}) = 5(\mathbf{0}) + 3 = 3 \neq 0$, f cannot be a linear transformation.

Example 4.1.7 Show that mapping $f: R^2 \to R^3$ defined by

$$f\left(\begin{bmatrix} x_1 \\ x_2 \end{bmatrix}\right) = \begin{bmatrix} x_1 \\ x_2 \\ 1 \end{bmatrix}$$

is not a linear transformation.

Solution Since

$$f(\mathbf{0}) = f\left(\begin{bmatrix} 0 \\ 0 \end{bmatrix}\right) = \begin{bmatrix} 0 \\ 0 \\ 1 \end{bmatrix} \neq \begin{bmatrix} 0 \\ 0 \\ 0 \end{bmatrix} = \mathbf{0}$$

f is not a linear transformation. ■

Warning It is possible that $f(\mathbf{0}) = \mathbf{0}$ when f is not a linear transformation.

Example 4.1.8 Show that the mapping $f: R^2 \to R^2$ defined by

$$f\left(\begin{bmatrix} x \\ y \end{bmatrix}\right) = \begin{bmatrix} x^2 \\ y^2 \end{bmatrix}$$

is not a linear transformation.

Solution First we note that $f(\mathbf{0}) = \mathbf{0}$. To show that f is not a linear transformation, we attempt to verify property 2 of the definition.

To this end,

$$f\left(a\begin{bmatrix} x \\ y \end{bmatrix}\right) = \begin{bmatrix} a^2x^2 \\ a^2y^2 \end{bmatrix}$$

whereas

$$af\left(\begin{bmatrix} x \\ y \end{bmatrix}\right) = \begin{bmatrix} ax^2 \\ ay^2 \end{bmatrix}$$

Property 2 requires that $a^2x^2 = ax^2$ for any real number x, and any scalar a. Choosing $a = 2$ (any choice of a other than 1 will do) and $x = 1$, $y = 0$ (any choice of y will do), we see

$$f\left(2\begin{bmatrix}1\\0\end{bmatrix}\right) = \begin{bmatrix}4\\0\end{bmatrix}$$

does not equal

$$2f\left(\begin{bmatrix}1\\0\end{bmatrix}\right) = \begin{bmatrix}2\\0\end{bmatrix}$$

Since property 2 is not satisfied, this transformation is not a linear transformation. ■

Example 4.1.9 Let the mapping $f: R^3 \to R^2$ be defined by

$$f\left(\begin{bmatrix}x_1\\x_2\\x_3\end{bmatrix}\right) = \begin{bmatrix}x_1x_2\\x_3\end{bmatrix}$$

Show that f is not a linear transformation, although $f(\mathbf{0}) = \mathbf{0}$.

Solution We show that f is not a linear transformation using a counterexample for property 2. Let

$$\mathbf{x} = \begin{bmatrix}1\\1\\1\end{bmatrix}$$

Then

$$f(\mathbf{x}) = f\left(\begin{bmatrix}1\\1\\1\end{bmatrix}\right) = \begin{bmatrix}1\\1\end{bmatrix}$$

Also,

$$3\mathbf{x} = \begin{bmatrix}3\\3\\3\end{bmatrix}$$

which means

$$f(3\mathbf{x}) = f\left(\begin{bmatrix}3\\3\\3\end{bmatrix}\right) = \begin{bmatrix}9\\3\end{bmatrix}$$

But

$$3f(\mathbf{x}) = 3\begin{bmatrix}1\\1\end{bmatrix} = \begin{bmatrix}3\\3\end{bmatrix}$$

and thus f is not linear.

Clearly,

$$f(\mathbf{0}) = f\left(\begin{bmatrix} 0 \\ 0 \\ 0 \end{bmatrix}\right) = \begin{bmatrix} 0 \\ 0 \end{bmatrix} = \mathbf{0}$$ ■

Example 4.1.10 Let A be an $m \times n$ matrix and $\mathbf{x}$ and $\mathbf{y}$ be arbitrary vectors in R^n. Then using properties of matrix multiplication,

$$A(\mathbf{x} + \mathbf{y}) = A\mathbf{x} + A\mathbf{y}$$

$$A(a\mathbf{x}) = aA\mathbf{x}$$

Thus matrix multiplication by A is a linear transformation from R^n to R^m $(A : R^n \to R^m)$. ■

This is the type of linear transformation with which we will work almost entirely. This is made clear in Section 4.3.

Example 4.1.11 Let $f : R^2 \to R^3$ be a linear transformation that satisfies

$$f\left(\begin{bmatrix} 1 \\ 0 \end{bmatrix}\right) = \begin{bmatrix} -2 \\ -1 \\ 3 \end{bmatrix} \quad \text{and} \quad f\left(\begin{bmatrix} 0 \\ 1 \end{bmatrix}\right) = \begin{bmatrix} 4 \\ 0 \\ -3 \end{bmatrix}$$

Find

$$f\left(\begin{bmatrix} 5 \\ 7 \end{bmatrix}\right)$$

Solution

$$f\left(\begin{bmatrix} 5 \\ 7 \end{bmatrix}\right) = f\left(5\begin{bmatrix} 1 \\ 0 \end{bmatrix} + 7\begin{bmatrix} 0 \\ 1 \end{bmatrix}\right)$$

$$= 5f\left(\begin{bmatrix} 1 \\ 0 \end{bmatrix}\right) + 7f\left(\begin{bmatrix} 0 \\ 1 \end{bmatrix}\right)$$

$$= 5\begin{bmatrix} -2 \\ -1 \\ 3 \end{bmatrix} + 7\begin{bmatrix} 4 \\ 0 \\ -3 \end{bmatrix}$$

$$= \begin{bmatrix} 18 \\ -5 \\ -6 \end{bmatrix}$$ ■

Example 4.1.11 illustrates the fact that a linear transformation is determined by its action on the elements of some basis of the domain space. This is made precise in Theorem 4.1.12.

Theorem 4.1.12 *Let $f : U \to V$ be a linear transformation and let $A = \{\mathbf{u}_1, \ldots, \mathbf{u}_n\}$ be a basis of U.*

1. If $\mathbf{u} \in U$, and if $\mathbf{u} = a_1\mathbf{u}_1 + \cdots + a_n\mathbf{u}_n$,

$$f(\mathbf{u}) = a_1 f(\mathbf{u}_1) + \cdots + a_n f(\mathbf{u}_n).$$

2. If $g : U \to V$ is a linear transformation and for $i = 1, \ldots, n$ we have $f(\mathbf{u}_i) = g(\mathbf{u}_i)$, then $f(\mathbf{u}) = g(\mathbf{u})$ for all $\mathbf{u} \in U$.

3. Let $\mathbf{v}_1, \ldots, \mathbf{v}_n$ be vectors in V. Then there is a unique linear transformation $g : U \to V$ such that $g(\mathbf{u}_i) = \mathbf{v}_i$ for $i = 1, \ldots, n$.

REMARK In Chapter 3 we saw that the elements of a basis of a vector space U were the building blocks of U in the sense that each element of U could be uniquely expressed as a linear combination of the basis elements. Theorem 4.1.12 strengthens this result in the sense that any linear transformation $f : U \to V$ is uniquely determined by the set of vectors $\{f(\mathbf{u}_1), \ldots, f(\mathbf{u}_n)\}$ in V.

Proof.

1. Let $\mathbf{u} = a_1\mathbf{u}_1 + \cdots + a_n\mathbf{u}_n$; then since f is linear,

$$\begin{aligned} f(\mathbf{u}) &= f(a_1\mathbf{u}_1 + \cdots + a_n\mathbf{u}_n) \\ &= f(a_1\mathbf{u}_1) + \cdots + f(a_n\mathbf{u}_n) \\ &= a_1 f(\mathbf{u}_1) + \cdots + a_n f(\mathbf{u}_n) \end{aligned}$$

2. Recall for two functions f and g to be equal, they must satisfy $f(\mathbf{u}) = g(\mathbf{u})$ for all $\mathbf{u} \in U$. Let $\mathbf{u} = a_1\mathbf{u}_1 + \cdots + a_n\mathbf{u}_n$ be any vector in U. Then

$$\begin{aligned} f(\mathbf{u}) &= a_1 f(\mathbf{u}_1) + \cdots + a_n f(\mathbf{u}_n) \\ &= a_1 g(\mathbf{u}_1) + \cdots + a_n g(\mathbf{u}_n) \\ &= g(\mathbf{u}) \end{aligned}$$

3. Define $g : U \to V$ as follows: If $\mathbf{u} \in U$, then there are unique real numbers $a_1, \ldots, a_n$ for which $\mathbf{u} = a_1\mathbf{u}_1 + \cdots + a_n\mathbf{u}_n$. Then set $g(\mathbf{u}) = a_1\mathbf{v}_1 + \cdots + a_n\mathbf{v}_n$. Then $g(\mathbf{u}_i) = \mathbf{v}_i$ for $i = 1, \ldots, n$, by its construction. That g is unique is shown by part 2 of the theorem.

We illustrate this theorem with one final example.

Example 4.1.13 Let $\{1, x, x^2\}$ be a basis of P_2 and let $f : P_2 \to P_3$ be the unique linear transformation (Theorem 4.1.12 part 3) defined by $f(1) = x$, $f(x) = x^3$, and $f(x^2) = x^2$. Calculate $f(a_0 + a_1x + a_2x^2)$.

Solution $f(a_0 + a_1x + a_2x^2) = a_0 f(1) + a_1 f(x) + a_2 f(x^2) = a_0x + a_1x^3 + a_2x^2$, which is more commonly written as $a_0x + a_2x^2 + a_1x^3$. ■

SUMMARY

Terms to know:

Linear transformation

Theorems to know:

If $f: U \to V$ is a linear transformation, then $f(\mathbf{0}) = \mathbf{0}$.

A linear transformation $f: U \to V$ is determined uniquely by $\{f(\mathbf{u}_1), \ldots, f(\mathbf{u}_n)\}$ where $\{\mathbf{u}_1, \ldots, \mathbf{u}_n\}$ is a basis of U.

EXERCISE SET 4.1

Let the mapping $f: R^3 \to R^2$ be defined as follows. Determine whether f is a linear transformation.

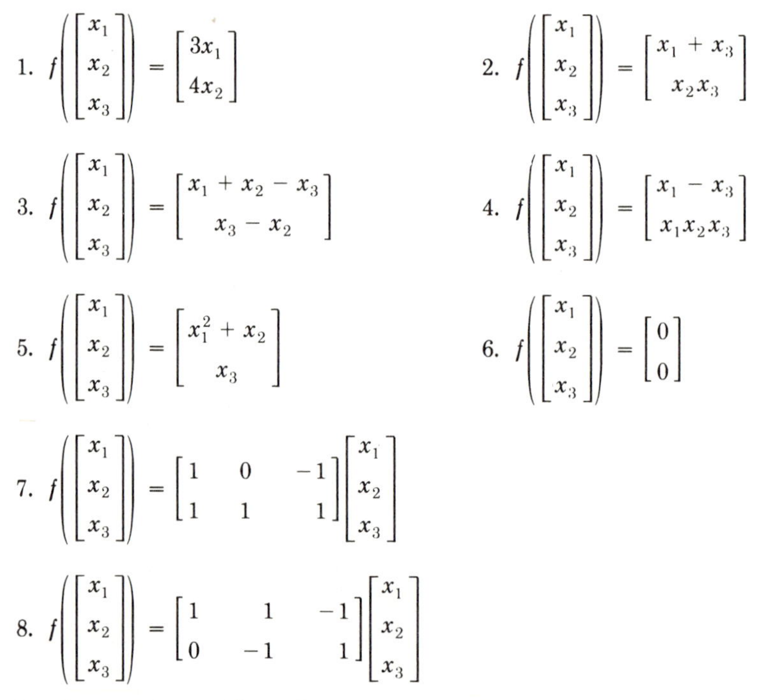

1. $f\left(\begin{bmatrix} x_1 \\ x_2 \\ x_3 \end{bmatrix}\right) = \begin{bmatrix} 3x_1 \\ 4x_2 \end{bmatrix}$

2. $f\left(\begin{bmatrix} x_1 \\ x_2 \\ x_3 \end{bmatrix}\right) = \begin{bmatrix} x_1 + x_3 \\ x_2 x_3 \end{bmatrix}$

3. $f\left(\begin{bmatrix} x_1 \\ x_2 \\ x_3 \end{bmatrix}\right) = \begin{bmatrix} x_1 + x_2 - x_3 \\ x_3 - x_2 \end{bmatrix}$

4. $f\left(\begin{bmatrix} x_1 \\ x_2 \\ x_3 \end{bmatrix}\right) = \begin{bmatrix} x_1 - x_3 \\ x_1 x_2 x_3 \end{bmatrix}$

5. $f\left(\begin{bmatrix} x_1 \\ x_2 \\ x_3 \end{bmatrix}\right) = \begin{bmatrix} x_1^2 + x_2 \\ x_3 \end{bmatrix}$

6. $f\left(\begin{bmatrix} x_1 \\ x_2 \\ x_3 \end{bmatrix}\right) = \begin{bmatrix} 0 \\ 0 \end{bmatrix}$

7. $f\left(\begin{bmatrix} x_1 \\ x_2 \\ x_3 \end{bmatrix}\right) = \begin{bmatrix} 1 & 0 & -1 \\ 1 & 1 & 1 \end{bmatrix}\begin{bmatrix} x_1 \\ x_2 \\ x_3 \end{bmatrix}$

8. $f\left(\begin{bmatrix} x_1 \\ x_2 \\ x_3 \end{bmatrix}\right) = \begin{bmatrix} 1 & 1 & -1 \\ 0 & -1 & 1 \end{bmatrix}\begin{bmatrix} x_1 \\ x_2 \\ x_3 \end{bmatrix}$

Let the mapping $f: R^3 \to R^3$ be defined as follows. Determine whether f is a linear transformation.

9. $f\left(\begin{bmatrix} x_1 \\ x_2 \\ x_3 \end{bmatrix}\right) = \begin{bmatrix} x_1 + 1 \\ x_2 + 1 \\ 0 \end{bmatrix}$

10. $f\left(\begin{bmatrix} x_1 \\ x_2 \\ x_3 \end{bmatrix}\right) = \begin{bmatrix} x_2 \\ x_1 \\ x_3 \end{bmatrix}$

11. $f\left(\begin{bmatrix} x_1 \\ x_2 \\ x_3 \end{bmatrix}\right) = \begin{bmatrix} -x_1 \\ -x_2 \\ -x_3 \end{bmatrix}$

12. $f\left(\begin{bmatrix} x_1 \\ x_2 \\ x_3 \end{bmatrix}\right) = \begin{bmatrix} x_1 \\ -x_2 \\ x_3 \end{bmatrix}$

Let the mapping $f : P_3 \to P_1$ be defined as follows. Determine whether f is a linear transformation.

13. $f(\mathbf{p}) = (p_1 - p_0) + (p_3 p_2)x$
14. $f(\mathbf{p}) = 3p_1 + p_3 x$
15. $f(\mathbf{p}) = (p_3 + p_0) + (p_1 - p_2)x$
16. $f(\mathbf{p}) = p_0 + p_1 x$
17. $f(\mathbf{p}) = p_1 p_2 - p_0 p_3 x$
18. $f(\mathbf{p}) = x - p_0$
19. $f(\mathbf{p}) = (p_0 + p_1 + p_2 + p_3) + (p_1 + p_2 + p_3)x$
20. $f(\mathbf{p}) = 3p_3 x + 2p_2$
21. $f(\mathbf{p}) = 3p_2 + p_0 + 1$
22. $f(\mathbf{p}) = p_1 + 2p_2 x$
23. Explain why the two rules given in the definition of a linear transformation are equivalent to the single condition

$$f(a\mathbf{u}_1 + b\mathbf{u}_2) = af(\mathbf{u}_1) + bf(\mathbf{u}_2)$$

for arbitrary vectors $\mathbf{u}_1$ and $\mathbf{u}_2$ and numbers a and b.

24. Verify that every $m \times n$ matrix M gives a linear transformation from the vector space R^n to the vector space R^m defined by $f(\mathbf{x}) = M\mathbf{x}$.
25. Verify that if D means to differentiate, then D is a linear transformation defined from the space of real differentiable functions to the space of functions from R to R.
26. Verify that if A means to antidifferentiate with constant of integration equal to 0, then A is a linear transformation defined on the space of integrable functions.
27. Let S and T be linear transformations on R^2 defined by

$$S\left(\begin{bmatrix} x \\ y \end{bmatrix}\right) = \begin{bmatrix} y \\ x \end{bmatrix} \quad \text{and} \quad T\left(\begin{bmatrix} x \\ y \end{bmatrix}\right) = \begin{bmatrix} x \\ 0 \end{bmatrix}$$

Describe S and T geometrically and give rules such as those defining S and T for each of the transformations $S + T$, TS, ST, S^2, and T^2, where

$$TS\left(\begin{bmatrix} x \\ y \end{bmatrix}\right) = T\left(S\begin{bmatrix} x \\ y \end{bmatrix}\right)$$

SECTION 4.2 VECTOR COORDINATES

In this and the next section, we discover that for all practical purposes, all n-dimensional vector spaces look like R^n and all linear transformations look like matrices. Since matrix multiplication and vector operations are relatively easy to perform, the fact that all vector spaces look like R^n becomes important. The phrase *looks like* is made precise later in this section.

Definition 4.2.1 Let V be an n-dimensional vector space with basis $B = \{\mathbf{b}_1, \mathbf{b}_2, \ldots, \mathbf{b}_n\}$. For any $\mathbf{v} \in V$, the **coordinate vector** $[\mathbf{v}]_B$ of $\mathbf{v}$ is

$$[\mathbf{v}]_B = \begin{bmatrix} x_1 \\ x_2 \\ \vdots \\ x_n \end{bmatrix}$$

where $\mathbf{v} = x_1\mathbf{b}_1 + x_2\mathbf{b}_2 + \cdots + x_n\mathbf{b}_n$.

Note that if $V = R^n$ and B is the standard basis E, then we have

$$\mathbf{v} = \begin{bmatrix} x_1 \\ \vdots \\ x_n \end{bmatrix} = [\mathbf{v}]_E$$

In this sense, the idea is familiar.

Since B is a basis of V, the scalars $x_1, x_2, \ldots, x_n$ are unique. Hence we obtain the following result.

Theorem 4.2.2 *If* $\mathbf{u}, \mathbf{v} \in V$ *and* $[\mathbf{u}]_B = [\mathbf{v}]_B$, *then* $\mathbf{u} = \mathbf{v}$.

Proof. Let

$$[\mathbf{u}]_B = [\mathbf{v}]_B = \begin{bmatrix} x_1 \\ x_2 \\ \vdots \\ x_n \end{bmatrix}$$

Then

$$\mathbf{u} = x_1\mathbf{b}_1 + x_2\mathbf{b}_2 + \cdots + x_n\mathbf{b}_n = \mathbf{v}$$

Example 4.2.3 Let $V = P_3$ and $B = \{1 + x, x + x^2, x^2 + x^3, x^3\}$. For $\mathbf{p} = 3 + 2x + 4x^2 + 2x^3$, to find

$$[\mathbf{p}]_B = \begin{bmatrix} a_1 \\ a_2 \\ a_3 \\ a_4 \end{bmatrix}$$

we must solve

$$\begin{aligned} 3 + 2x + 4x^2 + 2x^3 &= a_1(1 + x) + a_2(x + x^2) \\ &\quad + a_3(x^2 + x^3) + a_4x^3 \\ &= a_1(1) + (a_1 + a_2)x + (a_2 + a_3)x^2 \\ &\quad + (a_3 + a_4)x^3 \end{aligned}$$

Property 2 requires that $a^2x^2 = ax^2$ for any real number x, and any scalar a. Choosing $a = 2$ (any choice of a other than 1 will do) and $x = 1$, $y = 0$ (any choice of y will do), we see

$$f\left(2\begin{bmatrix}1\\0\end{bmatrix}\right) = \begin{bmatrix}4\\0\end{bmatrix}$$

does not equal

$$2f\left(\begin{bmatrix}1\\0\end{bmatrix}\right) = \begin{bmatrix}2\\0\end{bmatrix}$$

Since property 2 is not satisfied, this transformation is not a linear transformation. ■

Example 4.1.9 Let the mapping $f: R^3 \to R^2$ be defined by

$$f\left(\begin{bmatrix}x_1\\x_2\\x_3\end{bmatrix}\right) = \begin{bmatrix}x_1x_2\\x_3\end{bmatrix}$$

Show that f is not a linear transformation, although $f(\mathbf{0}) = \mathbf{0}$.

Solution We show that f is not a linear transformation using a counterexample for property 2. Let

$$\mathbf{x} = \begin{bmatrix}1\\1\\1\end{bmatrix}$$

Then

$$f(\mathbf{x}) = f\left(\begin{bmatrix}1\\1\\1\end{bmatrix}\right) = \begin{bmatrix}1\\1\end{bmatrix}$$

Also,

$$3\mathbf{x} = \begin{bmatrix}3\\3\\3\end{bmatrix}$$

which means

$$f(3\mathbf{x}) = f\left(\begin{bmatrix}3\\3\\3\end{bmatrix}\right) = \begin{bmatrix}9\\3\end{bmatrix}$$

But

$$3f(\mathbf{x}) = 3\begin{bmatrix}1\\1\end{bmatrix} = \begin{bmatrix}3\\3\end{bmatrix}$$

and thus f is not linear.

Clearly,

$$f(\mathbf{0}) = f\left(\begin{bmatrix} 0 \\ 0 \\ 0 \end{bmatrix}\right) = \begin{bmatrix} 0 \\ 0 \end{bmatrix} = \mathbf{0}$$ ■

Example 4.1.10 Let A be an $m \times n$ matrix and $\mathbf{x}$ and $\mathbf{y}$ be arbitrary vectors in R^n. Then using properties of matrix multiplication,

$$A(\mathbf{x} + \mathbf{y}) = A\mathbf{x} + A\mathbf{y}$$

$$A(a\mathbf{x}) = aA\mathbf{x}$$

Thus matrix multiplication by A is a linear transformation from R^n to R^m $(A : R^n \to R^m)$. ■

This is the type of linear transformation with which we will work almost entirely. This is made clear in Section 4.3.

Example 4.1.11 Let $f : R^2 \to R^3$ be a linear transformation that satisfies

$$f\left(\begin{bmatrix} 1 \\ 0 \end{bmatrix}\right) = \begin{bmatrix} -2 \\ -1 \\ 3 \end{bmatrix} \quad \text{and} \quad f\left(\begin{bmatrix} 0 \\ 1 \end{bmatrix}\right) = \begin{bmatrix} 4 \\ 0 \\ -3 \end{bmatrix}$$

Find

$$f\left(\begin{bmatrix} 5 \\ 7 \end{bmatrix}\right)$$

Solution

$$\begin{aligned} f\left(\begin{bmatrix} 5 \\ 7 \end{bmatrix}\right) &= f\left(5\begin{bmatrix} 1 \\ 0 \end{bmatrix} + 7\begin{bmatrix} 0 \\ 1 \end{bmatrix}\right) \\ &= 5f\left(\begin{bmatrix} 1 \\ 0 \end{bmatrix}\right) + 7f\left(\begin{bmatrix} 0 \\ 1 \end{bmatrix}\right) \\ &= 5\begin{bmatrix} -2 \\ -1 \\ 3 \end{bmatrix} + 7\begin{bmatrix} 4 \\ 0 \\ -3 \end{bmatrix} \\ &= \begin{bmatrix} 18 \\ -5 \\ -6 \end{bmatrix} \end{aligned}$$ ■

Example 4.1.11 illustrates the fact that a linear transformation is determined by its action on the elements of some basis of the domain space. This is made precise in Theorem 4.1.12.

Theorem 4.1.12 *Let $f : U \to V$ be a linear transformation and let $A = \{\mathbf{u}_1, \ldots, \mathbf{u}_n\}$ be a basis of U.*

1. *If $\mathbf{u} \in U$, and if $\mathbf{u} = a_1\mathbf{u}_1 + \cdots + a_n\mathbf{u}_n$,*

$$f(\mathbf{u}) = a_1 f(\mathbf{u}_1) + \cdots + a_n f(\mathbf{u}_n).$$

2. *If $g : U \to V$ is a linear transformation and for $i = 1, \ldots, n$ we have $f(\mathbf{u}_i) = g(\mathbf{u}_i)$, then $f(\mathbf{u}) = g(\mathbf{u})$ for all $\mathbf{u} \in U$.*
3. *Let $\mathbf{v}_1, \ldots, \mathbf{v}_n$ be vectors in V. Then there is a unique linear transformation $g : U \to V$ such that $g(\mathbf{u}_i) = \mathbf{v}_i$ for $i = 1, \ldots, n$.*

REMARK In Chapter 3 we saw that the elements of a basis of a vector space U were the building blocks of U in the sense that each element of U could be uniquely expressed as a linear combination of the basis elements. Theorem 4.1.12 strengthens this result in the sense that any linear transformation $f : U \to V$ is uniquely determined by the set of vectors $\{f(\mathbf{u}_1), \ldots, f(\mathbf{u}_n)\}$ in V.

Proof.

1. Let $\mathbf{u} = a_1\mathbf{u}_1 + \cdots + a_n\mathbf{u}_n$; then since f is linear,

$$\begin{aligned} f(\mathbf{u}) &= f(a_1\mathbf{u}_1 + \cdots + a_n\mathbf{u}_n) \\ &= f(a_1\mathbf{u}_1) + \cdots + f(a_n\mathbf{u}_n) \\ &= a_1 f(\mathbf{u}_1) + \cdots + a_n f(\mathbf{u}_n) \end{aligned}$$

2. Recall for two functions f and g to be equal, they must satisfy $f(\mathbf{u}) = g(\mathbf{u})$ for all $\mathbf{u} \in U$. Let $\mathbf{u} = a_1\mathbf{u}_1 + \cdots + a_n\mathbf{u}_n$ be any vector in U. Then

$$\begin{aligned} f(\mathbf{u}) &= a_1 f(\mathbf{u}_1) + \cdots + a_n f(\mathbf{u}_n) \\ &= a_1 g(\mathbf{u}_1) + \cdots + a_n g(\mathbf{u}_n) \\ &= g(\mathbf{u}) \end{aligned}$$

3. Define $g : U \to V$ as follows: If $\mathbf{u} \in U$, then there are unique real numbers $a_1, \ldots, a_n$ for which $\mathbf{u} = a_1\mathbf{u}_1 + \cdots + a_n\mathbf{u}_n$. Then set $g(\mathbf{u}) = a_1\mathbf{v}_1 + \cdots + a_n\mathbf{v}_n$. Then $g(\mathbf{u}_i) = \mathbf{v}_i$ for $i = 1, \ldots, n$, by its construction. That g is unique is shown by part 2 of the theorem.

We illustrate this theorem with one final example.

Example 4.1.13 Let $\{1, x, x^2\}$ be a basis of P_2 and let $f : P_2 \to P_3$ be the unique linear transformation (Theorem 4.1.12 part 3) defined by $f(1) = x$, $f(x) = x^3$, and $f(x^2) = x^2$. Calculate $f(a_0 + a_1x + a_2x^2)$.

Solution $f(a_0 + a_1x + a_2x^2) = a_0 f(1) + a_1 f(x) + a_2 f(x^2) = a_0x + a_1x^3 + a_2x^2$, which is more commonly written as $a_0x + a_2x^2 + a_1x^3$. ■

SUMMARY

Terms to know:

Linear transformation

Theorems to know:

If $f : U \to V$ is a linear transformation, then $f(\mathbf{0}) = \mathbf{0}$.

A linear transformation $f : U \to V$ is determined uniquely by $\{f(\mathbf{u}_1), \ldots, f(\mathbf{u}_n)\}$ where $\{\mathbf{u}_1, \ldots, \mathbf{u}_n\}$ is a basis of U.

EXERCISE SET 4.1

Let the mapping $f : R^3 \to R^2$ be defined as follows. Determine whether f is a linear transformation.

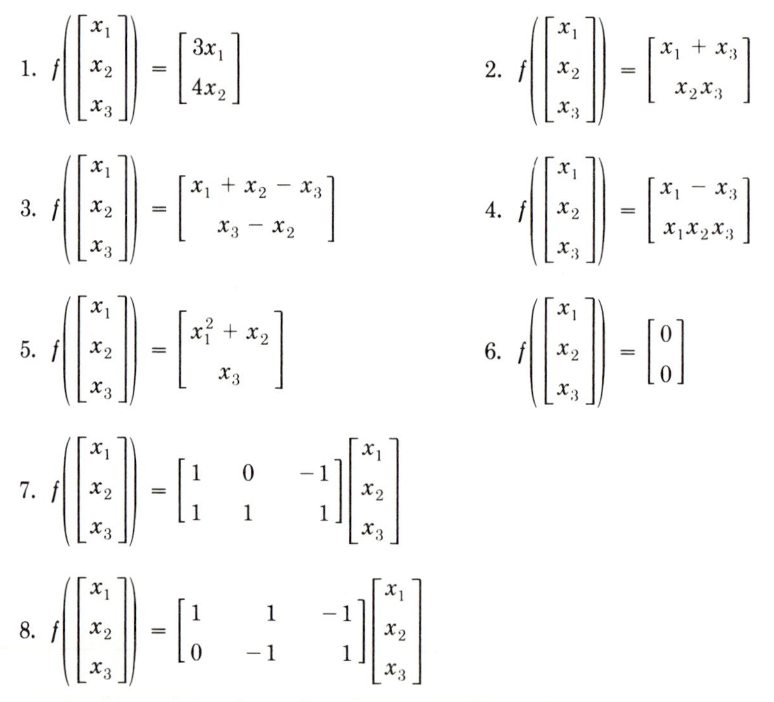

1. $f\left(\begin{bmatrix} x_1 \\ x_2 \\ x_3 \end{bmatrix}\right) = \begin{bmatrix} 3x_1 \\ 4x_2 \end{bmatrix}$

2. $f\left(\begin{bmatrix} x_1 \\ x_2 \\ x_3 \end{bmatrix}\right) = \begin{bmatrix} x_1 + x_3 \\ x_2 x_3 \end{bmatrix}$

3. $f\left(\begin{bmatrix} x_1 \\ x_2 \\ x_3 \end{bmatrix}\right) = \begin{bmatrix} x_1 + x_2 - x_3 \\ x_3 - x_2 \end{bmatrix}$

4. $f\left(\begin{bmatrix} x_1 \\ x_2 \\ x_3 \end{bmatrix}\right) = \begin{bmatrix} x_1 - x_3 \\ x_1 x_2 x_3 \end{bmatrix}$

5. $f\left(\begin{bmatrix} x_1 \\ x_2 \\ x_3 \end{bmatrix}\right) = \begin{bmatrix} x_1^2 + x_2 \\ x_3 \end{bmatrix}$

6. $f\left(\begin{bmatrix} x_1 \\ x_2 \\ x_3 \end{bmatrix}\right) = \begin{bmatrix} 0 \\ 0 \end{bmatrix}$

7. $f\left(\begin{bmatrix} x_1 \\ x_2 \\ x_3 \end{bmatrix}\right) = \begin{bmatrix} 1 & 0 & -1 \\ 1 & 1 & 1 \end{bmatrix}\begin{bmatrix} x_1 \\ x_2 \\ x_3 \end{bmatrix}$

8. $f\left(\begin{bmatrix} x_1 \\ x_2 \\ x_3 \end{bmatrix}\right) = \begin{bmatrix} 1 & 1 & -1 \\ 0 & -1 & 1 \end{bmatrix}\begin{bmatrix} x_1 \\ x_2 \\ x_3 \end{bmatrix}$

Let the mapping $f : R^3 \to R^3$ be defined as follows. Determine whether f is a linear transformation.

9. $f\left(\begin{bmatrix} x_1 \\ x_2 \\ x_3 \end{bmatrix}\right) = \begin{bmatrix} x_1 + 1 \\ x_2 + 1 \\ 0 \end{bmatrix}$

10. $f\left(\begin{bmatrix} x_1 \\ x_2 \\ x_3 \end{bmatrix}\right) = \begin{bmatrix} x_2 \\ x_1 \\ x_3 \end{bmatrix}$

11. $f\left(\begin{bmatrix} x_1 \\ x_2 \\ x_3 \end{bmatrix}\right) = \begin{bmatrix} -x_1 \\ -x_2 \\ -x_3 \end{bmatrix}$

12. $f\left(\begin{bmatrix} x_1 \\ x_2 \\ x_3 \end{bmatrix}\right) = \begin{bmatrix} x_1 \\ -x_2 \\ x_3 \end{bmatrix}$

Let the mapping $f: P_3 \to P_1$ be defined as follows. Determine whether f is a linear transformation.

13. $f(\mathbf{p}) = (p_1 - p_0) + (p_3 p_2)x$
14. $f(\mathbf{p}) = 3p_1 + p_3 x$
15. $f(\mathbf{p}) = (p_3 + p_0) + (p_1 - p_2)x$
16. $f(\mathbf{p}) = p_0 + p_1 x$
17. $f(\mathbf{p}) = p_1 p_2 - p_0 p_3 x$
18. $f(\mathbf{p}) = x - p_0$
19. $f(\mathbf{p}) = (p_0 + p_1 + p_2 + p_3) + (p_1 + p_2 + p_3)x$
20. $f(\mathbf{p}) = 3p_3 x + 2p_2$
21. $f(\mathbf{p}) = 3p_2 + p_0 + 1$
22. $f(\mathbf{p}) = p_1 + 2p_2 x$
23. Explain why the two rules given in the definition of a linear transformation are equivalent to the single condition

$$f(a\mathbf{u}_1 + b\mathbf{u}_2) = af(\mathbf{u}_1) + bf(\mathbf{u}_2)$$

for arbitrary vectors $\mathbf{u}_1$ and $\mathbf{u}_2$ and numbers a and b.

24. Verify that every $m \times n$ matrix M gives a linear transformation from the vector space R^n to the vector space R^m defined by $f(\mathbf{x}) = M\mathbf{x}$.
25. Verify that if D means to differentiate, then D is a linear transformation defined from the space of real differentiable functions to the space of functions from R to R.
26. Verify that if A means to antidifferentiate with constant of integration equal to 0, then A is a linear transformation defined on the space of integrable functions.
27. Let S and T be linear transformations on R^2 defined by

$$S\left(\begin{bmatrix} x \\ y \end{bmatrix}\right) = \begin{bmatrix} y \\ x \end{bmatrix} \quad \text{and} \quad T\left(\begin{bmatrix} x \\ y \end{bmatrix}\right) = \begin{bmatrix} x \\ 0 \end{bmatrix}$$

Describe S and T geometrically and give rules such as those defining S and T for each of the transformations $S + T$, TS, ST, S^2, and T^2, where

$$TS\left(\begin{bmatrix} x \\ y \end{bmatrix}\right) = T\left(S\begin{bmatrix} x \\ y \end{bmatrix}\right)$$

SECTION 4.2 VECTOR COORDINATES

In this and the next section, we discover that for all practical purposes, all n-dimensional vector spaces look like R^n and all linear transformations look like matrices. Since matrix multiplication and vector operations are relatively easy to perform, the fact that all vector spaces look like R^n becomes important. The phrase *looks like* is made precise later in this section.

Definition 4.2.1 Let V be an n-dimensional vector space with basis $B = \{\mathbf{b}_1, \mathbf{b}_2, \ldots, \mathbf{b}_n\}$. For any $\mathbf{v} \in V$, the **coordinate vector** $[\mathbf{v}]_B$ of $\mathbf{v}$ is

$$[\mathbf{v}]_B = \begin{bmatrix} x_1 \\ x_2 \\ \vdots \\ x_n \end{bmatrix}$$

where $\mathbf{v} = x_1\mathbf{b}_1 + x_2\mathbf{b}_2 + \cdots + x_n\mathbf{b}_n$.

Note that if $V = R^n$ and B is the standard basis E, then we have

$$\mathbf{v} = \begin{bmatrix} x_1 \\ \vdots \\ x_n \end{bmatrix} = [\mathbf{v}]_E$$

In this sense, the idea is familiar.

Since B is a basis of V, the scalars $x_1, x_2, \ldots, x_n$ are unique. Hence we obtain the following result.

Theorem 4.2.2 *If* $\mathbf{u}, \mathbf{v} \in V$ *and* $[\mathbf{u}]_B = [\mathbf{v}]_B$, *then* $\mathbf{u} = \mathbf{v}$.

Proof. Let

$$[\mathbf{u}]_B = [\mathbf{v}]_B = \begin{bmatrix} x_1 \\ x_2 \\ \vdots \\ x_n \end{bmatrix}$$

Then

$$\mathbf{u} = x_1\mathbf{b}_1 + x_2\mathbf{b}_2 + \cdots + x_n\mathbf{b}_n = \mathbf{v}$$

Example 4.2.3 Let $V = P_3$ and $B = \{1 + x, x + x^2, x^2 + x^3, x^3\}$. For $\mathbf{p} = 3 + 2x + 4x^2 + 2x^3$, to find

$$[\mathbf{p}]_B = \begin{bmatrix} a_1 \\ a_2 \\ a_3 \\ a_4 \end{bmatrix}$$

we must solve

$$\begin{aligned} 3 + 2x + 4x^2 + 2x^3 &= a_1(1 + x) + a_2(x + x^2) \\ &\quad + a_3(x^2 + x^3) + a_4x^3 \\ &= a_1(1) + (a_1 + a_2)x + (a_2 + a_3)x^2 \\ &\quad + (a_3 + a_4)x^3 \end{aligned}$$

for a_1, a_2, a_3, and a_4. By equating coefficients, we must solve the system of equations

$$\begin{aligned} a_1 \quad\quad\quad\quad &= 3 \\ a_1 + a_2 \quad\quad &= 2 \\ a_2 + a_3 \quad &= 4 \\ a_3 + a_4 &= 2 \end{aligned}$$

which has solution $a_1 = 3$, $a_2 = -1$, $a_3 = 5$, $a_4 = -3$, and so

$$[\mathbf{p}]_B = \begin{bmatrix} 3 \\ -1 \\ 5 \\ -3 \end{bmatrix}$$

Trivially, if B_0, is the basis $\{1, x, x^2, x^3\}$, then

$$[\mathbf{p}]_{B_0} = \begin{bmatrix} 3 \\ 2 \\ 4 \\ 2 \end{bmatrix}$$

Example 4.2.4 Let

$$V = R^3 \quad \text{and} \quad B = \left\{ \begin{bmatrix} 1 \\ 1 \\ 0 \end{bmatrix}, \begin{bmatrix} 0 \\ 1 \\ 1 \end{bmatrix}, \begin{bmatrix} 1 \\ 0 \\ 1 \end{bmatrix} \right\}$$

For

$$\mathbf{v} = \begin{bmatrix} 7 \\ 1 \\ 2 \end{bmatrix}, \qquad [\mathbf{v}]_B = \begin{bmatrix} a_1 \\ a_2 \\ a_3 \end{bmatrix}$$

where

$$\mathbf{v} = \begin{bmatrix} 7 \\ 1 \\ 2 \end{bmatrix} = a_1 \begin{bmatrix} 1 \\ 1 \\ 0 \end{bmatrix} + a_2 \begin{bmatrix} 0 \\ 1 \\ 1 \end{bmatrix} + a_3 \begin{bmatrix} 1 \\ 0 \\ 1 \end{bmatrix}$$

Again we must solve a system of equations, namely,

$$\begin{aligned} a_1 \quad\quad + a_3 &= 7 \\ a_1 + a_2 \quad\quad &= 1 \\ a_2 + a_3 &= 2 \end{aligned}$$

Since the solution to this system is $a_1 = 3$, $a_2 = -2$, and $a_3 = 4$,

$$[\mathbf{v}]_B = \begin{bmatrix} 7 \\ 1 \\ 2 \end{bmatrix}_B = \begin{bmatrix} 3 \\ -2 \\ 4 \end{bmatrix}$$

■

Example 4.2.5 Let $V = R_{2\times 2}$ with basis

$$B = \left\{ \begin{bmatrix} 1 & 0 \\ 0 & 1 \end{bmatrix}, \begin{bmatrix} 0 & 1 \\ 1 & 0 \end{bmatrix}, \begin{bmatrix} 1 & 1 \\ 0 & 0 \end{bmatrix}, \begin{bmatrix} 0 & 1 \\ 0 & 1 \end{bmatrix} \right\}$$

For

$$\mathbf{v} = \begin{bmatrix} 1 & 4 \\ 2 & -1 \end{bmatrix}, \qquad [\mathbf{v}]_B = \begin{bmatrix} a_1 \\ a_2 \\ a_3 \\ a_4 \end{bmatrix}$$

where

$$\begin{aligned} \mathbf{v} &= a_1\mathbf{b}_1 + a_2\mathbf{b}_2 + a_3\mathbf{b}_3 + a_4\mathbf{b}_4 \\ &= a_1\begin{bmatrix} 1 & 0 \\ 0 & 1 \end{bmatrix} + a_2\begin{bmatrix} 0 & 1 \\ 1 & 0 \end{bmatrix} + a_3\begin{bmatrix} 1 & 1 \\ 0 & 0 \end{bmatrix} + a_4\begin{bmatrix} 0 & 1 \\ 0 & 1 \end{bmatrix} \end{aligned}$$

The associated system of equations is

$$\begin{aligned} a_1 \quad + a_3 \qquad &= 1 \\ a_2 + a_3 + a_4 &= 4 \\ a_2 \qquad\qquad &= 2 \\ a_1 \qquad\quad + a_4 &= -1 \end{aligned}$$

with solution $a_1 = -1$, $a_2 = 2$, $a_3 = 2$, and $a_4 = 0$, so

$$[\mathbf{v}]_B = \begin{bmatrix} -1 \\ 2 \\ 2 \\ 0 \end{bmatrix}$$

Again, if

$$B_0 = \left\{ \begin{bmatrix} 1 & 0 \\ 0 & 0 \end{bmatrix}, \begin{bmatrix} 0 & 1 \\ 0 & 0 \end{bmatrix}, \begin{bmatrix} 0 & 0 \\ 1 & 0 \end{bmatrix}, \begin{bmatrix} 0 & 0 \\ 0 & 1 \end{bmatrix} \right\}$$

then

$$[\mathbf{v}]_{B_0} = \begin{bmatrix} 1 \\ 4 \\ 2 \\ -1 \end{bmatrix}$$

■

Theorem 4.2.6 *Let V be a vector space, $\mathbf{u}, \mathbf{v} \in V$, and B be a basis for V. Then*

1. $[\mathbf{u} + \mathbf{v}]_B = [\mathbf{u}]_B + [\mathbf{v}]_B$ *for any* $\mathbf{u}, \mathbf{v} \in V$.
2. $[r\mathbf{u}]_B = r[\mathbf{u}]_B$ *for any* $r \in R$.

Thus the mapping $\mathbf{u} \to [\mathbf{u}]_B$ is a linear transformation from V to R^n.

Proof. Let

$$[\mathbf{u}]_B = \begin{bmatrix} x_1 \\ x_2 \\ \vdots \\ x_n \end{bmatrix} \quad \text{and} \quad [\mathbf{v}]_B = \begin{bmatrix} y_1 \\ y_2 \\ \vdots \\ y_n \end{bmatrix}$$

Then

$$\mathbf{u} = x_1\mathbf{b}_1 + x_2\mathbf{b}_2 + \cdots + x_n\mathbf{b}_n \quad \text{and} \quad \mathbf{v} = y_1\mathbf{b}_1 + y_2\mathbf{b}_2 + \cdots + y_n\mathbf{b}_n$$

1. Since $\mathbf{u} + \mathbf{v} = (x_1 + y_1)\mathbf{b}_1 + (x_2 + y_2)\mathbf{b}_2 + \cdots + (x_n + y_n)\mathbf{b}_n$, we have

$$[\mathbf{u} + \mathbf{v}]_B = \begin{bmatrix} x_1 + y_1 \\ x_2 + y_2 \\ \vdots \\ x_n + y_n \end{bmatrix} = [\mathbf{u}]_B + [\mathbf{v}]_B$$

2. For any $r \in R$, $r\mathbf{u} = (rx_1)\mathbf{b}_1 + (rx_2)\mathbf{b}_2 + \cdots + (rx_n)\mathbf{b}_n$, so

$$[r\mathbf{u}]_B = \begin{bmatrix} rx_1 \\ rx_2 \\ \vdots \\ rx_n \end{bmatrix} = r[\mathbf{u}]_B$$

We illustrate the proof of Theorem 4.2.6 in the following example.

Example 4.2.7 Let $V = R_{2\times 2}$ with the same basis B as in Example 4.2.5 and let

$$\mathbf{u} = \begin{bmatrix} -1 & 2 \\ 0 & 1 \end{bmatrix}$$

so that

$$[\mathbf{u}]_B = \begin{bmatrix} -1 \\ 0 \\ 0 \\ 2 \end{bmatrix}$$

Let $\mathbf{v}$ be the matrix

$$\begin{bmatrix} 1 & 4 \\ 2 & -1 \end{bmatrix}$$

as in Example 4.2.5.

First,

$$\mathbf{u} + \mathbf{v} = \begin{bmatrix} -1 & 2 \\ 0 & 1 \end{bmatrix} + \begin{bmatrix} 1 & 4 \\ 2 & -1 \end{bmatrix} = \begin{bmatrix} 0 & 6 \\ 2 & 0 \end{bmatrix}$$

To find

$$[\mathbf{u} + \mathbf{v}]_B = \begin{bmatrix} z_1 \\ z_2 \\ z_3 \\ z_4 \end{bmatrix}$$

we solve the system of equations

$$\begin{aligned} z_1 \quad + z_3 \quad &= 0 \\ z_2 + z_3 + z_4 &= 6 \\ z_3 \quad &= 2 \\ z_1 \quad\quad + z_4 &= 0 \end{aligned}$$

obtaining $z_1 = -2$, $z_2 = 2$, $z_3 = 2$, $z_4 = 2$. And, indeed,

$$[\mathbf{u}]_B + [\mathbf{v}]_B = \begin{bmatrix} -1 \\ 0 \\ 0 \\ 2 \end{bmatrix} + \begin{bmatrix} -1 \\ 2 \\ 2 \\ 0 \end{bmatrix} = \begin{bmatrix} -2 \\ 2 \\ 2 \\ 2 \end{bmatrix} = [\mathbf{u} + \mathbf{v}]_B$$

Likewise, letting $r = 3$,

$$r\mathbf{u} = 3\begin{bmatrix} -1 & 2 \\ 0 & 1 \end{bmatrix} = \begin{bmatrix} -3 & 6 \\ 0 & 3 \end{bmatrix}$$

To find

$$[r\mathbf{u}]_B = \begin{bmatrix} w_1 \\ w_2 \\ w_3 \\ w_4 \end{bmatrix}$$

we solve the system of equations

$$\begin{aligned} w_1 \quad + w_3 \quad &= -3 \\ w_2 + w_3 + w_4 &= \quad 6 \\ w_3 \quad &= \quad 0 \\ w_1 \quad\quad + w_4 &= \quad 3 \end{aligned}$$

obtaining $w_1 = -3$, $w_2 = 0$, $w_3 = 0$, and $w_4 = 6$, or

$$[r\mathbf{u}]_B = \begin{bmatrix} -3 \\ 0 \\ 0 \\ 6 \end{bmatrix} = 3\begin{bmatrix} -1 \\ 0 \\ 0 \\ 2 \end{bmatrix} = r[\mathbf{u}]_B$$ ■

In the beginning of this section we said that if V is a vector space of dimension n with basis B, then V looks like R^n. We have shown that the function $f: V \to R^n$ defined by $f(\mathbf{v}) = [\mathbf{v}]_B$ is a linear transformation (Theorem 4.2.6) and that if $[\mathbf{u}]_B = [\mathbf{v}]_B$ (i.e., $f(\mathbf{u}) = f(\mathbf{v})$), then $\mathbf{u} = \mathbf{v}$ (Theorem 4.2.2). Such a function is called an **isomorphism**. Thus we say that the two vector spaces V and R^n are isomorphic if $\dim(V) = n$ and there is a linear transformation $f: V \to R^n$ for which $f(\mathbf{u}) = f(\mathbf{v})$ requires that $\mathbf{u} = \mathbf{v}$.

Isomorphisms are fundamental to mathematics. An isomorphism is essentially a strategy that enables us to study new structures in terms of structures that are familiar and more well known.

SUMMARY

Terms to know:

Coordinate vector

Theorems to know:

If B is a basis of a vector space V and $[\mathbf{u}]_B = [\mathbf{v}]_B$, then $\mathbf{u} = \mathbf{v}$.

Let V be a vector space with basis B. Then for $\mathbf{u}, \mathbf{v} \in V$, and $r \in R$,

1. $[\mathbf{u} + \mathbf{v}]_B = [\mathbf{u}]_B + [\mathbf{v}]_B$
2. $[r\mathbf{u}]_B = r[\mathbf{u}]_B$

Methods to know:

Determine the coordinate vector of $\mathbf{v}$ given a basis B of V.

EXERCISE SET 4.2

For each exercise, a vector space V having basis B and a vector $\mathbf{v} \in V$ are given. Find $[\mathbf{v}]_B$.

1. $V = P_2$, $B = \{1, x, x^2\}$

a. $\mathbf{v} = 3 + 4x + 2x^2$ b. $\mathbf{v} = -1 + 2x$

c. $\mathbf{v} = -2 + x^2$ d. $\mathbf{v} = x^2 - 2x - 1$

e. $\mathbf{v} = -3x^2$ f. $\mathbf{v} = 1$

g. $\mathbf{v} = x$ h. $\mathbf{v} = x^2$

2. Repeat Exercise 1 with the new basis $B = \{1, 1 + x, x + x^2\}$.

3. $V = R_{2\times 2}$, $B = \left\{\begin{bmatrix} 1 & 2 \\ 2 & 1 \end{bmatrix}, \begin{bmatrix} 1 & 0 \\ 0 & 1 \end{bmatrix}, \begin{bmatrix} 1 & 1 \\ 1 & 0 \end{bmatrix}, \begin{bmatrix} 1 & -1 \\ 1 & 0 \end{bmatrix}\right\}$

a. $\mathbf{v} = \begin{bmatrix} 3 & 4 \\ 0 & -2 \end{bmatrix}$ b. $\mathbf{v} = \begin{bmatrix} -1 & 0 \\ 0 & 1 \end{bmatrix}$ c. $\mathbf{v} = \begin{bmatrix} 2 & -1 \\ -1 & 1 \end{bmatrix}$

d. $\mathbf{v} = \begin{bmatrix} -2 & \frac{1}{2} \\ \sqrt{2} & 1 \end{bmatrix}$ e. $\mathbf{v} = \begin{bmatrix} -1 & 1 \\ 1 & 1 \end{bmatrix}$ f. $\mathbf{v} = \begin{bmatrix} 1 & 0 \\ 0 & 0 \end{bmatrix}$

g. $\mathbf{v} = \begin{bmatrix} 0 & 1 \\ 0 & 0 \end{bmatrix}$ h. $\mathbf{v} = \begin{bmatrix} 0 & 0 \\ 1 & 0 \end{bmatrix}$ i. $\mathbf{v} = \begin{bmatrix} 0 & 0 \\ 0 & 1 \end{bmatrix}$

4. $V = R^3$, $B = \left\{\begin{bmatrix} 1 \\ 0 \\ 1 \end{bmatrix}, \begin{bmatrix} 1 \\ 1 \\ 0 \end{bmatrix}, \begin{bmatrix} 0 \\ 1 \\ 1 \end{bmatrix}\right\}$

a. $\mathbf{v} = \begin{bmatrix} 1 \\ 2 \\ 1 \end{bmatrix}$ b. $\mathbf{v} = \begin{bmatrix} -2 \\ 1 \\ -1 \end{bmatrix}$ c. $\mathbf{v} = \begin{bmatrix} -7 \\ 1 \\ 2 \end{bmatrix}$

d. $\mathbf{v} = \begin{bmatrix} 4 \\ -1 \\ 6 \end{bmatrix}$ e. $\mathbf{v} = \begin{bmatrix} 1 \\ 0 \\ 0 \end{bmatrix}$ f. $\mathbf{v} = \begin{bmatrix} 0 \\ 1 \\ 0 \end{bmatrix}$ g. $\mathbf{v} = \begin{bmatrix} 0 \\ 0 \\ 1 \end{bmatrix}$

5. Let V be R^n and $B = \{\mathbf{b}_1, \ldots, \mathbf{b}_n\}$ be a basis of R^n. For $\mathbf{u}, \mathbf{v} \in V$, with

$$[\mathbf{u}]_B = \begin{bmatrix} u_1 \\ \vdots \\ u_n \end{bmatrix} \quad \text{and} \quad [\mathbf{v}]_B = \begin{bmatrix} v_1 \\ \vdots \\ v_n \end{bmatrix}$$

show that,

$$[\mathbf{u}]_B \cdot [\mathbf{v}]_B = u_1 v_1 \mathbf{b}_1 \cdot \mathbf{b}_1 + u_1 v_2 \mathbf{b}_1 \cdot \mathbf{b}_2 + \cdots + u_k v_j \mathbf{b}_k \cdot \mathbf{b}_j + \cdots + u_n v_n \mathbf{b}_n \cdot \mathbf{b}_n$$

where $\mathbf{b}_i \cdot \mathbf{b}_j$ is the scalar product in R^n.

6. A vector space U has basis $\{\mathbf{b}_1, \mathbf{b}_2, \mathbf{b}_3, \mathbf{b}_4\}$. If for $\mathbf{x}$ in U, $\mathbf{x} = 2\mathbf{b}_2 - \mathbf{b}_4$, what are the coordinates with respect to this basis?

7. Let V be R^n and B be a basis of R^n. For $\mathbf{u}, \mathbf{v}, \mathbf{w} \in V$ and c_1 and c_2 any real numbers, show each of the following.

   a. $[\mathbf{u}]_B + [\mathbf{v}]_B = [\mathbf{v}]_B + [\mathbf{u}]_B$

   b. $[\mathbf{u}]_B + ([\mathbf{v}_B] + [\mathbf{w}]_B) = ([\mathbf{u}]_B + [\mathbf{v}]_B) + [\mathbf{w}]_B$

   c. $c_1([\mathbf{u}]_B + [\mathbf{v}]_B) = c_1[\mathbf{u}]_B + c_1[\mathbf{v}]_B$

   d. $(c_1 + c_2)[\mathbf{u}]_B = c_1[\mathbf{u}]_B + c_2[\mathbf{u}]_B$

   e. $(c_1c_2)[\mathbf{u}]_B = c_1(c_2[\mathbf{u}]_B)$

SECTION 4.3 MATRIX REPRESENTATION OF A LINEAR TRANSFORMATION

Let $f: U \to V$ be a linear transformation from vector space U to vector space V. Associated with this linear transformation is a matrix M, which can be used in computations more effectively than f. How we find M is the topic of this section.

First, we construct the matrix M associated with a linear transformation $f: R^n \to R^m$ using standard bases. Then this procedure is generalized for arbitrary bases and arbitrary vector spaces.

Example 4.3.1 Suppose $f: R^3 \to R^2$ is a linear transformation defined by

$$f\left(\begin{bmatrix} x_1 \\ x_2 \\ x_3 \end{bmatrix}\right) = \begin{bmatrix} x_1 - x_3 \\ x_2 + x_3 \end{bmatrix}$$

Using the standard basis for R^3,

$$f\left(\begin{bmatrix} 1 \\ 0 \\ 0 \end{bmatrix}\right) = \begin{bmatrix} 1 \\ 0 \end{bmatrix}, \qquad f\left(\begin{bmatrix} 0 \\ 1 \\ 0 \end{bmatrix}\right) = \begin{bmatrix} 0 \\ 1 \end{bmatrix},$$

and

$$f\left(\begin{bmatrix} 0 \\ 0 \\ 1 \end{bmatrix}\right) = \begin{bmatrix} -1 \\ 1 \end{bmatrix}$$

The matrix M is defined to be the 2×3 matrix

$$\begin{bmatrix} 1 & 0 & -1 \\ 0 & 1 & 1 \end{bmatrix}$$

whose ith column is $f(\mathbf{e}_i)$. As a check, notice that

$$\begin{bmatrix} 1 & 0 & -1 \\ 0 & 1 & 1 \end{bmatrix}\begin{bmatrix} x_1 \\ x_2 \\ x_3 \end{bmatrix} = \begin{bmatrix} x_1 - x_3 \\ x_2 + x_3 \end{bmatrix} = f\left(\begin{bmatrix} x_1 \\ x_2 \\ x_3 \end{bmatrix}\right)$$ ■

In general, consider the linear transformation $f: R^n \to R^m$. Let $E_n = \{\mathbf{e}_1, \mathbf{e}_2, \ldots, \mathbf{e}_n\}$ be the standard basis for R^n and $E_m = \{\mathbf{e}_1, \mathbf{e}_2, \ldots, \mathbf{e}_m\}$ be the standard basis for R^m. Then

$$f(\mathbf{e}_1) = f\left(\begin{bmatrix} 1 \\ 0 \\ \vdots \\ 0 \end{bmatrix}\right) = \begin{bmatrix} a_{11} \\ a_{21} \\ \vdots \\ a_{m1} \end{bmatrix}, \qquad f(\mathbf{e}_2) = f\left(\begin{bmatrix} 0 \\ 1 \\ \vdots \\ 0 \end{bmatrix}\right) = \begin{bmatrix} a_{12} \\ a_{22} \\ \vdots \\ a_{m2} \end{bmatrix}, \ldots,$$

and

$$f(\mathbf{e}_n) = f\left(\begin{bmatrix} 0 \\ 0 \\ \vdots \\ 1 \end{bmatrix}\right) = \begin{bmatrix} a_{1n} \\ a_{2n} \\ \vdots \\ a_{mn} \end{bmatrix}$$

The matrix M associated with f is the $m \times n$ matrix

$$\begin{bmatrix} a_{11} & a_{12} & \cdots & a_{1n} \\ a_{21} & a_{22} & \cdots & a_{2n} \\ \vdots & \vdots & & \vdots \\ a_{m1} & a_{m2} & \cdots & a_{mn} \end{bmatrix}$$

Example 4.3.2 Define the linear transformation $f: R^2 \to R^3$ by

$$f\left(\begin{bmatrix} x_1 \\ x_2 \end{bmatrix}\right) = \begin{bmatrix} 3x_1 - x_2 \\ x_2 \\ x_1 + x_2 \end{bmatrix}$$

Then

$$f\left(\begin{bmatrix} 1 \\ 0 \end{bmatrix}\right) = \begin{bmatrix} 3 \\ 0 \\ 1 \end{bmatrix} \quad \text{and} \quad f\left(\begin{bmatrix} 0 \\ 1 \end{bmatrix}\right) = \begin{bmatrix} -1 \\ 1 \\ 1 \end{bmatrix}$$

The matrix associated with f is

$$M = \begin{bmatrix} 3 & -1 \\ 0 & 1 \\ 1 & 1 \end{bmatrix}$$

Check:

$$M\begin{bmatrix} x_1 \\ x_2 \end{bmatrix} = \begin{bmatrix} 3 & -1 \\ 0 & 1 \\ 1 & 1 \end{bmatrix}\begin{bmatrix} x_1 \\ x_2 \end{bmatrix} = \begin{bmatrix} 3x_1 - x_2 \\ x_2 \\ x_1 + x_2 \end{bmatrix} = f\left(\begin{bmatrix} x_1 \\ x_2 \end{bmatrix}\right) \qquad \blacksquare$$

The key to constructing M associated with the linear transformation $f: R^n \to R^m$ is (1) f operates on each basis vector in R^n, resulting in (2) a vector in R^m, which is (3) a linear combination of the basis vectors in R^m which is (4) a column of the matrix M.

In Example 4.3.1,

$$f\left(\begin{bmatrix} x_1 \\ x_2 \\ x_3 \end{bmatrix}\right) = \begin{bmatrix} x_1 - x_3 \\ x_2 + x_3 \end{bmatrix}$$

and

$$f\left(\begin{bmatrix} 1 \\ 0 \\ 0 \end{bmatrix}\right) = \begin{bmatrix} 1 \\ 0 \end{bmatrix} = 1\begin{bmatrix} 1 \\ 0 \end{bmatrix} + 0\begin{bmatrix} 0 \\ 1 \end{bmatrix}$$

$$f\left(\begin{bmatrix} 0 \\ 1 \\ 0 \end{bmatrix}\right) = \begin{bmatrix} 0 \\ 1 \end{bmatrix} = 0\begin{bmatrix} 1 \\ 0 \end{bmatrix} + 1\begin{bmatrix} 0 \\ 1 \end{bmatrix}$$

$$f\left(\begin{bmatrix} 0 \\ 0 \\ 1 \end{bmatrix}\right) = \begin{bmatrix} -1 \\ 1 \end{bmatrix} = -1\begin{bmatrix} 1 \\ 0 \end{bmatrix} + 1\begin{bmatrix} 0 \\ 1 \end{bmatrix}$$

In Example 4.3.2,

$$f\left(\begin{bmatrix} x_1 \\ x_2 \end{bmatrix}\right) = \begin{bmatrix} 3x_1 - x_2 \\ x_2 \\ x_1 + x_2 \end{bmatrix}$$

and

$$f\left(\begin{bmatrix} 1 \\ 0 \end{bmatrix}\right) = \begin{bmatrix} 3 \\ 0 \\ 1 \end{bmatrix} = 3\begin{bmatrix} 1 \\ 0 \\ 0 \end{bmatrix} + 0\begin{bmatrix} 0 \\ 1 \\ 0 \end{bmatrix} + 1\begin{bmatrix} 0 \\ 0 \\ 1 \end{bmatrix}$$

$$f\left(\begin{bmatrix} 0 \\ 1 \end{bmatrix}\right) = \begin{bmatrix} -1 \\ 1 \\ 1 \end{bmatrix} = -1\begin{bmatrix} 1 \\ 0 \\ 0 \end{bmatrix} + 1\begin{bmatrix} 0 \\ 1 \\ 0 \end{bmatrix} + 1\begin{bmatrix} 0 \\ 0 \\ 1 \end{bmatrix}$$

In Example 4.3.3, we use the same transformation as in Example 4.3.2 but with different bases.

Example 4.3.3 Define $f: R^2 \to R^3$ by

$$f\left(\begin{bmatrix} x_1 \\ x_2 \end{bmatrix}\right) = \begin{bmatrix} 3x_1 - x_2 \\ x_2 \\ x_1 + x_2 \end{bmatrix}$$

Using the basis

$$A = \left\{ \begin{bmatrix} -1 \\ 1 \end{bmatrix}, \begin{bmatrix} 2 \\ 1 \end{bmatrix} \right\}$$

for R^2 and

$$B = \left\{ \begin{bmatrix} 1 \\ -1 \\ 1 \end{bmatrix}, \begin{bmatrix} -1 \\ 1 \\ 1 \end{bmatrix}, \begin{bmatrix} 1 \\ 0 \\ 1 \end{bmatrix} \right\}$$

for R^3, determine the matrix M associated with f and the chosen bases for R^2 and R^3.

Solution Following the above procedure, we first let f operate on the basis vectors of A, obtaining

$$f\left(\begin{bmatrix} -1 \\ 1 \end{bmatrix}\right) = \begin{bmatrix} -4 \\ 1 \\ 0 \end{bmatrix} \quad \text{and} \quad f\left(\begin{bmatrix} 2 \\ 1 \end{bmatrix}\right) = \begin{bmatrix} 5 \\ 1 \\ 3 \end{bmatrix}$$

Next we express each of these vectors as a linear combination of the basis vectors of R^3, by writing

$$\begin{bmatrix} -4 \\ 1 \\ 0 \end{bmatrix} = s_1 \begin{bmatrix} 1 \\ -1 \\ 1 \end{bmatrix} + s_2 \begin{bmatrix} -1 \\ 1 \\ 1 \end{bmatrix} + s_3 \begin{bmatrix} 1 \\ 0 \\ 1 \end{bmatrix}$$

and

$$\begin{bmatrix} 5 \\ 1 \\ 3 \end{bmatrix} = t_1 \begin{bmatrix} 1 \\ -1 \\ 1 \end{bmatrix} + t_2 \begin{bmatrix} -1 \\ 1 \\ 1 \end{bmatrix} + t_3 \begin{bmatrix} 1 \\ 0 \\ 1 \end{bmatrix}$$

The first equation is equivalent to the system

$$\begin{aligned} s_1 - s_2 + s_3 &= -4 \\ -s_1 + s_2 \phantom{{}+s_3} &= 1 \\ s_1 + s_2 + s_3 &= 0 \end{aligned}$$

having the solution $s_1 = 1$, $s_2 = 2$, and $s_3 = -3$. The second equation is equivalent to the system

$$\begin{aligned} t_1 - t_2 + t_3 &= 5 \\ -t_1 + t_2 \phantom{{}+t_3} &= 1 \\ t_1 + t_2 + t_3 &= 3 \end{aligned}$$

which has the solution $t_1 = -2$, $t_2 = -1$, $t_3 = 6$. The first column of matrix M is

$$\begin{bmatrix} s_1 \\ s_2 \\ s_3 \end{bmatrix}$$

and the second column is

$$\begin{bmatrix} t_1 \\ t_2 \\ t_3 \end{bmatrix}$$

That is,

$$M = \begin{bmatrix} 1 & -2 \\ 2 & -1 \\ -3 & 6 \end{bmatrix}$$

Thus if

$$\mathbf{x} = 3\begin{bmatrix} -1 \\ 1 \end{bmatrix} + 4\begin{bmatrix} 2 \\ 1 \end{bmatrix} = \begin{bmatrix} 5 \\ 7 \end{bmatrix} \quad \text{then} \quad f(\mathbf{x}) = \begin{bmatrix} 8 \\ 7 \\ 12 \end{bmatrix}$$

whereas

$$[f(\mathbf{x})]_B = M[\mathbf{x}]_A = \begin{bmatrix} 1 & -2 \\ 2 & -1 \\ -3 & 6 \end{bmatrix}\begin{bmatrix} 3 \\ 4 \end{bmatrix} = \begin{bmatrix} -5 \\ 2 \\ 15 \end{bmatrix}$$

since

$$-5\begin{bmatrix} 1 \\ -1 \\ 1 \end{bmatrix} + 2\begin{bmatrix} -1 \\ 1 \\ 1 \end{bmatrix} + 15\begin{bmatrix} 1 \\ 0 \\ 1 \end{bmatrix} = \begin{bmatrix} 8 \\ 7 \\ 12 \end{bmatrix}$$

■

Definition 4.3.4 Let $f: U \to V$ be a linear transformation from vector space U to V. Let $A = \{\mathbf{a}_1, \mathbf{a}_2, \ldots, \mathbf{a}_n\}$ be a basis for U and $B = \{\mathbf{b}_1, \mathbf{b}_2, \ldots, \mathbf{b}_m\}$ be a basis for V. Then if

$$\begin{aligned} f(\mathbf{a}_1) &= \alpha_{11}\mathbf{b}_1 + \alpha_{21}\mathbf{b}_2 + \cdots + \alpha_{m1}\mathbf{b}_m \\ f(\mathbf{a}_2) &= \alpha_{12}\mathbf{b}_1 + \alpha_{22}\mathbf{b}_2 + \cdots + \alpha_{m2}\mathbf{b}_m \\ &\vdots \\ f(\mathbf{a}_n) &= \alpha_{1n}\mathbf{b}_1 + \alpha_{2n}\mathbf{b}_2 + \cdots + \alpha_{mn}\mathbf{b}_m \end{aligned} \tag{4.3.1}$$

the **associated matrix** of f is the $m \times n$ matrix

$$M = \begin{bmatrix} \alpha_{11} & \alpha_{12} & \cdots & \alpha_{1n} \\ \alpha_{21} & \alpha_{22} & \cdots & \alpha_{2n} \\ \vdots & \vdots & & \vdots \\ \alpha_{m1} & \alpha_{m2} & \cdots & \alpha_{mn} \end{bmatrix}$$

where the ith column of M is the coordinate vector of $f(\mathbf{a}_i)$ with respect to the basis B.

Note that M is the transpose of the coefficient matrix of the system in (4.3.1).

Example 4.3.5 Let $U = P_2$ and $V = R_{2\times 2}$. If $A = \{1, x, x^2\}$, and

$$B = \left\{ \begin{bmatrix} 1 & 0 \\ 0 & 1 \end{bmatrix}, \begin{bmatrix} 0 & 1 \\ 0 & 1 \end{bmatrix}, \begin{bmatrix} 1 & 1 \\ 1 & 0 \end{bmatrix}, \begin{bmatrix} 0 & 0 \\ 0 & 1 \end{bmatrix} \right\}$$

find the matrix M of the linear transformation

$$f(p_0 + p_1 x + p_2 x^2) = \begin{bmatrix} p_0 + p_1 & p_1 + p_2 \\ p_0 + p_2 & p_2 \end{bmatrix}$$

Solution We have

$$f(1) = \begin{bmatrix} 1 & 0 \\ 1 & 0 \end{bmatrix}$$

$$= s_1 \begin{bmatrix} 1 & 0 \\ 0 & 1 \end{bmatrix} + s_2 \begin{bmatrix} 0 & 1 \\ 0 & 1 \end{bmatrix} + s_3 \begin{bmatrix} 1 & 1 \\ 1 & 0 \end{bmatrix} + s_4 \begin{bmatrix} 0 & 0 \\ 0 & 1 \end{bmatrix}$$

$$f(x) = \begin{bmatrix} 1 & 1 \\ 0 & 0 \end{bmatrix}$$

$$= t_1 \begin{bmatrix} 1 & 0 \\ 0 & 1 \end{bmatrix} + t_2 \begin{bmatrix} 0 & 1 \\ 0 & 1 \end{bmatrix} + t_3 \begin{bmatrix} 1 & 1 \\ 1 & 0 \end{bmatrix} + t_4 \begin{bmatrix} 0 & 0 \\ 0 & 1 \end{bmatrix}$$

and

$$f(x^2) = \begin{bmatrix} 0 & 1 \\ 1 & 1 \end{bmatrix}$$

$$= r_1 \begin{bmatrix} 1 & 0 \\ 0 & 1 \end{bmatrix} + r_2 \begin{bmatrix} 0 & 1 \\ 0 & 1 \end{bmatrix} + r_3 \begin{bmatrix} 1 & 1 \\ 1 & 0 \end{bmatrix} + r_4 \begin{bmatrix} 0 & 0 \\ 0 & 1 \end{bmatrix}$$

The corresponding systems of equations are

$$\begin{aligned} s_1 \quad\quad + s_3 \quad\quad &= 1 \\ s_2 + s_3 \quad\quad &= 0 \\ s_3 \quad\quad &= 1 \\ s_1 + s_2 \quad\quad + s_4 &= 0 \end{aligned}$$

$$\begin{aligned} t_1 \quad\quad + t_3 \quad\quad &= 1 \\ t_2 + t_3 \quad\quad &= 1 \\ t_3 \quad\quad &= 0 \\ t_1 + t_2 \quad\quad + t_4 &= 0 \end{aligned}$$

and

$$\begin{aligned} r_1 \quad\quad + r_3 \quad\quad &= 0 \\ r_2 + r_3 \quad\quad &= 1 \\ r_3 \quad\quad &= 1 \\ r_1 + r_2 \quad\quad + r_4 &= 1 \end{aligned}$$

and the corresponding solutions are

$$\begin{array}{llll} s_1 = 0, & s_2 = -1, & s_3 = 1, & s_4 = 1 \\ t_1 = 1, & t_2 = 1, & t_3 = 0, & t_4 = -2 \\ r_1 = -1, & r_2 = 0, & r_3 = 1, & r_4 = 2 \end{array}$$

The 4×3 matrix is

$$M = \begin{bmatrix} 0 & 1 & -1 \\ -1 & 1 & 0 \\ 1 & 0 & 1 \\ 1 & -2 & 2 \end{bmatrix}$$

Note that if $\mathbf{p} = 3 + 4x - x^2$, then

$$[\mathbf{p}]_A = \begin{bmatrix} 3 \\ 4 \\ -1 \end{bmatrix}$$

and

$$M[\mathbf{p}]_A = \begin{bmatrix} 0 & 1 & -1 \\ -1 & 1 & 0 \\ 1 & 0 & 1 \\ 1 & -2 & 2 \end{bmatrix} \begin{bmatrix} 3 \\ 4 \\ -1 \end{bmatrix} = \begin{bmatrix} 5 \\ 1 \\ 2 \\ -7 \end{bmatrix} = [f(\mathbf{p})]_B$$

since

$$5\begin{bmatrix} 1 & 0 \\ 0 & 1 \end{bmatrix} + 1\begin{bmatrix} 0 & 1 \\ 0 & 1 \end{bmatrix} + 2\begin{bmatrix} 1 & 1 \\ 1 & 0 \end{bmatrix} - 7\begin{bmatrix} 0 & 0 \\ 0 & 1 \end{bmatrix} = \begin{bmatrix} 7 & 3 \\ 2 & -1 \end{bmatrix}$$

$$= f(3 + 4x - x^2) \quad \blacksquare$$

One idea should be stressed:

> Given f, U, V, A, and B, the matrix M depends on all of f, U, V, A, and B. In particular, changing A or B will change M.

Example 4.3.6 In Example 4.3.5, let $A = \{1 + x, x + x^2, x^2\}$. Then

$$f(1 + x) = \begin{bmatrix} 2 & 1 \\ 1 & 0 \end{bmatrix}$$

$$= s_1\begin{bmatrix} 1 & 0 \\ 0 & 1 \end{bmatrix} + s_2\begin{bmatrix} 0 & 1 \\ 0 & 1 \end{bmatrix} + s_3\begin{bmatrix} 1 & 1 \\ 1 & 0 \end{bmatrix} + s_4\begin{bmatrix} 0 & 0 \\ 0 & 1 \end{bmatrix}$$

$$f(x + x^2) = \begin{bmatrix} 1 & 2 \\ 1 & 1 \end{bmatrix}$$

$$= t_1\begin{bmatrix} 1 & 0 \\ 0 & 1 \end{bmatrix} + t_2\begin{bmatrix} 0 & 1 \\ 0 & 1 \end{bmatrix} + t_3\begin{bmatrix} 1 & 1 \\ 1 & 0 \end{bmatrix} + t_4\begin{bmatrix} 0 & 0 \\ 0 & 1 \end{bmatrix}$$

and

$$f(x^2) = \begin{bmatrix} 0 & 1 \\ 1 & 1 \end{bmatrix}$$

$$= r_1\begin{bmatrix} 1 & 0 \\ 0 & 1 \end{bmatrix} + r_2\begin{bmatrix} 0 & 1 \\ 0 & 1 \end{bmatrix} + r_3\begin{bmatrix} 1 & 1 \\ 1 & 0 \end{bmatrix} + r_4\begin{bmatrix} 0 & 0 \\ 0 & 1 \end{bmatrix}$$

Solving these systems yields

$$\begin{array}{llll} s_1 = 1, & s_2 = 0, & s_3 = 1, & s_4 = -1 \\ t_1 = 0, & t_2 = 1, & t_3 = 1, & t_4 = 0 \\ r_1 = -1, & r_2 = 0, & r_3 = 1, & r_4 = 2 \end{array}$$

and so

$$M = \begin{bmatrix} 1 & 0 & -1 \\ 0 & 1 & 0 \\ 1 & 1 & 1 \\ -1 & 0 & 2 \end{bmatrix}$$

which is different from the M in Example 4.3.5. As in Example 4.3.5, if $\mathbf{p} = 3 + 4x - x^2$; then

$$[\mathbf{p}]_A = \begin{bmatrix} 3 \\ 1 \\ -2 \end{bmatrix}$$

and

$$[f(\mathbf{p})]_B = \begin{bmatrix} 1 & 0 & -1 \\ 0 & 1 & 0 \\ 1 & 1 & 1 \\ -1 & 0 & 2 \end{bmatrix}\begin{bmatrix} 3 \\ 1 \\ -2 \end{bmatrix} = \begin{bmatrix} 5 \\ 1 \\ 2 \\ -7 \end{bmatrix}$$

the same answer obtained in Example 4.3.5. ■

Example 4.3.7 In Example 4.3.5, let

$$B = \left\{\begin{bmatrix}1 & 0\\0 & 0\end{bmatrix}, \begin{bmatrix}0 & 1\\0 & 0\end{bmatrix}, \begin{bmatrix}0 & 0\\1 & 0\end{bmatrix}, \begin{bmatrix}0 & 0\\0 & 1\end{bmatrix}\right\}$$

be a basis for $R_{2\times 2}$. Then

$$f(1) = \begin{bmatrix}1 & 0\\1 & 0\end{bmatrix} = 1\begin{bmatrix}1 & 0\\0 & 0\end{bmatrix} + 0\begin{bmatrix}0 & 1\\0 & 0\end{bmatrix} + 1\begin{bmatrix}0 & 0\\1 & 0\end{bmatrix} + 0\begin{bmatrix}0 & 0\\0 & 1\end{bmatrix}$$

$$f(x) = \begin{bmatrix}1 & 1\\0 & 0\end{bmatrix} = 1\begin{bmatrix}1 & 0\\0 & 0\end{bmatrix} + 1\begin{bmatrix}0 & 1\\0 & 0\end{bmatrix} + 0\begin{bmatrix}0 & 0\\1 & 0\end{bmatrix} + 0\begin{bmatrix}0 & 0\\0 & 1\end{bmatrix}$$

$$f(x^2) = \begin{bmatrix}0 & 1\\1 & 1\end{bmatrix} = 0\begin{bmatrix}1 & 0\\0 & 0\end{bmatrix} + 1\begin{bmatrix}0 & 1\\0 & 0\end{bmatrix} + 1\begin{bmatrix}0 & 0\\1 & 0\end{bmatrix} + 1\begin{bmatrix}0 & 0\\0 & 1\end{bmatrix}$$

and consequently,

$$M = \begin{bmatrix}1 & 1 & 0\\0 & 1 & 1\\1 & 0 & 1\\0 & 0 & 1\end{bmatrix}$$

which is different than the Ms obtained in Examples 4.3.5 and 4.3.6. ■

In Chapter 6, we show, for special linear transformations, how A and B can be chosen to guarantee that M is a diagonal matrix. We summarize the preceding discussion and examples in Theorem 4.3.8.

Theorem 4.3.8 *Let $f : U \to V$ be a linear transformation between vector spaces U and V with $A = \{\mathbf{a}_1, \mathbf{a}_2, \ldots, \mathbf{a}_n\}$ a basis for U and $B = \{\mathbf{b}_1, \mathbf{b}_2, \ldots, \mathbf{b}_m\}$ a basis for V. For each $\mathbf{u} \in U$, form $[\mathbf{u}]_A$. Since $f(\mathbf{u}) \in V$, we may form $[f(\mathbf{u})]_B$. The matrix M associated with f, U, V, A, and B has the property $[f(\mathbf{u})]_B = M[\mathbf{u}]_A$.*

Proof. Let $[\mathbf{u}]_A = [u_1, \ldots, u_n]^T$. Since

$$M = \begin{bmatrix}\alpha_{11} & \alpha_{12} & \cdots & \alpha_{1n}\\ \alpha_{21} & \alpha_{22} & \cdots & \alpha_{2n}\\ \vdots & \vdots & & \vdots\\ \alpha_{m1} & \alpha_{2n} & \cdots & \alpha_{mn}\end{bmatrix}$$

then

$$M[\mathbf{u}]_A = \begin{bmatrix}\alpha_{11}u_1 + \alpha_{12}u_2 + \alpha_{1n}u_n\\ \alpha_{21}u_1 + \alpha_{22}u_2 + \alpha_{2n}u_n\\ \vdots\\ \alpha_{m1}u_1 + \alpha_{m2}u_2 + \alpha_{mn}u_n\end{bmatrix}$$

However, since $[\mathbf{u}]_A = [u_1, \ldots, u_n]^T$, $\mathbf{u} = u_1\mathbf{a}_1 + \cdots + u_n\mathbf{a}_n$, and $f(\mathbf{u}) = u_1 f(\mathbf{a}_1) + \cdots + u_n f(\mathbf{a}_n)$, whereas

$$\begin{aligned} f(\mathbf{a}_1) &= \alpha_{11}\mathbf{b}_1 + \alpha_{21}\mathbf{b}_2 + \cdots + \alpha_{m1}\mathbf{b}_m \\ f(\mathbf{a}_2) &= \alpha_{12}\mathbf{b}_1 + \alpha_{22}\mathbf{b}_2 + \cdots + \alpha_{m2}\mathbf{b}_m \\ &\vdots \\ f(\mathbf{a}_n) &= \alpha_{1n}\mathbf{b}_1 + \alpha_{2n}\mathbf{b}_2 + \cdots + \alpha_{mn}\mathbf{b}_m \end{aligned}$$

Therefore,

$$\begin{aligned} f(\mathbf{u}) &= u_1(\alpha_{11}\mathbf{b}_1 + \alpha_{21}\mathbf{b}_2 + \cdots + \alpha_{m1}\mathbf{b}_m) \\ &\quad + u_2(\alpha_{12}\mathbf{b}_1 + \alpha_{22}\mathbf{b}_2 + \cdots + \alpha_{m2}\mathbf{b}_m) + \cdots \\ &\quad + u_n(\alpha_{1n}\mathbf{b}_1 + \alpha_{2n}\mathbf{b}_2 + \cdots + \alpha_{mn}\mathbf{b}_m) \\ &= (u_1\alpha_{11} + \cdots + u_n\alpha_{1n})\mathbf{b}_1 \\ &\quad + (u_1\alpha_{21} + \cdots + u_n\alpha_{2n})\mathbf{b}_2 + \cdots \\ &\quad + (u_1\alpha_{m1} + \cdots + u_n\alpha_{mn})\mathbf{b}_m \end{aligned}$$

This says that

$$\begin{aligned} [f(\mathbf{u})]_B &= \begin{bmatrix} u_1\alpha_{11} + \cdots + u_n\alpha_{1n} \\ u_1\alpha_{21} + \cdots + u_n\alpha_{2n} \\ \vdots \\ u_1\alpha_{m1} + \cdots + u_n\alpha_{mn} \end{bmatrix} \\ &= \begin{bmatrix} \alpha_{11}u_1 + \cdots + \alpha_{1n}u_1 \\ \alpha_{21}u_1 + \cdots + \alpha_{2n}u_n \\ \vdots \\ \alpha_{m1}u_1 + \cdots + \alpha_{mn}u_n \end{bmatrix} \\ &= M[\mathbf{u}]_A \end{aligned}$$

which is the desired result.

Figure 4.1 describes the relationship demonstrated by Theorem 4.3.8. As a special case, if $U = R^m$ and A is the standard basis of R^n, then the ith column of M is $[f(\mathbf{e}_i)]_B$.

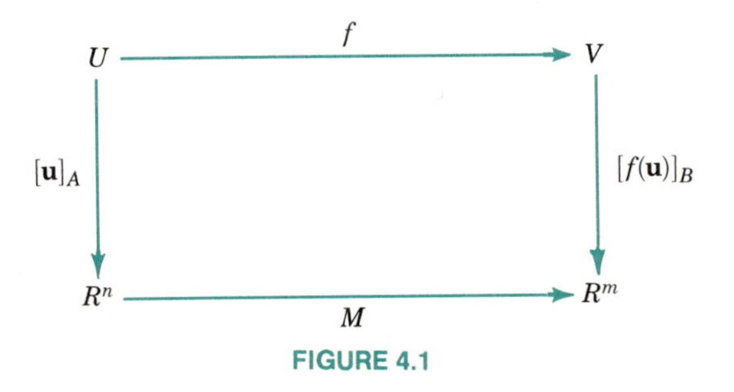

FIGURE 4.1

SUMMARY

Terms to know:

Associated matrix of a linear transformation.

Methods to know:

Given a linear transformation $f: U \to V$ and a basis A of U and basis B of V, find the associated matrix of f.

EXERCISE SET 4.3

1. Let the basis be the standard basis for each of the following linear transformations $f: R^n \to R^m$. Find the corresponding matrices.

a. $f: R^3 \to R^3$, where $f\left(\begin{bmatrix} x_1 \\ x_2 \\ x_3 \end{bmatrix}\right) = \begin{bmatrix} 2x_2 + 3x_3 \\ 2x_1 + x_2 + x_3 \end{bmatrix}$

b. $f: R^2 \to R^4$, where $f\left(\begin{bmatrix} x_1 \\ x_2 \end{bmatrix}\right) = \begin{bmatrix} x_1 + x_2 \\ x_1 - x_2 \\ 2x_1 \\ 4x_2 \end{bmatrix}$

c. $f: R^4 \to R^2$, where $f\left(\begin{bmatrix} x_1 \\ x_2 \\ x_3 \\ x_4 \end{bmatrix}\right) = \begin{bmatrix} x_1 + x_3 \\ x_2 + x_4 \end{bmatrix}$

d. $f: R^5 \to R^3$, where $f\left(\begin{bmatrix} x_1 \\ x_2 \\ x_3 \\ x_4 \\ x_5 \end{bmatrix}\right) = \begin{bmatrix} x_1 \\ x_2 \\ x_3 \end{bmatrix}$

2. Find the matrices corresponding to the following linear transformations.

a. $f: R^3 \to R^3$, where

$$f\left(\begin{bmatrix} x_1 \\ x_2 \\ x_3 \end{bmatrix}\right) = \begin{bmatrix} 2x_2 + 3x_3 \\ 0 \\ 2x_1 - x_2 + x_3 \end{bmatrix}$$

where the basis for R^3 is

$$\left\{\begin{bmatrix} 1 \\ 1 \\ 1 \end{bmatrix}, \begin{bmatrix} -1 \\ 1 \\ 0 \end{bmatrix}, \begin{bmatrix} 1 \\ 0 \\ -1 \end{bmatrix}\right\}$$

b. $f: R^2 \to R^4$, where

$$f\left(\begin{bmatrix} x_1 \\ x_2 \end{bmatrix}\right) = \begin{bmatrix} x_1 + x_2 \\ x_1 - x_2 \\ 2x_1 \\ 4x_2 \end{bmatrix}$$

where the basis for R^2 is

$$\left\{\begin{bmatrix} 1 \\ 1 \end{bmatrix}, \begin{bmatrix} -1 \\ 2 \end{bmatrix}\right\}$$

and for R^4 is

$$\left\{\begin{bmatrix} 0 \\ 1 \\ 1 \\ 0 \end{bmatrix}, \begin{bmatrix} 1 \\ 1 \\ 0 \\ 0 \end{bmatrix}, \begin{bmatrix} -1 \\ 0 \\ -1 \\ 0 \end{bmatrix}, \begin{bmatrix} 1 \\ 1 \\ 1 \\ 1 \end{bmatrix}\right\}$$

c. $f: R^4 \to R^2$, where

$$f\left(\begin{bmatrix} x_1 \\ x_2 \\ x_3 \\ x_4 \end{bmatrix}\right) = \begin{bmatrix} x_1 + x_3 \\ x_2 + x_4 \end{bmatrix}$$

using the bases given in part (b).

3. Let

$$A = \left\{\begin{bmatrix} 1 \\ 2 \end{bmatrix}, \begin{bmatrix} -2 \\ 1 \end{bmatrix}\right\} \quad \text{and} \quad B = \left\{\begin{bmatrix} 1 \\ 1 \end{bmatrix}, \begin{bmatrix} 3 \\ 1 \end{bmatrix}\right\}$$

be bases for R^2.

a. Let

$$f\left(\begin{bmatrix} 1 \\ 2 \end{bmatrix}\right) = \begin{bmatrix} 3 \\ 4 \end{bmatrix} \quad \text{and} \quad f\left(\begin{bmatrix} -2 \\ 1 \end{bmatrix}\right) = \begin{bmatrix} -1 \\ -4 \end{bmatrix}$$

Find the matrix M associated with this transformation and the bases A and B.

b. Now let

$$B_1 = \left\{\begin{bmatrix} 1 \\ 0 \end{bmatrix}, \begin{bmatrix} -2 \\ 3 \end{bmatrix}\right\}$$

and find the matrix M associated with f and the bases A and B_1.

4. Given the basis $A = \{1, x - 1, x^2 - x\}$ of P_2 and the basis

$$B = \left\{\begin{bmatrix} 1 \\ 0 \\ 1 \end{bmatrix}, \begin{bmatrix} 0 \\ 1 \\ 0 \end{bmatrix}, \begin{bmatrix} -1 \\ 0 \\ 1 \end{bmatrix}\right\}$$

of R^3, find the matrix M with respect to A and B representing each given linear transformation $f: P_2 \to R^3$.

a. $f(p_0 + p_1x + p_2x^2) = \begin{bmatrix} p_1 \\ p_0 \\ p_2 \end{bmatrix}$

b. $f(p_0 + p_1x + p_2x^2) = \begin{bmatrix} -p_0 \\ 0 \\ p_1 \end{bmatrix}$

c. $f(p_0 + p_1x + p_2x^2) = \begin{bmatrix} p_2 \\ p_0 \\ p_1 \end{bmatrix}$

d. $f(p_0 + p_1x + p_2x^2) = \begin{bmatrix} p_2 - p_1 \\ p_1 - p_0 \\ p_2 \end{bmatrix}$

e. $f(p_0 + p_1x + p_2x^2) = \begin{bmatrix} p_0 \\ 0 \\ p_2 \end{bmatrix}$

5. Given the basis $A = \{1, x - 1, x^2 - x\}$ of P_2 and the basis

$$B = \left\{ \begin{bmatrix} 1 & 0 \\ 0 & 1 \end{bmatrix}, \begin{bmatrix} 0 & 1 \\ 1 & 0 \end{bmatrix}, \begin{bmatrix} 0 & 0 \\ 1 & 0 \end{bmatrix}, \begin{bmatrix} 0 & 0 \\ 1 & 1 \end{bmatrix} \right\}$$

of $R_{2\times 2}$, find the matrix M associated with A and B representing each given linear transformations $f: P_2 \to R_{2\times 2}$.

a. $f(p_0 + p_1x + p_2x^2) = \begin{bmatrix} p_0 & p_1 \\ 0 & p_2 \end{bmatrix}$

b. $f(p_0 + p_1x + p_2x^2) = \begin{bmatrix} p_1 - p_2 & p_0 \\ p_1 & p_2 \end{bmatrix}$

c. $f(p_0 + p_1x + p_2x^2) = \begin{bmatrix} p_0 & p_1 \\ p_2 & 0 \end{bmatrix}$

d. $f(p_0 + p_1x + p_2x^2) = \begin{bmatrix} p_0 & 0 \\ 0 & p_2 \end{bmatrix}$

e. $f(p_0 + p_1x + p_2x^2) = \begin{bmatrix} 0 & 0 \\ p_2 & 0 \end{bmatrix}$

SECTION 4.4 CHANGE OF BASIS

In the previous section, we considered a linear transformation $f: U \to V$ and its matrix representation M with respect to a basis A of U and a basis B of V. In this situation,

$$[f(\mathbf{u})]_B = M[\mathbf{u}]_A$$

In this section, we change one or both of the bases A and B and determine the effect this change has on M. In particular, we determine a method to change these bases without completely recalculating M.

We begin by considering the identity linear transformation $i_U : U \to U$ given by $i_U(\mathbf{u}) = \mathbf{u}$. If A_1 and A_2 are bases of U, we consider the matrix P for which $P[\mathbf{u}]_1 = [\mathbf{u}]_2 = [i(\mathbf{u})]_2$, as shown in Figure 4.2, where $[\mathbf{u}]_i$ is the representation of $\mathbf{u}$ with respect to basis A_i $(i = 1, 2)$. (Note that we are denoting $[\mathbf{u}]_{A_i}$ by $[\mathbf{u}]_i$.)

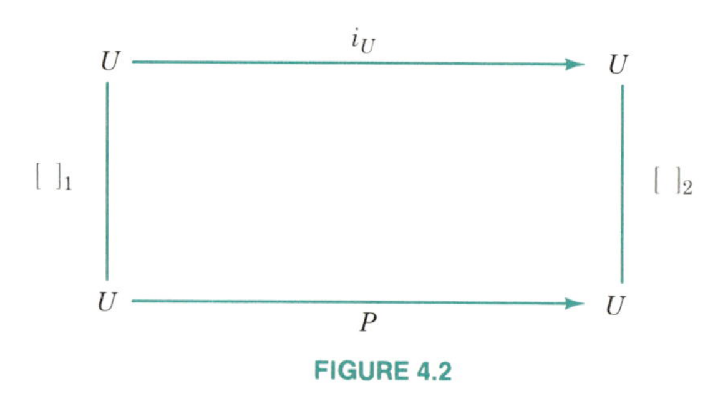

FIGURE 4.2

Definition 4.4.1 P is called the **change of basis** matrix from A_1 to A_2.

Example 4.4.2 Let $U = R^3$, $A_1 = \{\mathbf{e}_1, \mathbf{e}_2, \mathbf{e}_3\}$ and

$$A_2 = \left\{ \begin{bmatrix} 1 \\ 0 \\ 1 \end{bmatrix}, \begin{bmatrix} 0 \\ 1 \\ 1 \end{bmatrix}, \begin{bmatrix} 1 \\ 1 \\ 0 \end{bmatrix} \right\}$$

Find the A_1 to A_2 change of basis matrix P.

Solution Recall from the preceding section that

$$\begin{bmatrix} p_{11} \\ p_{21} \\ p_{31} \end{bmatrix} = \begin{bmatrix} 1 \\ 0 \\ 0 \end{bmatrix}_{A_2}, \quad \begin{bmatrix} p_{12} \\ p_{22} \\ p_{32} \end{bmatrix} = \begin{bmatrix} 0 \\ 1 \\ 0 \end{bmatrix}_{A_2}, \quad \text{and} \quad \begin{bmatrix} p_{13} \\ p_{23} \\ p_{33} \end{bmatrix} = \begin{bmatrix} 0 \\ 0 \\ 1 \end{bmatrix}_{A_2}$$

thus

$$\begin{bmatrix} 1 \\ 0 \\ 0 \end{bmatrix} = p_{11} \begin{bmatrix} 1 \\ 0 \\ 1 \end{bmatrix} + p_{21} \begin{bmatrix} 0 \\ 1 \\ 1 \end{bmatrix} + p_{31} \begin{bmatrix} 1 \\ 1 \\ 0 \end{bmatrix}$$

The augmented matrix of this system is

$$\left[\begin{array}{ccc|c} 1 & 0 & 1 & 1 \\ 0 & 1 & 1 & 0 \\ 1 & 1 & 0 & 0 \end{array}\right]$$

Likewise, to find $\begin{bmatrix} 0 \\ 1 \\ 0 \end{bmatrix}_{A_2}$ we solve the system whose augmented matrix is

$$\left[\begin{array}{ccc|c} 1 & 0 & 1 & 0 \\ 0 & 1 & 1 & 1 \\ 1 & 1 & 0 & 0 \end{array}\right]$$

To find $\begin{bmatrix} 0 \\ 0 \\ 1 \end{bmatrix}_{A_2}$ we consider the augmented matrix

$$\left[\begin{array}{ccc|c} 1 & 0 & 1 & 0 \\ 0 & 1 & 1 & 0 \\ 1 & 1 & 0 & 1 \end{array}\right]$$

We can solve all three systems at once as follows: Consider the matrix

$$A = \left[\begin{array}{ccc|ccc} 1 & 0 & 1 & 1 & 0 & 0 \\ 0 & 1 & 1 & 0 & 1 & 0 \\ 1 & 1 & 0 & 0 & 0 & 1 \end{array}\right]$$

Reduce A to $A^1 = [E|P]$ where E is the reduced row echelon form of

$$B = \begin{bmatrix} 1 & 0 & 1 \\ 0 & 1 & 1 \\ 0 & 1 & 1 \end{bmatrix}$$

Since the columns of B form the basis A_2, $E = I$. Thus

$$\left[\begin{array}{ccc|ccc} 1 & 0 & 1 & 1 & 0 & 0 \\ 0 & 1 & 1 & 0 & 1 & 0 \\ 1 & 1 & 0 & 0 & 0 & 1 \end{array}\right]$$

becomes

$$A^1 = \left[\begin{array}{ccc|ccc} 1 & 0 & 0 & \frac{1}{2} & -\frac{1}{2} & \frac{1}{2} \\ 0 & 1 & 0 & -\frac{1}{2} & \frac{1}{2} & \frac{1}{2} \\ 0 & 0 & 1 & \frac{1}{2} & \frac{1}{2} & -\frac{1}{2} \end{array}\right]$$

and the A_1 to A_2 change of basis matrix is

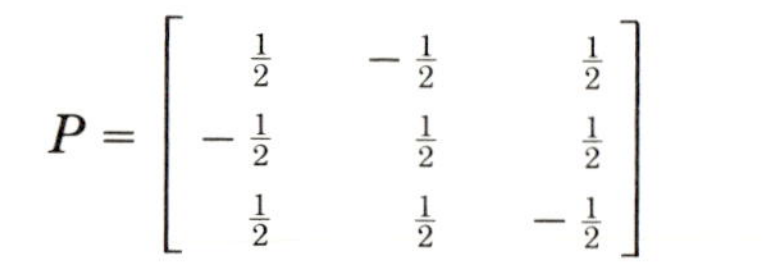

$$P = \begin{bmatrix} \frac{1}{2} & -\frac{1}{2} & \frac{1}{2} \\ -\frac{1}{2} & \frac{1}{2} & \frac{1}{2} \\ \frac{1}{2} & \frac{1}{2} & -\frac{1}{2} \end{bmatrix}$$

Notice that in the matrix A, the first 3 columns form the basis A_2 while the last 3 columns form the basis A_1. ■

Example 4.4.3 Let $V = P_2$ with basis $A_1 = \{1, 1 + x, 1 + x + x^2\}$ and $A_2 = \{1 - x, 1 + x^2, x - x^2\}$. Find the A_1 to A_2 change of basis matrix P.

Solution We want a matrix P such that

$$P[1]_{A_1} = [1]_{A_2}$$

$$P[1 + x]_{A_1} = [1 + x]_{A_2}$$

and

$$P[1 + x + x^2]_{A_1} = [1 + x + x^2]_{A_2}$$

We let

$$[1]_{A_2} = \begin{bmatrix} a \\ b \\ c \end{bmatrix}, \quad [1 + x]_{A_2} = \begin{bmatrix} d \\ e \\ f \end{bmatrix}, \quad \text{and} \quad [1 + x + x^2]_{A_2} = \begin{bmatrix} g \\ h \\ i \end{bmatrix}$$

Then

$$P = \begin{bmatrix} a & d & g \\ b & e & h \\ c & f & i \end{bmatrix}$$

Thus we solve three systems of equations:

1.
$$\begin{aligned} 1 &= a(1 - x) + b(1 + x^2) + c(x - x^2) \\ &= (a + b)1 + (c - a)x + (b - c)x^2 \end{aligned} \qquad (4.4.1)$$

or

$$\begin{aligned} a + b &= 1 \\ -a + c &= 0 \\ b - c &= 0 \end{aligned}$$

2. $$1 + x = d(1 - x) + e(1 + x^2) + f(x - x^2) \tag{4.4.2}$$

or

$$\begin{aligned} 1 &= \quad d + e \\ 1 &= -d \qquad + f \\ 0 &= \qquad\quad e - f \end{aligned}$$

and

3. $$1 + x + x^2 = g(1 - x) + h(1 + x^2) + i(x - x^2) \tag{4.4.3}$$

or

$$\begin{aligned} 1 &= \quad g + h \\ 1 &= -g \qquad + i \\ 1 &= \qquad\quad h - i \end{aligned}$$

Thus we reduce

$$A = \left[\begin{array}{ccc|ccc} 1 & 1 & 0 & 1 & 1 & 1 \\ -1 & 0 & 1 & 0 & 1 & 1 \\ 0 & 1 & -1 & 0 & 0 & 1 \end{array}\right]$$

to

$$A' = \left[\begin{array}{ccc|ccc} 1 & 0 & 0 & \frac{1}{2} & 0 & -\frac{1}{2} \\ 0 & 1 & 0 & \frac{1}{2} & 1 & \frac{3}{2} \\ 0 & 0 & 1 & \frac{1}{2} & 1 & \frac{1}{2} \end{array}\right] = [I|P]$$

Thus the change of basis matrix is

$$P = \begin{bmatrix} \frac{1}{2} & 0 & -\frac{1}{2} \\ \frac{1}{2} & 1 & \frac{3}{2} \\ \frac{1}{2} & 1 & \frac{1}{2} \end{bmatrix}$$ ■

In general, if V has bases $A_1 = \{\mathbf{a}_1, \ldots, \mathbf{a}_n\}$ and $A_2 = \{\mathbf{b}_1, \ldots, \mathbf{b}_n\}$, then we form

$$A = [[\mathbf{b}_1] \quad [\mathbf{b}_2] \quad \cdots \quad [\mathbf{b}_n] | [\mathbf{a}_1] \quad [\mathbf{a}_2] \quad \cdots \quad [\mathbf{a}_n]]$$

where $[\mathbf{u}]$ is the representation of $\mathbf{u}$ with respect to the *standard basis* of V. We row reduce A to obtain $A' = [I|P]$. Then P is the A_1 to A_2 change of basis matrix.

Example 4.4.4 Let $V = R_{2\times 2}$ with basis A_1 the standard basis and

$$A_2 = \left\{\begin{bmatrix} 1 & 0 \\ 0 & 1 \end{bmatrix}, \begin{bmatrix} 1 & -1 \\ 0 & 0 \end{bmatrix}, \begin{bmatrix} 0 & 1 \\ 0 & -1 \end{bmatrix}, \begin{bmatrix} 0 & 0 \\ 1 & 1 \end{bmatrix}\right\}$$

Find the A_1 to A_2 change of basis matrix P.

Solution We form

$$A = \left[\begin{array}{cccc|cccc} 1 & 1 & 0 & 0 & 1 & 0 & 0 & 0 \\ 0 & -1 & 1 & 0 & 0 & 1 & 0 & 0 \\ 0 & 0 & 0 & 1 & 0 & 0 & 1 & 0 \\ 1 & 0 & -1 & 1 & 0 & 0 & 0 & 1 \end{array}\right]$$

and obtain

$$P = \begin{bmatrix} \frac{1}{2} & \frac{1}{2} & -\frac{1}{2} & \frac{1}{2} \\ \frac{1}{2} & -\frac{1}{2} & \frac{1}{2} & -\frac{1}{2} \\ \frac{1}{2} & \frac{1}{2} & \frac{1}{2} & -\frac{1}{2} \\ 0 & 0 & 1 & 0 \end{bmatrix}$$

In this case

$$P = \begin{bmatrix} 1 & 1 & 0 & 0 \\ 0 & -1 & 1 & 0 \\ 0 & 0 & 0 & 1 \\ 1 & 0 & -1 & 1 \end{bmatrix}^{-1}$$

■

In Example 4.4.2,

$$P = \begin{bmatrix} 1 & 0 & 1 \\ 0 & 1 & 1 \\ 1 & 1 & 0 \end{bmatrix}^{-1}$$

also.

REMARK In general, if A_1 is the standard basis, the A_1 to A_2 change of basis matrix is the inverse of the matrix

$$C = [[\mathbf{b}_1] \quad [\mathbf{b}_2] \cdots [\mathbf{b}_n]]$$

which is obtained by listing as its column the vectors of A_2.

Example 4.4.5 Let V, A_1, A_2 be as in Example 4.4.4. Then the A_2 to A_1 change of basis matrix is

$$P = \begin{bmatrix} 1 & 1 & 0 & 0 \\ 0 & -1 & 1 & 0 \\ 0 & 0 & 0 & 1 \\ 1 & 0 & -1 & 1 \end{bmatrix}$$

Thus P is obtained by listing the vectors of A_2 as the columns of P. (Note that we switched the roles of A_1 and A_2 from Example 4.4.4.) ■

Warning This shortcut works only when A_2 is the standard basis of V.

In summary, let A_1 and A_2 be bases of the vector space V. If A_1 is the standard basis, then the A_1 to A_2 change of basis matrix is

$$P = Q^{-1}$$

where Q has as its ith column the ith vector of A_2. If A_2 is the standard basis, the A_1 to A_2 change of basis matrix is $P = Q$ where Q has as its ith column the ith vector of A_1.

Theorem 4.4.6 *Let V be a vector space with the basis A_1 and A_2. Let P be the A_1 to A_2 change of basis matrix. Then P^{-1} is the A_2 to A_1 change of basis matrix.*

Proof. Let Q denote the A_2 to A_1 change of basis matrix. Then for $\mathbf{v} \in V$,

$$P[\mathbf{v}]_{A_1} = [\mathbf{v}]_{A_2} \quad \text{and} \quad Q[\mathbf{v}]_{A_2} = [\mathbf{v}]_{A_1}$$

Thus

$$QP[\mathbf{v}]_{A_1} = Q[\mathbf{v}]_{A_2} = [\mathbf{v}]_{A_1}$$

Therefore, if $\mathbf{v} = \mathbf{a}_i$, and

$$A_1 = \{\mathbf{a}_1, \mathbf{a}_2, \ldots, \mathbf{a}_n\}, \quad \text{then } [\mathbf{v}]_{A_1} = \mathbf{e}_i \quad \text{and} \quad QP\mathbf{e}_i = \mathbf{e}_i$$

for $i = 1, 2, \ldots, n$ and so $QP = I_n$. Likewise,

$$PQ[\mathbf{v}]_{A_2} = P[\mathbf{v}]_{A_1} = [\mathbf{v}]_{A_2}$$

so that if $\mathbf{v} = \mathbf{b}_i$, and

$$A_2 = \{\mathbf{b}_1, \mathbf{b}_2, \ldots, \mathbf{b}_n\}, \quad \text{then } [\mathbf{v}]_{A_2} = \mathbf{e}_i \quad \text{and} \quad PQ\mathbf{e}_i = \mathbf{e}_i$$

for $i = 1, 2, \ldots, n$ and so $Q = P^{-1}$.

Now we return to the matrix representation of $f: U \to V$. Let A_1 and A_2 be bases of U and B_1 and B_2 be bases of V. Let M_1 represent f with respect to A_1 and B_1, whereas M_2 represents f with respect to A_2 and B_2. Then

$$[f(\mathbf{u})]_{B_2} = M_2[\mathbf{u}]_{A_2} = M_2P[\mathbf{u}]_{A_1}$$

where P is the A_1 to A_2 change of basis matrix. However, $[f(\mathbf{u})]_{B_2} =$

$Q[f(\mathbf{u})]_{B_1}$, where Q is the B_1 to B_2 change of basis matrix. Thus

$$[f(\mathbf{u})]_{B_2} = Q[f(\mathbf{u})]_{B_1} = QM_1[\mathbf{u}]_{A_1} = M_2P[\mathbf{u}]_{A_1}.$$

Thus

$$QM_1 = M_2P$$

or

$$M_1 = Q^{-1}M_2P$$

Definition 4.4.7 Let A and B be $n \times n$ matrices. A and B are **similar** if and only if there exists an $n \times n$ invertible matrix P with

$$A = P^{-1}BP$$

Theorem 4.4.8 *Let A and B be $m \times n$ matrices and P and Q be invertible matrices for which $A = Q^{-1}BP$. Then A and B represent the same linear transformation $f: R^n \to R^m$.*

Proof. Let A_1 be the standard basis of R^n and B_1 be the basis of R^n whose vectors are the columns of P. Then P is the B_1 to A_1 change of basis matrix. Likewise, let A_2 be the standard basis of R^m and B_2 be the basis of R^m whose vectors are the columns of Q. Then Q is the B_2 to A_2 change of basis matrix.

Let $f: R^n \to R^m$ be the linear transformation given by $f(\mathbf{u}) = B\mathbf{u}$. Then B is the matrix of f with respect to the standard bases A_1 and A_2 and $[f(\mathbf{u})]_{A_2} = B[\mathbf{u}]_{A_1}$. Then

$$[f(\mathbf{u})]_{A_2} = Q[f(\mathbf{u})]_{B_2}$$

and

$$P[\mathbf{u}]_{B_1} = [\mathbf{u}]_{A_1}$$

so that

$$BP[\mathbf{u}]_{B_1} = B[\mathbf{u}]_{A_1} = [f(\mathbf{u})]_{A_2} = Q[f(\mathbf{u})]_{B_2}$$

or

$$Q^{-1}BP[\mathbf{u}]_{B_1} = [f(\mathbf{u})]_{B_2}$$

Thus $A = Q^{-1}BP$ is the matrix representing f with respect to the bases B_1 and B_2. Thus A and B represent the same linear transformation.

Perhaps the situation in the proof of Theorem 4.4.8 can be clarified by Figure 4.3.

We illustrate this theorem by the Example 4.4.9.

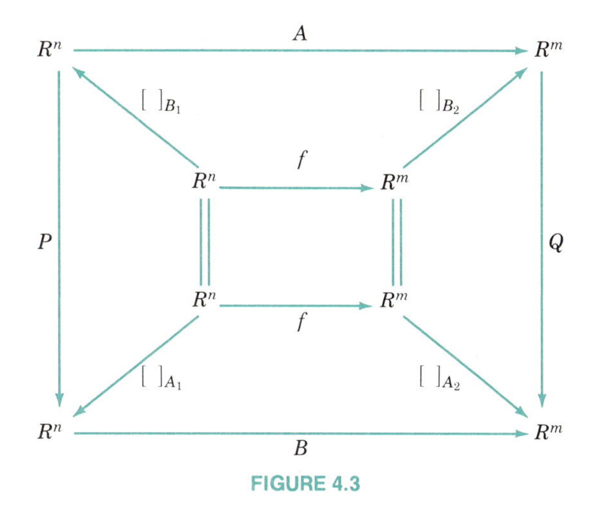

FIGURE 4.3

Example 4.4.9 Let $f: R^3 \to R^2$ be the linear transformation given by

$$f\left(\begin{bmatrix} x_1 \\ x_2 \\ x_3 \end{bmatrix}\right) = \begin{bmatrix} 3x_1 \\ x_1 + x_2 \end{bmatrix}$$

Let R^3 have bases

$$A_1 = \left\{\begin{bmatrix} 1 \\ 0 \\ 1 \end{bmatrix}, \begin{bmatrix} 1 \\ 0 \\ -1 \end{bmatrix}, \begin{bmatrix} 0 \\ -1 \\ 1 \end{bmatrix}\right\} \quad \text{and} \quad B_1 = \left\{\begin{bmatrix} 1 \\ -1 \\ 0 \end{bmatrix}, \begin{bmatrix} -1 \\ 0 \\ 1 \end{bmatrix}, \begin{bmatrix} 0 \\ 1 \\ 0 \end{bmatrix}\right\}$$

Moreover, let R^2 have bases

$$A_2 = \left\{\begin{bmatrix} 1 \\ 1 \end{bmatrix}, \begin{bmatrix} 1 \\ -1 \end{bmatrix}\right\} \quad \text{and} \quad B_2 = \left\{\begin{bmatrix} 3 \\ 1 \end{bmatrix}, \begin{bmatrix} 1 \\ -2 \end{bmatrix}\right\}$$

a. Find the matrix B representing f with respect to A_1 and A_2.
b. Find the matrix A representing f with respect to B_1 and B_2.
c. Find the B_1 to A_1 change of basis matrix P.
d. Find the B_2 to A_2 change of basis matrix Q.
e. Verify that $Q^{-1}BP = A$.

Solution a. $f(\mathbf{a}_1) = f\left(\begin{bmatrix} 1 \\ 0 \\ 1 \end{bmatrix}\right) = \begin{bmatrix} 3 \\ 1 \end{bmatrix} = a\begin{bmatrix} 1 \\ 1 \end{bmatrix} + b\begin{bmatrix} 1 \\ -1 \end{bmatrix}$

Thus

$$a + b = 3$$
$$a - b = 1$$

and $a = 2$, $b = 1$, so

$$[f(\mathbf{a}_1)]_{A_2} = \begin{bmatrix} 2 \\ 1 \end{bmatrix}$$

Similarly

$$f(\mathbf{a}_2) = f\left(\begin{bmatrix} 1 \\ 0 \\ -1 \end{bmatrix}\right) = \begin{bmatrix} 3 \\ 1 \end{bmatrix} = a\begin{bmatrix} 1 \\ 1 \end{bmatrix} + b\begin{bmatrix} 1 \\ -1 \end{bmatrix}$$

so

$$[f(\mathbf{a}_2)]_{A_2} = \begin{bmatrix} 2 \\ 1 \end{bmatrix}$$

Now

$$f(\mathbf{a}_3) = f\left(\begin{bmatrix} 0 \\ -1 \\ 1 \end{bmatrix}\right) = \begin{bmatrix} 0 \\ -1 \end{bmatrix} = a\begin{bmatrix} 1 \\ 1 \end{bmatrix} + b\begin{bmatrix} 1 \\ -1 \end{bmatrix}$$

thus

$$\begin{aligned} a + b &= 0 \\ a - b &= -1 \end{aligned}$$

and $a = -\frac{1}{2}$, $b = \frac{1}{2}$, so

$$[f(\mathbf{a}_3)]_{A_2} = \begin{bmatrix} -\frac{1}{2} \\ \frac{1}{2} \end{bmatrix}$$

Hence

$$B = \begin{bmatrix} 2 & 2 & -\frac{1}{2} \\ 1 & 1 & \frac{1}{2} \end{bmatrix}$$

b. $f(\mathbf{b}_1) = f\left(\begin{bmatrix} 1 \\ -1 \\ 0 \end{bmatrix}\right) = \begin{bmatrix} 3 \\ 0 \end{bmatrix} = a\begin{bmatrix} 3 \\ 1 \end{bmatrix} + b\begin{bmatrix} 1 \\ -2 \end{bmatrix}$

Thus

$$\begin{aligned} 3a + b &= 3 \\ a - 2b &= 0 \end{aligned}$$

and $a = \frac{6}{7}$, $b = \frac{3}{7}$, so

$$[f(\mathbf{b}_1)]_{B_2} = \begin{bmatrix} \frac{6}{7} \\ \frac{3}{7} \end{bmatrix}$$

Similarly

$$f(\mathbf{b}_2) = f\left(\begin{bmatrix} -1 \\ 0 \\ 1 \end{bmatrix}\right) = \begin{bmatrix} -3 \\ -1 \end{bmatrix} = a\begin{bmatrix} 3 \\ 1 \end{bmatrix} + b\begin{bmatrix} 1 \\ -2 \end{bmatrix}$$

Hence

$$\begin{aligned} 3a + b &= -3 \\ a - 2b &= -1 \end{aligned}$$

and $a = -1,\ b = 0$, so

$$[f(\mathbf{b}_2)]_{B_2} = \begin{bmatrix} -1 \\ 0 \end{bmatrix}$$

$$f(\mathbf{b}_3) = f\left(\begin{bmatrix} 0 \\ 1 \\ 0 \end{bmatrix}\right) = \begin{bmatrix} 0 \\ 1 \end{bmatrix} = a\begin{bmatrix} 3 \\ 1 \end{bmatrix} + b\begin{bmatrix} 1 \\ -2 \end{bmatrix}$$

Therefore,

$$\begin{aligned} 3a + b &= 0 \\ a - 2b &= 1 \end{aligned}$$

and $a = \frac{1}{7},\ b = -\frac{3}{7}$, so

$$[f(\mathbf{b}_3)]_{B_2} = \begin{bmatrix} \frac{1}{7} \\ -\frac{3}{7} \end{bmatrix}$$

Therefore,

$$A = \begin{bmatrix} \frac{6}{7} & -1 & \frac{1}{7} \\ \frac{3}{7} & 0 & -\frac{3}{7} \end{bmatrix}$$

c. To find P, we reduce

$$\left[\begin{array}{ccc|ccc} 1 & 1 & 0 & 1 & -1 & 0 \\ 0 & 0 & -1 & -1 & 0 & 1 \\ 1 & -1 & 1 & 0 & 1 & 0 \end{array}\right]$$

to obtain

$$\left[\begin{array}{ccc|ccc} 1 & 0 & 0 & 0 & 0 & \frac{1}{2} \\ 0 & 1 & 0 & 1 & -1 & -\frac{1}{2} \\ 0 & 0 & 1 & 1 & 0 & -1 \end{array}\right]$$

Thus

$$P = \begin{bmatrix} 0 & 0 & \frac{1}{2} \\ 1 & -1 & -\frac{1}{2} \\ 1 & 0 & -1 \end{bmatrix}$$

d. To find Q, we reduce

$$\left[\begin{array}{cc|cc} 1 & 1 & 3 & 1 \\ 1 & -1 & 1 & -2 \end{array}\right]$$

to obtain

$$\left[\begin{array}{cc|cc} 1 & 0 & 2 & -\frac{1}{2} \\ 0 & 1 & 1 & \frac{3}{2} \end{array}\right]$$

or

$$Q = \begin{bmatrix} 2 & -\frac{1}{2} \\ 1 & \frac{3}{2} \end{bmatrix}$$

e. First calculate Q^{-1}, obtaining

$$\begin{bmatrix} \frac{3}{7} & \frac{1}{7} \\ -\frac{2}{7} & \frac{4}{7} \end{bmatrix}$$

Thus

$$Q^{-1}BP = \begin{bmatrix} \frac{3}{7} & \frac{1}{7} \\ -\frac{2}{7} & \frac{4}{7} \end{bmatrix} \begin{bmatrix} 2 & 2 & -\frac{1}{2} \\ 1 & 1 & \frac{1}{2} \end{bmatrix} \begin{bmatrix} 0 & 0 & \frac{1}{2} \\ 1 & -1 & -\frac{1}{2} \\ 1 & 0 & -1 \end{bmatrix}$$

$$= \begin{bmatrix} 1 & 1 & -\frac{1}{7} \\ 0 & 0 & \frac{3}{7} \end{bmatrix} \begin{bmatrix} 0 & 0 & \frac{1}{2} \\ 1 & -1 & -\frac{1}{2} \\ 1 & 0 & -1 \end{bmatrix} = \begin{bmatrix} \frac{6}{7} & -1 & \frac{1}{7} \\ \frac{3}{7} & 0 & -\frac{3}{7} \end{bmatrix}$$

$$= A$$

■

SUMMARY

Terms to know:

Change of basis matrix
Similar matrices

Theorems to know:

If P is the A_1 to A_2 change of basis matrix, then P^{-1} is the A_2 to A_1 change of basis matrix.

If A and B are $m \times n$ matrices for which $A = Q^{-1}BP$ where P and Q are invertible, then A and B represent the same linear transformation $f: R^n \to R^m$.

EXERCISE SET 4.4

In Exercises 1–10, for the given vector space V and bases A and B of V, find the A to B change of basis matrix P.

1. $V = R^3$, $A = \left\{\begin{bmatrix}1\\-1\\1\end{bmatrix}, \begin{bmatrix}1\\0\\1\end{bmatrix}, \begin{bmatrix}1\\3\\2\end{bmatrix}\right\}$, $B = \left\{\begin{bmatrix}1\\1\\1\end{bmatrix}, \begin{bmatrix}1\\2\\3\end{bmatrix}, \begin{bmatrix}0\\1\\0\end{bmatrix}\right\}$

2. $V = P_2$, $A = \{x, 1 + x, x + x^2\}$, $B = \{x - x^2, 1 + x, x^2 - 1\}$

3. $V = P_2$, $A = \{x - x^2, 1 + x, x^2 - 1\}$, $B = \{x, x + x^2, 1 + x\}$

4. $V = P_2$, $A = \{x^2 + 1, x^2 + x, x + 1\}$, $B = \{1 + x, x^2 + 1, x^2 + x\}$

5. $V = R^3$, $A = \left\{\begin{bmatrix}1\\-1\\1\end{bmatrix}, \begin{bmatrix}0\\2\\1\end{bmatrix}, \begin{bmatrix}1\\0\\-1\end{bmatrix}\right\}$, $B = \left\{\begin{bmatrix}-1\\1\\0\end{bmatrix}, \begin{bmatrix}1\\-3\\2\end{bmatrix}, \begin{bmatrix}1\\0\\1\end{bmatrix}\right\}$

6. $V = R^2$, $A = \left\{\begin{bmatrix}1\\0\end{bmatrix}, \begin{bmatrix}0\\1\end{bmatrix}\right\}$, $B = \left\{\begin{bmatrix}2\\1\end{bmatrix}, \begin{bmatrix}-3\\4\end{bmatrix}\right\}$

7. $V = R^3$, $A = \left\{\begin{bmatrix}-3\\0\\-3\end{bmatrix}, \begin{bmatrix}-3\\2\\-1\end{bmatrix}, \begin{bmatrix}1\\6\\-1\end{bmatrix}\right\}$, $B_2 = \left\{\begin{bmatrix}3\\1\\-5\end{bmatrix}, \begin{bmatrix}1\\1\\-3\end{bmatrix}, \begin{bmatrix}-1\\0\\2\end{bmatrix}\right\}$

8. $V = R^3$, $A = \left\{\begin{bmatrix}-6\\-6\\0\end{bmatrix}, \begin{bmatrix}-2\\-6\\4\end{bmatrix}, \begin{bmatrix}-2\\-3\\7\end{bmatrix}\right\}$, $B = \left\{\begin{bmatrix}2\\1\\1\end{bmatrix}, \begin{bmatrix}2\\-1\\1\end{bmatrix}, \begin{bmatrix}1\\2\\1\end{bmatrix}\right\}$

9. $V = R^3$, $A = \left\{\begin{bmatrix}1\\-1\\1\end{bmatrix}, \begin{bmatrix}1\\0\\1\end{bmatrix}, \begin{bmatrix}1\\3\\2\end{bmatrix}\right\}$, $B = \left\{\begin{bmatrix}3\\1\\-5\end{bmatrix}, \begin{bmatrix}1\\1\\-3\end{bmatrix}, \begin{bmatrix}-1\\0\\2\end{bmatrix}\right\}$

10. $V = R_{2\times 2}$, $A = \left\{\begin{bmatrix}1 & 1\\0 & 1\end{bmatrix}, \begin{bmatrix}1 & 1\\1 & 0\end{bmatrix}, \begin{bmatrix}1 & 0\\1 & 1\end{bmatrix}, \begin{bmatrix}0 & 1\\1 & 1\end{bmatrix}\right\}$,

$B = \left\{\begin{bmatrix}1 & 0\\0 & 1\end{bmatrix}, \begin{bmatrix}0 & 1\\1 & 0\end{bmatrix}, \begin{bmatrix}0 & 0\\0 & 1\end{bmatrix}, \begin{bmatrix}0 & -1\\0 & 0\end{bmatrix}\right\}$

11. If the invertible matrices A and B are similar, prove that A^{-1} and B^{-1} are similar.

12. If matrices A and B are similar, prove that A^2 and B^2 are similar.

13. If matrices A and B are similar, prove that A^n and B^n are similar for any integer n.

14. Let $f: R^2 \to R^3$ be given by

$$f\left(\begin{bmatrix} x_1 \\ x_2 \end{bmatrix}\right) = \begin{bmatrix} 3x_1 - x_2 \\ x_2 \\ x_1 + x_2 \end{bmatrix}$$

In Example 4.3.3, we found the matrix M of f with respect to bases

$$A_1 = \left\{\begin{bmatrix} -1 \\ 1 \end{bmatrix}, \begin{bmatrix} 2 \\ 1 \end{bmatrix}\right\} \quad \text{and} \quad B_1 = \left\{\begin{bmatrix} 1 \\ -1 \\ 1 \end{bmatrix}, \begin{bmatrix} -1 \\ 1 \\ 1 \end{bmatrix}, \begin{bmatrix} 1 \\ 0 \\ 1 \end{bmatrix}\right\}$$

Let

$$A_2 = \left\{\begin{bmatrix} 0 \\ -1 \end{bmatrix}, \begin{bmatrix} 1 \\ 1 \end{bmatrix}\right\} \quad \text{and} \quad B_2 = \left\{\begin{bmatrix} 1 \\ 1 \\ 1 \end{bmatrix}, \begin{bmatrix} 3 \\ 1 \\ 2 \end{bmatrix}, \begin{bmatrix} 0 \\ 1 \\ 0 \end{bmatrix}\right\}$$

a. Find the matrix representation N of f with respect to A_2 and B_2.

b. Find the A_1 to A_2 change of basis matrix P.

c. Find the B_1 to B_2 change of basis matrix Q.

d. Verify that $Q^{-1}NP = M$.

15. In Example 4.3.5, we found the matrix M representing the linear transformation $f: P_2 \to R_{2\times 2}$ given by

$$f(p_0 + p_1x + p_2x^2) = \begin{bmatrix} p_0 + p_1 & p_1 + p_2 \\ p_0 + p_2 & p_2 \end{bmatrix}$$

with respect to the bases $A_1 = \{1, x, x^2\}$ of P_2 and

$$B_1 = \left\{\begin{bmatrix} 1 & 0 \\ 0 & 1 \end{bmatrix}, \begin{bmatrix} 0 & 1 \\ 0 & 1 \end{bmatrix}, \begin{bmatrix} 1 & 1 \\ 1 & 0 \end{bmatrix}, \begin{bmatrix} 0 & 0 \\ 0 & 1 \end{bmatrix}\right\}$$

of $R_{2\times 2}$. Let $A_2 = \{1 - x + x^2, 1 + x^2, 1 + 3x + 2x^2\}$ be a basis of P_2 and

$$B_2 = \left\{\begin{bmatrix} 1 & 1 \\ 0 & 1 \end{bmatrix}, \begin{bmatrix} 1 & 1 \\ 1 & 0 \end{bmatrix}, \begin{bmatrix} 1 & 0 \\ 1 & 1 \end{bmatrix}, \begin{bmatrix} 0 & 1 \\ 1 & 1 \end{bmatrix}\right\}$$

be a basis of $R_{2\times 2}$.

a. Find the matrix N representing f with respect to A_2 and B_2.

b. Find the A_1 to A_2 change of basis matrix P.

c. Find the B_1 to B_2 change of basis matrix Q.

d. Verify that $Q^{-1}NP = M$.

16. What does Theorem 4.4.8 say about similar matrices?

17. If A is an $n \times n$ matrix and $A = P^{-1}BP$, what can be said about the linear transformations represented by A and B?

SECTION 4.5 THE IMAGE OF A LINEAR TRANSFORMATION

There are two important vector spaces associated with a linear transformation $f: U \to V$. One, a subspace of V, is considered in this section and the other, a subspace of U, is taken up in the next section.

Definition 4.5.1 The **image** of a linear transformation $f: U \to V$, denoted by im(f), is the set of all vectors $\mathbf{v}$ in V for which the equation $f(\mathbf{u}) = \mathbf{v}$ has a solution in U. Using set notation,

$$\text{im}(f) = \{\mathbf{v} \in V \mid \mathbf{v} = f(\mathbf{u}) \text{ for some } \mathbf{u} \in U\}$$

Example 4.5.2 Let $f: R^3 \to R^2$ be defined by

$$f\left(\begin{bmatrix} x_1 \\ x_2 \\ x_3 \end{bmatrix}\right) = \begin{bmatrix} x_1 - x_3 \\ 2x_2 - x_3 \end{bmatrix}$$

Does

$$\begin{bmatrix} 2 \\ -3 \end{bmatrix}$$

belong to im(f)?

Solution In order for

$$\begin{bmatrix} 2 \\ -3 \end{bmatrix}$$

to belong to the im(f), there needs to be at least one

$$\begin{bmatrix} x_1 \\ x_2 \\ x_3 \end{bmatrix}$$

in R^3 such that

$$f\left(\begin{bmatrix} x_1 \\ x_2 \\ x_3 \end{bmatrix}\right) = \begin{bmatrix} x_1 - x_3 \\ 2x_2 - x_3 \end{bmatrix} = \begin{bmatrix} 2 \\ -3 \end{bmatrix}$$

In other words, the corresponding system

$$\begin{aligned} x_1 \qquad - x_3 &= 2 \\ 2x_2 - x_3 &= -3 \end{aligned}$$

needs to be solvable for x_1, x_2, and x_3. The vector

$$\begin{bmatrix} 2 \\ -3 \end{bmatrix}$$

belongs to im(f) because one solution of the system is $x_1 = 3$, $x_2 = -1$, $x_3 = 1$, since

$$f\left(\begin{bmatrix} 3 \\ -1 \\ 1 \end{bmatrix}\right) = \begin{bmatrix} 3 - (1) \\ 2(-1) - 1 \end{bmatrix} = \begin{bmatrix} 2 \\ -3 \end{bmatrix}$$

■

Example 4.5.3 The matrix

$$A = \begin{bmatrix} 3 & 1 \\ 4 & 2 \\ 1 & 0 \end{bmatrix}$$

gives a linear transformation $f: R^2 \to R^3$ by $f(\mathbf{x}) = A\mathbf{x}$. The vector

$$\begin{bmatrix} 3 \\ 4 \\ 1 \end{bmatrix} \in \text{im}(f)$$

since the vector

$$\begin{bmatrix} 1 \\ 0 \end{bmatrix}$$

is in R^2 and

$$A\begin{bmatrix} 1 \\ 0 \end{bmatrix} = \begin{bmatrix} 3 & 1 \\ 4 & 2 \\ 1 & 0 \end{bmatrix}\begin{bmatrix} 1 \\ 0 \end{bmatrix} = \begin{bmatrix} 3 \\ 4 \\ 1 \end{bmatrix}$$

On the other hand, the vector

$$\begin{bmatrix} 4 \\ 6 \\ 2 \end{bmatrix} \notin \text{im}(f)$$

since there is no vector

$$\begin{bmatrix} x_1 \\ x_2 \end{bmatrix}$$

in R^2 such that

$$A\begin{bmatrix} x_1 \\ x_2 \end{bmatrix} = \begin{bmatrix} 3 & 1 \\ 4 & 2 \\ 1 & 0 \end{bmatrix}\begin{bmatrix} x_1 \\ x_2 \end{bmatrix} = \begin{bmatrix} 3x_1 + x_2 \\ 4x_1 + 2x_2 \\ x_1 \end{bmatrix} = \begin{bmatrix} 4 \\ 6 \\ 2 \end{bmatrix}$$

The last row says $x_1 = 2$, and for this value of x_1, $3x_1 + x_2 = 4$ yields $x_2 = -2$. However, using the second row, $4(2) + 2(-2) \neq 6$. ■

Theorem 4.5.4 tells us that $\text{im}(f)$ is a subspace of U.

Theorem 4.5.4 *If $f: U \to V$ is a linear transformation, then* $\text{im}(f)$ *is a subspace of V.*

Proof. First, $\mathbf{0} \in V$ is in $\text{im}(f)$, since $\mathbf{0} \in U$ and $f(\mathbf{0}) = \mathbf{0}$. Let $\mathbf{v}_1, \mathbf{v}_2 \in \text{im}(f)$; then for some $\mathbf{u}_1 \in U$, $f(\mathbf{u}_1) = \mathbf{v}_1$. Likewise, for some $\mathbf{u}_2 \in U$, $f(\mathbf{u}_2) = \mathbf{v}_2$. To show that $\text{im}(f)$ is a subspace of V, we must show for arbitrary $a, b \in R$ that $a\mathbf{v}_1 + b\mathbf{v}_2 \in \text{im}(f)$. But f is a linear transformation, so that

$$a\mathbf{v}_1 + b\mathbf{v}_2 = af(\mathbf{u}_1) + bf(\mathbf{u}_2) = f(a\mathbf{u}_1) + f(b\mathbf{u}_2) = f(a\mathbf{u}_1 + b\mathbf{u}_2)$$

Since $a\mathbf{u}_1 + b\mathbf{u}_2 \in U$, then $f(a\mathbf{u}_1 + b\mathbf{u}_2) = a\mathbf{v}_1 + b\mathbf{v}_2 \in \text{im}(f)$. Therefore, $\text{im}(f)$ is a subspace of V.

How do we find $\text{im}(f)$ for arbitrary linear transformations $f: U \to V$? Since $\text{im}(f)$ is a subspace of V, it has a basis. We will find a basis for $\text{im}(f)$. First choose a basis for A for U and B for V. (Any basis will do!) Then we represent f with respect to A and B by the matrix M. If $\dim(U) = n$ and $\dim(V) = m$, then M is an $m \times n$ matrix.

To find $\text{im}(f)$, we find those vectors $\mathbf{y} \in R^m$ for which $M\mathbf{x} = \mathbf{y}$ can be solved for $\mathbf{x}$. Then $\mathbf{v} \in \text{im}(f)$ if and only if $M\mathbf{x} = \mathbf{y}$ where $\mathbf{y} = [\mathbf{v}]_B$. As is common in mathematics, we have changed the problem, "Find $\text{im}(f)$" to the more familiar problem "For which $\mathbf{y} \in R^m$ does $M\mathbf{x} = \mathbf{y}$ have a solution?"

If $M\mathbf{x} = \mathbf{y}$, where

$$\mathbf{x} = \begin{bmatrix} x_1 \\ x_2 \\ \vdots \\ x_n \end{bmatrix} \quad \text{and} \quad \mathbf{y} = \begin{bmatrix} y_1 \\ y_2 \\ \vdots \\ y_m \end{bmatrix}$$

then $M\mathbf{x} = \mathbf{y}$ can be written as

$$\begin{bmatrix} a_{11} & a_{12} & \cdots & a_{1n} \\ a_{21} & a_{22} & \cdots & a_{2n} \\ \vdots & \vdots & & \vdots \\ a_{m1} & a_{m2} & \cdots & a_{mn} \end{bmatrix} \begin{bmatrix} x_1 \\ x_2 \\ \vdots \\ x_n \end{bmatrix} = \begin{bmatrix} y_1 \\ y_2 \\ \vdots \\ y_m \end{bmatrix}$$

$$\begin{bmatrix} a_{11}x_1 + a_{12}x_2 + \cdots + a_{1n}x_n \\ a_{21}x_1 + a_{22}x_2 + \cdots + a_{2n}x_n \\ \vdots \qquad \vdots \qquad\qquad \vdots \\ a_{m1}x_1 + a_{m2}x_2 + \cdots + a_{mn}x_n \end{bmatrix} = \begin{bmatrix} y_1 \\ y_2 \\ \vdots \\ y_m \end{bmatrix}$$

or as

$$x_1\begin{bmatrix} a_{11} \\ a_{21} \\ \vdots \\ a_{m1} \end{bmatrix} + x_2\begin{bmatrix} a_{12} \\ a_{22} \\ \vdots \\ a_{m2} \end{bmatrix} + \cdots + x_n\begin{bmatrix} a_{1n} \\ a_{2n} \\ \vdots \\ a_{mn} \end{bmatrix} = \begin{bmatrix} y_1 \\ y_2 \\ \vdots \\ y_m \end{bmatrix}$$

The vectors in the left-hand side of the last equation are the column vectors of the matrix M. Thus $\mathbf{y} \in \operatorname{im}(f)$ if and only if $[\mathbf{y}]_B$ is a linear combination of the columns of M. This leads to the following definition.

Definition 4.5.5 The subspace

$$\left\langle \begin{bmatrix} a_{11} \\ a_{21} \\ \vdots \\ a_{m1} \end{bmatrix}, \begin{bmatrix} a_{12} \\ a_{22} \\ \vdots \\ a_{m2} \end{bmatrix}, \ldots, \begin{bmatrix} a_{1n} \\ a_{2n} \\ \vdots \\ a_{mn} \end{bmatrix} \right\rangle$$

of R^m is called the **column space of the matrix**

$$M = \begin{bmatrix} a_{11} & a_{12} & \cdots & a_{1n} \\ a_{21} & a_{22} & \cdots & a_{2n} \\ \vdots & \vdots & & \vdots \\ a_{m1} & a_{m2} & \cdots & a_{mn} \end{bmatrix}$$

and is denoted by $\mathrm{CS}(M)$.

The following result is merely a restatement of the last few paragraphs.

Theorem 4.5.6 *Let M be an $m \times n$ matrix and $\mathbf{y} \in R^m$. Then $M\mathbf{x} = \mathbf{y}$ has a solution $\mathbf{x} \in R^n$ if and only if $\mathbf{y} \in \mathrm{CS}(M)$.*

Example 4.5.7 Let

$$M = \begin{bmatrix} 1 & 3 & 1 \\ 2 & 1 & 4 \end{bmatrix} \quad \text{and} \quad \mathbf{y} = \begin{bmatrix} 1 \\ 5 \end{bmatrix}$$

Is $\mathbf{y} \in \mathrm{CS}(M)$?

Solution We are asking whether $M\mathbf{x} = \mathbf{y}$ can be solved for

$$\mathbf{x} = \begin{bmatrix} x_1 \\ x_2 \\ x_3 \end{bmatrix}$$

But, using the methods of Chapter 1, we see that

$$\mathbf{x} = \begin{bmatrix} -\frac{1}{2} \\ 0 \\ \frac{3}{2} \end{bmatrix}$$

is a solution, and so the answer is *yes*. ■

Similarly, we define the row space of a matrix M as follows.

Definition 4.5.8 The **row space of an $m \times n$ matrix** M is the subspace of R^n spanned by the rows of M and is denoted by $\mathrm{RS}(M)$.

For the matrix

$$M = \begin{bmatrix} 1 & 3 & 1 \\ 2 & 1 & 4 \end{bmatrix}, \qquad \mathrm{CS}(M) = \left\langle \begin{bmatrix} 1 \\ 2 \end{bmatrix}, \begin{bmatrix} 3 \\ 1 \end{bmatrix}, \begin{bmatrix} 1 \\ 4 \end{bmatrix} \right\rangle$$

Notice that

$$M^T = \begin{bmatrix} 1 & 2 \\ 3 & 1 \\ 1 & 4 \end{bmatrix}$$

and that $\mathrm{RS}(M^T) = \mathrm{CS}(M)$.

This holds in general.

Theorem 4.5.9 *Given an $m \times n$ matrix M, then* $\mathrm{CS}(M) = \mathrm{RS}(M^T)$.

The proof follows from the fact that the rows of M^T are the columns of M. Thus, Theorem 4.5.9 says that to find a basis for $\mathrm{CS}(M)$, we can merely find a basis for $\mathrm{RS}(M^T)$. And, in general, given a linear transformation $f: U \to V$, to find a basis for $\mathrm{im}(f)$, we can proceed as follows:

1. Choose a basis A for U.
2. Choose a basis B for V.
3. Find the matrix M representing f with respect to A and B.

trarily chosen and $x_1 = x_3$ and $x_2 = -x_3$. Consequently,

$$\mathrm{NS}(f) = \left\{ \begin{bmatrix} x_1 \\ x_2 \\ x_3 \end{bmatrix} \in R^3 \,\middle|\, x_1 = x_3,\ x_2 = -x_3 \text{ and } x_3 \text{ arbitrary} \right\}$$

$$= \left\{ \begin{bmatrix} x_3 \\ -x_3 \\ x_3 \end{bmatrix} \,\middle|\, x_3 \text{ arbitrary} \right\}$$

$$= \left\{ x_3 \begin{bmatrix} 1 \\ -1 \\ 1 \end{bmatrix} \,\middle|\, x_3 \text{ arbitrary} \right\} = \left\langle \begin{bmatrix} 1 \\ -1 \\ 1 \end{bmatrix} \right\rangle$$

c. Since $\mathrm{NS}(f)$ is the set of all vectors in R such that

$$f(x) = 5x = 0$$

we have $x \in \mathrm{NS}(f)$ if and only if $5x = 0$, or $x = 0$. Thus

$$\mathrm{NS}(f) = \{x \in R \mid x = 0\} = \{\mathbf{0}\}$$

d. We have

$$f\left(\begin{bmatrix} a & b \\ c & d \end{bmatrix}\right) = \begin{bmatrix} b+c & a+d \\ a+c & b+d \end{bmatrix} = \begin{bmatrix} 0 & 0 \\ 0 & 0 \end{bmatrix}$$

and so $b = -c$, $a = -d$, $a = -c$ and $b = -d$. Thus

$$\begin{bmatrix} a & b \\ c & d \end{bmatrix} = \begin{bmatrix} a & a \\ -a & -a \end{bmatrix} = a\begin{bmatrix} 1 & 1 \\ -1 & -1 \end{bmatrix}$$

and

$$\mathrm{NS}(f) = \left\langle \begin{bmatrix} 1 & 1 \\ -1 & -1 \end{bmatrix} \right\rangle$$

e. Since

$$f(p_0 + p_1x + p_2x^2) = \begin{bmatrix} p_0 + p_1 & p_0 + p_2 \\ p_1 + p_2 & p_2 \end{bmatrix} = \begin{bmatrix} 0 & 0 \\ 0 & 0 \end{bmatrix}$$

we have $p_0 + p_1 = 0$, $p_0 + p_2 = 0$, $p_1 + p_2 = 0$, and $p_2 = 0$. Solving this system, we have $p_2 = p_1 = p_0 = 0$. Thus $\mathrm{NS}(f) = \{p_0 + p_1x + p_2x^2 \in P_2 \mid p_0 = p_1 = p_2 = 0\} = \{\mathbf{0}\}$. ■

The null space of an $m \times n$ matrix A is given by

$$\mathrm{NS}(A) = \{\mathbf{x} \in R^n \mid A\mathbf{x} = \mathbf{0}\}$$

Thus NS(A) is the set of all solutions of the homogeneous system of equations

$$\begin{aligned} a_{11}x_1 + a_{12}x_2 + \cdots + a_{1n}x_n &= 0 \\ a_{21}x_1 + a_{22}x_2 + \cdots + a_{2n}x_n &= 0 \\ \vdots \qquad \vdots \qquad\quad \vdots \\ a_{m1}x_1 + a_{m2}x_2 + \cdots + a_{mn}x_n &= 0 \end{aligned}$$

We have already verified that the solution set of a homogeneous system is a subspace of R^n. This result is contained in the next theorem.

Theorem 4.6.3 *Let $f: U \to V$ be a linear transformation. Then NS(f) is a subspace of U.*

Proof. First $\mathbf{0} \in$ NS(f), since $f(\mathbf{0}) = \mathbf{0}$; hence NS($f$) is not empty.

Let $\mathbf{u}_1, \mathbf{u}_2 \in$ NS(f) and $a, b \in R$. Then we must show that $a\mathbf{u}_1 + b\mathbf{u}_2 \in$ NS(f), that is, that $f(a\mathbf{u}_1 + b\mathbf{u}_2) = \mathbf{0}$. But since f is a linear transformation,

$$f(a\mathbf{u}_1 + b\mathbf{u}_2) = f(a\mathbf{u}_1) + f(b\mathbf{u}_2) = af(\mathbf{u}_1) + bf(\mathbf{u}_2) = a\mathbf{0} + b\mathbf{0} = \mathbf{0}$$

Therefore, $a\mathbf{u}_1 + b\mathbf{u}_2 \in$ NS(f).

How do we find a basis for the subspace NS(f)? We choose a basis A for U and a basis B for V. We represent f by the $m \times n$ matrix M, determine NS(M), and use that to find a basis for NS(f).

First, NS(M) is nonempty, since $M\mathbf{0} = \mathbf{0}$, so $\mathbf{0} \in$ NS(M). If this is the only vector in NS(M), then we are finished, since $\{\mathbf{0}\}$ has no basis. Are there NS(M) that contain nonzero vectors? Example 4.6.2(b) gave an example of a matrix M with NS(M) containing nonzero vectors. The existence of such M can also be seen from Theorem 1.3.6, in which we showed that if there are more unknowns than equations ($n > m$), then the system

$$\begin{aligned} a_{11}x_1 + \cdots + a_{1n}x_n &= 0 \\ a_{21}x_1 + \cdots + a_{2n}x_n &= 0 \\ \vdots \qquad\quad \vdots \\ a_{m1}x_1 + \cdots + a_{mn}x_n &= 0 \end{aligned}$$

always has nonzero solutions. To obtain such solutions, we use elementary row operations to reduce M to row echelon form, E. Setting the x_i corresponding to columns not containing a leading row entry (i.e., free variable) equal to nonzero numbers and solving for the remaining

x_i's (pivot variables) in terms of the pivoting free variables yields a nonzero vector $\mathbf{x} \in \text{NS}(M)$. We state this result as a theorem.

Theorem 4.6.4 *Let A be an $m \times n$ matrix. If $m < n$, then there are nonzero vectors in* $\text{NS}(A)$.

In fact, the system $A\mathbf{x} = \mathbf{0}$ has infinitely many solutions when $m < n$.

To determine a basis for $\text{NS}(A)$, we solve $A\mathbf{x} = \mathbf{0}$ as we did in Chapter 1.

Example 4.6.5 Let

$$A = \begin{bmatrix} 1 & 2 \\ -3 & -4 \\ 8 & 12 \end{bmatrix}$$

Find a basis for $\text{NS}(A)$.

Solution Since $m = 3$ and $n = 2$, $m > n$ and there may or may not be nonzero vectors in $\text{NS}(A)$. Applying row operations to A yields the echelon form

$$\left[\begin{array}{cc|c} 1 & 2 & 0 \\ 0 & 2 & 0 \\ 0 & 0 & 0 \end{array}\right]$$

Thus if

$$\mathbf{x} = \begin{bmatrix} x_1 \\ x_2 \end{bmatrix} \in \text{NS}(A)$$

then

$$x_1 + 2x_2 = 0$$
$$2x_2 = 0$$

or $x_1 = x_2 = 0$. $\text{NS}(A)$ contains only the zero vector

$$\mathbf{0} = \begin{bmatrix} 0 \\ 0 \end{bmatrix}$$ ■

Example 4.6.6 Let

$$A = \begin{bmatrix} 1 & -3 \\ -2 & 6 \\ -1 & 3 \end{bmatrix}$$

Find a basis for $\text{NS}(A)$.

Solution Again $m = 3$ and $n = 2$ $(m > n)$, so that we do not know whether NS(A) contains nonzero vectors. A sequence of row operations applied to

$$\left[\begin{array}{rr|r} 1 & -3 & 0 \\ -2 & 6 & 0 \\ -1 & 3 & 0 \end{array}\right]$$

results in

$$\left[\begin{array}{rr|r} 1 & -3 & 0 \\ 0 & 0 & 0 \\ 0 & 0 & 0 \end{array}\right]$$

If

$$\mathbf{x} = \begin{bmatrix} x_1 \\ x_2 \end{bmatrix} \in \text{NS}(A)$$

then $x_1 - 3x_2 = 0$. Thus $x_1 = 3x_2$, and so, $\mathbf{x} \in \text{NS}(A)$ if and only if

$$\mathbf{x} = \begin{bmatrix} x_1 \\ x_2 \end{bmatrix} = \begin{bmatrix} 3x_2 \\ x_2 \end{bmatrix} = x_2 \begin{bmatrix} 3 \\ 1 \end{bmatrix}$$

where x_2 is arbitrary. Thus

$$\left\{ \begin{bmatrix} 3 \\ 1 \end{bmatrix} \right\}$$

is a basis for NS(A). ■

Example 4.6.7 Let $f: R^4 \to R^3$ be a linear transformation given by

$$f\left(\begin{bmatrix} x_1 \\ x_2 \\ x_3 \\ x_4 \end{bmatrix}\right) = \begin{bmatrix} x_1 + x_2 + x_3 + x_4 \\ x_2 + x_4 \\ x_1 + x_3 + x_4 \end{bmatrix}$$

Find a basis for NS(f).

Solution Choosing the standard basis of R^4 and R^3 for A and B, f is represented by

$$A = \begin{bmatrix} 1 & 1 & 1 & 1 \\ 0 & 1 & 0 & 1 \\ 1 & 0 & 1 & 1 \end{bmatrix}$$

We wish to find NS(A).

This matrix has $m = 3$ rows and $n = 4$ columns $(n > m)$, so we know that there are nonzero vectors in NS(A).

Applying row transformations to

$$\left[\begin{array}{cccc|c} 1 & 1 & 1 & 1 & 0 \\ 0 & 1 & 0 & 1 & 0 \\ 1 & 0 & 1 & 1 & 0 \end{array}\right]$$

results in

$$\left[\begin{array}{cccc|c} 1 & 1 & 1 & 1 & 0 \\ 0 & 1 & 0 & 1 & 0 \\ 0 & 0 & 0 & 1 & 0 \end{array}\right]$$

If

$$\mathbf{x} = \begin{bmatrix} x_1 \\ x_2 \\ x_3 \\ x_4 \end{bmatrix} \in \mathrm{NS}(A)$$

then

$$\begin{aligned} x_1 + x_2 + x_3 + x_4 &= 0 \\ x_2 \qquad + x_4 &= 0 \\ x_4 &= 0 \end{aligned}$$

Solving this system yields $x_4 = 0$, $x_2 = 0$, and $x_1 = -x_3$ for x_3 arbitrary. Thus

$$\mathbf{x} = \begin{bmatrix} x_1 \\ x_2 \\ x_3 \\ x_4 \end{bmatrix} = \begin{bmatrix} -x_3 \\ 0 \\ x_3 \\ 0 \end{bmatrix} = x_3 \begin{bmatrix} -1 \\ 0 \\ 1 \\ 0 \end{bmatrix}$$

where x_3 is an arbitrary number. Since every nonzero vector in $\mathrm{NS}(A)$ is obtained by multiplying

$$\begin{bmatrix} -1 \\ 0 \\ 1 \\ 0 \end{bmatrix}$$

by a scalar, then

$$\left\{ \begin{bmatrix} -1 \\ 0 \\ 1 \\ 0 \end{bmatrix} \right\}$$

is a basis for $\mathrm{NS}(f)$. ■

REMARK We could also have solved this system for x_1, obtaining $x_4 = 0$, $x_2 = 0$, and $x_3 = -x_1$. These give the basis

$$\left\{\begin{bmatrix} 1 \\ 0 \\ -1 \\ 0 \end{bmatrix}\right\}$$

Example 4.6.8 Find the null space of

$$A = \begin{bmatrix} 1 & 1 & 0 & 1 & 1 \\ 1 & 0 & 1 & 0 & 1 \\ 0 & 1 & -1 & 1 & 1 \end{bmatrix}$$

and determine a basis for NS(A).

Solution Here $m = 3$ and $n = 5$. Thus $m < n$, and there are nonzero vectors in NS(A). We apply row operations to

$$\left[\begin{array}{ccccc|c} 1 & 1 & 0 & 1 & 1 & 0 \\ 1 & 0 & 1 & 0 & 1 & 0 \\ 0 & 1 & -1 & 1 & 1 & 0 \end{array}\right]$$

to obtain

$$\left[\begin{array}{ccccc|c} 1 & 1 & 0 & 1 & 1 & 0 \\ 0 & -1 & 1 & -1 & 0 & 0 \\ 0 & 0 & 0 & 0 & 1 & 0 \end{array}\right]$$

If

$$\mathbf{x} = \begin{bmatrix} x_1 \\ x_2 \\ x_3 \\ x_4 \\ x_5 \end{bmatrix} \in \mathrm{NS}(A)$$

then

$$\begin{aligned} x_1 + x_2 \qquad + x_4 + x_5 &= 0 \\ - x_2 + x_3 - x_4 \qquad &= 0 \\ x_5 &= 0 \end{aligned}$$

Since $x_5 = 0$, this system reduces to

$$\begin{aligned} x_1 + x_2 + \qquad x_4 &= 0 \\ - x_2 + x_3 - x_4 &= 0 \end{aligned}$$

Backsolving, we obtain

$$\begin{aligned} x_1 &= -x_3 \\ x_2 &= \quad x_3 - x_4 \end{aligned}$$

where x_3 and x_4 are arbitrarily chosen. Thus

$$\mathbf{x} = \begin{bmatrix} x_1 \\ x_2 \\ x_3 \\ x_4 \\ x_5 \end{bmatrix} = \begin{bmatrix} -x_3 \\ x_3 - x_4 \\ x_3 \\ x_4 \\ 0 \end{bmatrix} = \begin{bmatrix} -x_3 \\ x_3 \\ x_3 \\ 0 \\ 0 \end{bmatrix} + \begin{bmatrix} 0 \\ -x_4 \\ 0 \\ x_4 \\ 0 \end{bmatrix} = x_3 \begin{bmatrix} -1 \\ 1 \\ 1 \\ 0 \\ 0 \end{bmatrix} + x_4 \begin{bmatrix} 0 \\ -1 \\ 0 \\ -1 \\ 0 \end{bmatrix}$$

The set

$$\left\{ \begin{bmatrix} -1 \\ 1 \\ 1 \\ 0 \\ 0 \end{bmatrix}, \begin{bmatrix} 0 \\ -1 \\ 0 \\ 1 \\ 0 \end{bmatrix} \right\}$$

is linearly independent and spans NS(A), so it is a basis for NS(A). ■

If the null space of the $m \times n$ matrix A contains only the vector $\mathbf{0}$, then it has no basis. However, if the null space of A contains nonzero vectors, then its basis has, at most, n vectors, each of which is a solution of $A\mathbf{x} = \mathbf{0}$. Every solution vector of $A\mathbf{x} = \mathbf{0}$ can be written as a linear combination of the basis vectors.

Definition 4.6.9 Let $f : U \to V$ be a linear transformation. The **nullity of f** is the dimension of NS(f) (i.e., nullity(f) = dim[NS(f)]). For an $m \times n$ matrix A, the **nullity of A** is the dimension of NS(A).

Example 4.6.10 Consider the system

$$\begin{aligned} x_1 + 3x_2 \phantom{{}+ 2x_3} + 5x_4 + 6x_5 &= 0 \\ x_1 + 3x_2 + x_3 + 7x_4 + 4x_5 &= 0 \\ x_1 + 3x_2 + 2x_3 + 3x_4 + 10x_5 &= 0 \end{aligned}$$

Find a basis for null space of the coefficient matrix and express every solution of the system as a linear combination of the basis elements.

Solution Applying the row operations to

$$A = \left[\begin{array}{ccccc|c} 1 & 3 & 0 & 5 & 6 & 0 \\ 1 & 3 & 1 & 7 & 4 & 0 \\ 1 & 3 & 2 & 3 & 10 & 0 \end{array}\right]$$

results in the row echelon form

$$\left[\begin{array}{ccccc|c} 1 & 3 & 0 & 5 & 6 & 0 \\ 0 & 0 & 1 & 2 & -2 & 0 \\ 0 & 0 & 0 & 1 & -\frac{4}{3} & 0 \end{array}\right]$$

which means

$$\begin{aligned} x_4 &= \tfrac{4}{3}x_5 \\ x_3 &= -2x_4 + 2x_5 = -2\left(\tfrac{4}{3}x_5\right) + 2x_5 = -\tfrac{2}{3}x_5 \\ x_1 &= -3x_2 - 5x_4 - 6x_5 = -3x_2 - \tfrac{20}{3}x_5 - 6x_5 \\ &= -3x_2 - \tfrac{38}{3}x_5 \end{aligned}$$

or

$$\begin{bmatrix} x_1 \\ x_2 \\ x_3 \\ x_4 \\ x_5 \end{bmatrix} = \begin{bmatrix} -3x_2 - \frac{38}{3}x_5 \\ x_2 \\ -\frac{2}{3}x_5 \\ \frac{4}{3}x_5 \\ x_5 \end{bmatrix} = \begin{bmatrix} -3x_2 \\ x_2 \\ 0 \\ 0 \\ 0 \end{bmatrix} + \begin{bmatrix} -\frac{38}{3}x_5 \\ 0 \\ -\frac{2}{3}x_5 \\ \frac{4}{3}x_5 \\ x_5 \end{bmatrix}$$

$$= x_2 \begin{bmatrix} -3 \\ 1 \\ 0 \\ 0 \\ 0 \end{bmatrix} + x_5 \begin{bmatrix} -\frac{38}{3} \\ 0 \\ -\frac{2}{3} \\ \frac{4}{3} \\ 1 \end{bmatrix}$$

The null space of the coefficient matrix is spanned by the set of linearly independent vectors

$$\left\{ \begin{bmatrix} -3 \\ 1 \\ 0 \\ 0 \\ 0 \end{bmatrix}, \begin{bmatrix} -\frac{38}{3} \\ 0 \\ -\frac{2}{3} \\ \frac{4}{3} \\ 1 \end{bmatrix} \right\}$$

which means this set is a basis of NS(A). This also means the set of all possible solutions of the homogeneous system $A\mathbf{x} = \mathbf{0}$ is

$$\left\langle \begin{bmatrix} -3 \\ 1 \\ 0 \\ 0 \\ 0 \end{bmatrix}, \begin{bmatrix} -\frac{38}{3} \\ 0 \\ -\frac{2}{3} \\ \frac{4}{3} \\ 1 \end{bmatrix} \right\rangle = \text{NS}(A)$$

Thus the nullity of A is 2. ■

SUMMARY

Terms to know:

The null space of a linear transformation $f: U \to V$

Theorems to know:

Let $f: U \to V$ be a linear transformation; then NS(f) is a subspace of U.

Let A be an $m \times n$ matrix. If $m < n$, there are nonzero vectors in NS(A).

Methods to know:

Determine a basis of NS(f)

EXERCISE SET 4.6

Find a basis if it exists for the null space of the following matrices.

1. $\begin{bmatrix} 3 & -2 & -1 & -4 \\ 1 & 1 & -2 & -3 \end{bmatrix}$

2. $\begin{bmatrix} 1 & 9 & 7 \\ 2 & 1 & 4 \end{bmatrix}$

3. $\begin{bmatrix} 1 & 1 & 1 \\ 1 & 2 & 1 \\ 2 & 1 & 1 \end{bmatrix}$

4. $\begin{bmatrix} 0 & 1 \\ 1 & 0 \end{bmatrix}$

5. $[7 \quad 2 \quad 9]$

6. $\begin{bmatrix} 1 & 4 & 0 & 3 & 5 \\ 1 & 4 & 1 & 3 & 2 \\ 0 & 4 & 8 & 12 & 4 \end{bmatrix}$

7. $\begin{bmatrix} 1 & 3 & -2 & 0 \\ 2 & 6 & 0 & -1 \\ -1 & -3 & 4 & 1 \end{bmatrix}$

8. $\begin{bmatrix} 1 & 2 & 3 \\ 4 & 5 & 6 \\ 7 & 8 & 9 \end{bmatrix}$

9. $\begin{bmatrix} 1 & 3 & 15 \\ 2 & 1 & 10 \\ 1 & 0 & 3 \end{bmatrix}$

10. $\begin{bmatrix} 3 & 6 & 1 \\ 2 & 1 & 1 \\ -1 & 0 & 1 \end{bmatrix}$

11. $\begin{bmatrix} 1 & 0 & 0 \\ 1 & 1 & 0 \\ 1 & 1 & 1 \end{bmatrix}$

12. $\begin{bmatrix} 8 & 1 & 2 \\ 0 & 1 & 1 \end{bmatrix}$

13. $\begin{bmatrix} 2 & 1 \\ 4 & 3 \end{bmatrix}$

14. $\begin{bmatrix} 21 & -7 & 35 \\ 8 & 10 & -1 \\ -3 & 1 & -5 \end{bmatrix}$

15. $\begin{bmatrix} 2 & 1 & -1 \\ -6 & -3 & 3 \\ 2 & 1 & -1 \end{bmatrix}$

16. $\begin{bmatrix} 1 & 0 & 1 \\ -1 & 1 & -1 \\ 0 & 1 & 1 \end{bmatrix}$

17. $\begin{bmatrix} 1 & 1 & -1 \\ -1 & 1 & 1 \\ 1 & 1 & 1 \end{bmatrix}$

Find nontrivial solutions of the following systems of homogeneous equations whenever they exist. Also, find a basis for the solution space of each system.

18. $$\begin{aligned} x_1 + 2x_2 + x_3 &= 0 \\ x_2 + 3x_3 &= 0 \end{aligned}$$

19. $$\begin{aligned} x_1 - x_2 + x_3 &= 0 \\ x_1 + x_2 - x_3 &= 0 \\ -x_1 + x_2 + x_3 &= 0 \end{aligned}$$

20. $$\begin{aligned} x_1 - x_2 + x_3 &= 0 \\ x_2 - x_3 &= 0 \\ -x_1 + 2x_2 + 3x_3 &= 0 \\ x_1 - x_2 + x_3 &= 0 \end{aligned}$$

21. $$\begin{aligned} x_1 + \tfrac{1}{2}x_2 + \tfrac{2}{3}x_3 &= 0 \\ \tfrac{1}{2}x_1 - x_2 \qquad &= 0 \\ 2x_2 - 3x_3 &= 0 \end{aligned}$$

22. Prove that if A is an $m \times n$ matrix, NS(A) is a subspace of R^n.

23. Find a system of homogeneous linear equations in x_1, x_2, x_3, and x_4 with the complete solution

$$\left\{ s\begin{bmatrix} 1 \\ 0 \\ 1 \\ 0 \end{bmatrix} + t\begin{bmatrix} 0 \\ 1 \\ 0 \\ 1 \end{bmatrix} \middle| \; s, t \in R \right\}$$

24. Suppose $f: U \to V$ is a linear transformation. Let $A = \{\mathbf{a}_1, \mathbf{a}_2, \ldots, \mathbf{a}_n\}$ and $B = \{\mathbf{b}_1, \ldots, \mathbf{b}_m\}$ and M be the matrix representing f with respect to the basis A of U and the basis B of V.
    a. Show that if $f(\mathbf{u}) = \mathbf{0}$, then $M[\mathbf{u}]_A = [\mathbf{0}]_B$.
    b. Show that if $M[\mathbf{u}]_A = [\mathbf{0}]_B$, then $f(\mathbf{u}) = \mathbf{0}$.
    c. Conclude that $\{\mathbf{u}_1, \ldots, \mathbf{u}_k\}$ is a basis for NS(f) if, and only if, $\{[\mathbf{u}_1]_A, \ldots, [\mathbf{u}_k]_A\}$ is a basis for NS(M).

25. Prove that if $f: U \to V$ is a linear transformation then $f(\mathbf{0}) = \mathbf{0}$.

26. A linear transformation f is one-to-one if and only if $f(\mathbf{x}) \neq f(\mathbf{y})$ for $\mathbf{x} \neq \mathbf{y}$. Prove that if f is one-to-one, the null space of f is $\{\mathbf{0}\}$.

27. Let A be an $n \times n$ matrix. If $\mathbf{v}_1, \mathbf{v}_2, \ldots, \mathbf{v}_k$ are vectors in R^n and $\{A\mathbf{v}_1, A\mathbf{v}_2, \ldots, A\mathbf{v}_k\}$ is linearly independent, prove that $\{\mathbf{v}_1, \mathbf{v}_2, \ldots, \mathbf{v}_k\}$ is also linearly independent.

SECTION 4.7 SOLUTIONS TO SYSTEMS OF LINEAR EQUATIONS (REVISITED)

In Chapter 4 we were interested primarily in finding all solutions to a system of m equations in n unknowns. First we considered a matrix as the coefficient matrix of a system of linear equations. We saw examples of homogeneous systems of linear equations having exactly one solution or an infinite number of solutions. We also saw nonhomogeneous systems having no solutions, exactly one solution, or an infinite number of solutions. In this chapter, we have regarded matrices as representations of linear transformations from R^n to R^m. The connection between these two ways to consider matrices arose when noting that for a matrix A, the vector space NS(A) is the space of solutions to the homogeneous system $A\mathbf{x} = \mathbf{0}$.

The structure of the solution set of $A\mathbf{x} = \mathbf{b}$ is different from that of $A\mathbf{x} = \mathbf{0}$, since the first set is not a vector space. To see this, let $\mathbf{u}$ and $\mathbf{v}$ be solutions of $A\mathbf{x} = \mathbf{b}$ where $\mathbf{b} \neq \mathbf{0}$. Then

$$A(\mathbf{u} + \mathbf{v}) = A\mathbf{u} + A\mathbf{v} = \mathbf{b} + \mathbf{b} = 2\mathbf{b} \neq \mathbf{b}$$

and so $\mathbf{u} + \mathbf{v}$ is not a solution of $A\mathbf{x} = \mathbf{b}$. However, $\mathbf{u} - \mathbf{v}$ is a solution of $A\mathbf{x} = \mathbf{0}$, since

$$A(\mathbf{u} - \mathbf{v}) = A\mathbf{u} - A\mathbf{v} = \mathbf{b} - \mathbf{b} = \mathbf{0}$$

That is to say, the difference of any two solutions of $A\mathbf{x} = \mathbf{b}$ is always a solution of $A\mathbf{x} = \mathbf{0}$.

Theorem 4.7.1 *Let $\mathbf{y}$ be a solution of $A\mathbf{x} = \mathbf{b}$ and let $\{\mathbf{x}_1, \mathbf{x}_2, \ldots, \mathbf{x}_t\}$ $(t \leq n)$ be a basis of* NS(A). *Then every solution of $A\mathbf{x} = \mathbf{b}$ is of the form*

$$\mathbf{x} = \mathbf{y} + c_1\mathbf{x}_1 + c_2\mathbf{x}_2 + \cdots + c_t\mathbf{x}_t \tag{4.7.1}$$

Conversely, every vector of the form of (4.7.1) *is a solution of $A\mathbf{x} = \mathbf{b}$.*

Proof. Let $A\mathbf{x} = \mathbf{b}$. If $A\mathbf{y} = \mathbf{b}$ then $\mathbf{x} - \mathbf{y} \in$ NS(A), thus

$$\mathbf{x} - \mathbf{y} = c_1\mathbf{x}_1 + c_2\mathbf{x}_2 + \cdots + c_t\mathbf{x}_t$$

or

$$\mathbf{x} = \mathbf{y} + c_1\mathbf{x}_1 + c_2\mathbf{x}_2 + \cdots + c_t\mathbf{x}_t$$

Conversely, let $\mathbf{x} = \mathbf{y} + c_1\mathbf{x}_1 + c_2\mathbf{x}_2 + \cdots + c_t\mathbf{x}_t$ where $A\mathbf{y} = \mathbf{b}$. Then

$$\begin{aligned} A\mathbf{x} &= A(\mathbf{y} + c_1\mathbf{x}_1 + c_2\mathbf{x}_2 + \cdots + c_t\mathbf{x}_t) \\ &= A\mathbf{y} + c_1A\mathbf{x}_1 + c_2A\mathbf{x}_2 + \cdots + c_tA\mathbf{x}_t \\ &= \mathbf{b} + \mathbf{0} + \mathbf{0} + \cdots + \mathbf{0} = \mathbf{b} \end{aligned}$$

The solution $\mathbf{x}$ is called the **general solution** of $A\mathbf{x} = \mathbf{b}$, and $\mathbf{y}$ is called a **particular** solution of $A\mathbf{x} = \mathbf{b}$. To find the general solution of $A\mathbf{x} = \mathbf{b}$, we first find a particular solution $\mathbf{y}$ of $A\mathbf{x} = \mathbf{b}$, find a basis for the solution space of $A\mathbf{x} = \mathbf{0}$ [i.e., NS(A)], and then construct all sums of $\mathbf{y}$ with elements of NS(A).

Example 4.7.2 Find all solutions of the following system of equations:

$$\begin{aligned} 3x - 2y - z - 4w &= 2 \\ x + y - 2z - 3w &= 1 \end{aligned}$$

Solution The augmented matrix of this system is

$$\left[\begin{array}{cccc|c} 3 & -2 & -1 & -4 & 2 \\ 1 & 1 & -2 & -3 & 1 \end{array}\right]$$

Using the methods of Chapter 1, we arrive at

$$\left[\begin{array}{cccc|c} 1 & 1 & -2 & -3 & 1 \\ 0 & 1 & -1 & -1 & \frac{1}{5} \end{array}\right] \tag{4.7.2}$$

Thus

$$\begin{aligned} x + y - 2z - 3w &= 1 \\ y - z - w &= \tfrac{1}{5} \end{aligned}$$

Letting $z = w = 0$, we find that $y = \frac{1}{5}$ and $x = \frac{4}{5}$, so

$$\begin{bmatrix} \frac{4}{5} \\ \frac{1}{5} \\ 0 \\ 0 \end{bmatrix}$$

is a particular solution of the system.

Next we must find the null space of the coefficient matrix

$$A = \begin{bmatrix} 3 & -2 & -1 & -4 \\ 1 & 1 & -2 & -3 \end{bmatrix}$$

Using the methods of Chapter 1, we obtain

$$\left[\begin{array}{cccc|c} 1 & 1 & -2 & -3 & 0 \\ 0 & 1 & -1 & -1 & 0 \end{array}\right]$$

Notice that this is the augmented matrix of the system in (4.7.2) with 0s for the last column rather than

$$\begin{bmatrix} 1 \\ \frac{1}{5} \end{bmatrix}$$

Thus

$$\begin{aligned} x + y - 2z - 3w &= 0 \\ y - z - w &= 0 \end{aligned}$$

and it follows that

$$y = z + w \quad \text{and} \quad x = z + 2w, \qquad \text{for arbitrary } w \text{ and } z.$$

If

$$\begin{bmatrix} x \\ y \\ z \\ w \end{bmatrix} \in \text{NS}(A)$$

then

$$\begin{bmatrix} x \\ y \\ z \\ w \end{bmatrix} = \begin{bmatrix} z + 2w \\ z + w \\ z \\ w \end{bmatrix} = \begin{bmatrix} z \\ z \\ z \\ 0 \end{bmatrix} + \begin{bmatrix} 2w \\ w \\ 0 \\ w \end{bmatrix} = z\begin{bmatrix} 1 \\ 1 \\ 1 \\ 0 \end{bmatrix} + w\begin{bmatrix} 2 \\ 1 \\ 0 \\ 1 \end{bmatrix}$$

We have thus found that

$$\text{NS}(A) = \left\langle \begin{bmatrix} 1 \\ 1 \\ 1 \\ 0 \end{bmatrix}, \begin{bmatrix} 2 \\ 1 \\ 0 \\ 1 \end{bmatrix} \right\rangle$$

Consequently, any solution of the original system of equations has the form

$$\begin{bmatrix} \frac{4}{5} \\ \frac{1}{5} \\ 0 \\ 0 \end{bmatrix} + z\begin{bmatrix} 1 \\ 1 \\ 1 \\ 0 \end{bmatrix} + w\begin{bmatrix} 2 \\ 1 \\ 0 \\ 1 \end{bmatrix}$$

where z and w can be replaced by any numbers. ■

SUMMARY

Terms to know:

General solution of a system of linear equations

Particular solution of a system of linear equations

Theorems to know:

Let A be an $m \times n$ matrix, $\{\mathbf{x}_1, \ldots, \mathbf{x}_t\}$ $t \leq n$ be a basis for NS(A), and let $A\mathbf{y} = \mathbf{b}$. Then every solution $\mathbf{x}$ of $A\mathbf{x} = \mathbf{b}$ has the form

$$\mathbf{x} = \mathbf{y} + c_1\mathbf{x}_1 + \cdots + c_t\mathbf{x}_t$$

EXERCISE SET 4.7

Find all solutions for the following systems of equations using the method of this section.

1. $$\begin{aligned} x - y + z - w &= 2 \\ 3x + 4y - z + w &= 5 \end{aligned}$$

2. $$\begin{aligned} 7x + 3y - 2z + w &= 1 \\ 4x - 3y + z - w &= 2 \\ x + z - w &= 0 \end{aligned}$$

3. $$\begin{aligned} x + y + z &= 1 \\ -x + 2y + 4z &= 0 \end{aligned}$$

4. $$\begin{aligned} 2x_1 + 4x_2 - 5x_3 + 4x_4 &= 0 \\ 3x_1 - 4x_2 + 5x_3 - 7x_4 &= 2. \end{aligned}$$

5. Let **u** and **v** be solutions of $A\mathbf{x} = \mathbf{b}$. Explain why, in general, $r\mathbf{u} + s\mathbf{v}$ is not a solution. When would the sum be a solution?

6. Given the system

$$\begin{aligned} x_1 + 2x_2 \qquad + x_5 &= 5 \\ x_3 \quad + 3x_5 &= 0 \\ x_4 \qquad &= 3 \end{aligned}$$

Show that

a.

$$\begin{bmatrix} 5 - 2x_2 - x_5 \\ x_2 \\ -3x_5 \\ 3 \\ x_5 \end{bmatrix}$$

is a general solution,

b.

$$\begin{bmatrix} 3 - 3s - t \\ 1 + s + t \\ 3t - 3s \\ 3 \\ s - t \end{bmatrix}$$

is also a general solution,

c.

$$\begin{bmatrix} 5 - 3u + v - 2w \\ u - v + w \\ -3u - 3v \\ 3 \\ u + v \end{bmatrix}$$

is also a general solution.

d. Show that any 5×1 vector that can be obtained from the formula in (a) by assigning values to x_2 and x_5 can also be obtained from the formula in (b) by assigning values to s and t (and vice versa), and can also be obtained from the formula in (c) by assigning values to u, v, and w (and vice versa).

or

$$a_1\mathbf{v}_1 + \cdots + a_k\mathbf{v}_k - c_1\mathbf{u}_1 - \cdots - c_{n-k}\mathbf{u}_{n-k} = \mathbf{0}$$

Since $\{\mathbf{v}_1, \ldots, \mathbf{v}_k, \mathbf{u}_1, \ldots, \mathbf{u}_{n-k}\}$ is a basis for U, $a_1 = a_2 = \cdots = a_k = c_1 = \cdots = c_{n-k} = 0$. Hence $\{f(\mathbf{u}_1), \ldots, f(\mathbf{u}_{n-k})\}$ is a linearly independent set.

It may seem tiresome to consider $\text{RS}(A^T)$ when we determine the rank of the $m \times n$ matrix A. However, we do not have to transpose A to find $r(A)$, as shown by Theorem 4.8.5.

Theorem 4.8.5 *Let A be an $m \times n$ matrix. Then $r(A) = \dim(RS(A))$.*

Proof. Let $s = \dim[\text{RS}(A)]$. We can find s by reducing A to row echelon form E. To find $\dim[\text{NS}(A)]$, we reduce the augmented matrix $[A \mid 0]$ to $[E \mid 0]$. Then s is the number of nonzero rows of E. If the leading entry of row i occurs in column $L(i)$, we can rewrite the system $[E \mid 0]$ as

$$\begin{aligned}
b_{1L(1)}x_{L(1)} + \quad \cdots \quad &= 0 \\
b_{2L(2)}x_{L(2)} + \cdots \quad &= 0 \\
&\vdots \\
b_{sL(s)}x_{L(s)} + \cdots &= 0
\end{aligned}$$

We can backsolve to find $x_{L(1)}, x_{L(2)}, \ldots, x_{L(s)}$ in terms of the remaining $n - s$ x's.

Now for each i such that $1 \le i \le n$ and $i \ne L(j)$ for $j = 1, \ldots, s$, construct a vector of $\text{NS}(A)$ by first letting the ith component be 1 and the kth component be 0 for the $(n-1) - s$ k's not equal to $L(j)$, $j = 1, \ldots, s$ and backsolving to find the values of $x_{L(1)}, x_{L(2)}, \ldots, x_{L(s)}$. The $n - s$ vectors so produced span $\text{NS}(A)$ and are linearly independent, and thus they form a basis for $\text{NS}(A)$. Thus, $\dim[\text{NS}(A)] = n - s$, and so by Theorem 2, $r(A) = \dim[\text{CS}(A)] = \dim(U) - \dim[\text{NS}(A)] = n - (n - s) = s$.

Let us look at some consequences of Theorem 4.8.5.

1. Let A be an $m \times n$ matrix. If $r(A) = m$, then $\dim(\text{CS}(A)) = m$, but $\text{CS}(A)$ is a subspace of R^m, so $\text{CS}(A) = R^m$. Thus, for every $\mathbf{v}$ in R^m, we can write

$$\begin{aligned}
\mathbf{v} &= x_1\mathbf{A}_1 + x_2\mathbf{A}_2 + \cdots + x_n\mathbf{A}_n \\
&= A\begin{bmatrix} x_1 \\ \vdots \\ x_n \end{bmatrix}
\end{aligned}$$

where $\mathbf{A}_i$ is the ith column of A. We state this as a theorem.

Theorem 4.8.6 *Let A be an $m \times n$ matrix. If $r(A) = m$, for every* $\mathbf{b}$ *in* R^m, $A\mathbf{x} = \mathbf{b}$ *has a solution.*

2. If $NS(A) = \{\mathbf{0}\}$, then $r(A) = n$ since $r(A) + \dim(NS(A)) = n$. Thus if $A\mathbf{x} = \mathbf{b}$ has a solution, the solution is unique.
3. If $r(A) = n$, the columns of A are linearly independent.

Example 4.8.7 Are the columns of

$$A = \begin{bmatrix} -3 & 5 & 1 & 2 \\ 7 & 2 & 0 & -4 \\ -8 & 3 & 1 & 6 \end{bmatrix}$$

linearly independent?

Solution First observe that $CS(A)$ is a subspace of R^3, so $r(A) \leqq 3$; however, there are four columns, so the columns of A cannot be linearly independent. ■

Example 4.8.8 Are the columns of

$$A = \begin{bmatrix} 1 & 1 & 1 \\ -3 & 0 & 1 \\ 1 & 1 & 4 \\ 2 & 7 & 2 \\ 4 & 3 & 8 \end{bmatrix}$$

linearly independent?

Solution Using Gauss elimination reduces A to

$$\begin{bmatrix} 1 & 1 & 1 \\ 0 & 3 & 4 \\ 0 & 0 & 3 \\ 0 & 0 & 0 \\ 0 & 0 & 0 \end{bmatrix}$$

Thus $r(A) = 3$, so the columns are linearly independent because $\dim CS(A) = 3$ and there are 3 columns in A. ■

SUMMARY

Terms to know:

Rank of a linear transformation

Rank of a matrix

Theorems to know:

If $f : U \to V$ is a linear transformation, then

$$r(f) \leq \dim(U) \quad \text{and} \quad r(f) \leq \dim(V)$$
$$r(f) + \dim[\mathrm{NS}(f)] = \dim(U)$$

If A is an $m \times n$ matrix, then $r(A) = \dim[\mathrm{RS}(A)] = \dim[\mathrm{CS}(A)]$.

Methods to know:

Given a linear transformation $f : U \to V$, find $r(f)$.

EXERCISE SET 4.8

1. Given an $m \times n$ matrix A with rank r, find $\dim \mathrm{NS}(A)$.
   a. $m = 4,\ n = 3,\ r = 1$
   b. $m = 5,\ n = 10,\ r = 4$
   c. $m = 75,\ n = 100,\ r = 62$
   d. $m = 16,\ n = 16,\ r = 16$
   e. $m = 22,\ n = 13,\ r = 10$
2. We wish to find all solutions to $A\mathbf{x} = \mathbf{b}$. If A is an $m \times n$ matrix of rank r, find the number of *possible* solutions to $A\mathbf{x} = \mathbf{b}$.

   Hint: There are either zero, one, or an infinite number of solutions to $A\mathbf{x} = \mathbf{b}$.

   a. $m = 7,\ n = 6,\ r = 6$
   b. $m = 8,\ n = 9,\ r = 8$
   c. $m = 97,\ n = 97,\ r = 96$
   d. $m = 100,\ n = 100,\ r = 92$
   e. $m = 100{,}000,\ n = 100{,}000,\ r = 15$
3. Find all solutions in each case.

   a. $\begin{bmatrix} 1 & 2 & 3 \\ 4 & 5 & 6 \\ 1 & 6 & 1 \end{bmatrix} \mathbf{x} = \begin{bmatrix} 1 \\ 9 \\ 6 \end{bmatrix}$
   b. $\begin{bmatrix} 1 & -1 & 1 \\ -1 & 1 & -1 \\ -1 & 1 & -1 \end{bmatrix} \mathbf{x} = \begin{bmatrix} 0 \\ 0 \\ 0 \end{bmatrix}$

   c. $\begin{bmatrix} 1 & 9 & 0 \\ 0 & 9 & 1 \\ 9 & 0 & 1 \end{bmatrix} \begin{bmatrix} x_1 \\ x_2 \\ x_3 \end{bmatrix} = \begin{bmatrix} 7 \\ 3 \\ 1 \end{bmatrix}$
   d. $\begin{bmatrix} 0 & 1 & 2 \\ 2 & 1 & 0 \\ 1 & 0 & 2 \end{bmatrix} \begin{bmatrix} x_1 \\ x_2 \\ x_3 \end{bmatrix} = \begin{bmatrix} 6 \\ 2 \\ 1 \end{bmatrix}$

   e. $\begin{bmatrix} 1 & 0 \\ 2 & -3 \end{bmatrix} \mathbf{x} = \begin{bmatrix} 1 \\ 9 \end{bmatrix}$
4. If A is an $m \times n$ matrix of rank m, is it true that the equation $A\mathbf{x} = \mathbf{b}$ always has a solution? Why?
5. Prove that the null space of A^T is the set of all vectors that are orthogonal to every vector the image of A.
6. If A is an $m \times n$ matrix, prove that nullity $(A^T) = m - n +$ nullity (A).

These problems can best be solved using the accompanying computer software or a hand-held calculator. Determine the nullity and rank of the following matrices.

7. $\begin{bmatrix} 1.2 & 1.9 & 3.2 \\ 2.3 & 1.1 & 2.9 \\ 3.7 & 0 & 2.8 \\ 3.7 & 0 & 2.8 \end{bmatrix}$ 8. $\begin{bmatrix} -4.1 & 1.3 & 4.5 & -2.6 & 0.8 \\ 12.2 & -4.7 & 3.7 & 8.5 & -9.7 \\ 8.1 & -3.4 & 8.2 & 5.9 & -8.9 \\ 1.7 & -1.8 & 6.3 & 7.2 & 0.5 \end{bmatrix}$

9. $\begin{bmatrix} 132 & -64 & 89 \\ 112 & 267 & -44 \\ 842 & -531 & 342 \end{bmatrix}$

10. $\begin{bmatrix} 3.4 & -2.2 & 1.8 & -2.0 & 10 \\ 42 & 12 & -8 & 4 & -22 \\ 7 & 16 & 20 & 21 & 3 \\ -1.9 & -4.1 & 2.9 & 2.7 & 2 \\ 1 & -1 & 44 & 42 & -11 \end{bmatrix}$

SECTION 4.9 PROPERTIES OF NONSINGULAR MATRICES

In this section we discuss several properties of nonsingular (or invertible) matrices and also connect nonsingular matrices with the concepts of rank, column space, and row space of a matrix discussed in earlier sections. The first property is a restatement of uniqueness, which was verified in Chapter 1.

Theorem 4.9.1 *Let A be an $n \times n$ invertible matrix. Then A has a unique inverse.*

Loosely stated, the theorem says that if a matrix B looks like an inverse of A, then B is the inverse, A^{-1}, of A. In Chapter 1, this idea led to proofs of the next two theorems, assuming that A and B have inverses.

Theorem 4.9.2 $(A^{-1})^{-1} = A$

Theorem 4.9.3 $(AB)^{-1} = B^{-1}A^{-1}$ *[assuming A and B have inverses]*

In Chapter 1, you learned how to find the inverse of a matrix A. In the following theorem, we answer the question of when A has an inverse without necessarily finding the inverse of A.

Theorem 4.9.4 *Let A be an $n \times n$ matrix. The following conditions on A are equivalent:*

1. *A is invertible.*
2. *$r(A) = n$.*
3. *The columns of A are linearly independent.*
4. *The columns of A span R^n.*

5. *The rows of A are linearly independent.*
6. *The rows of A span R^n.*
7. *Given any $\mathbf{b} \in R^n$, $A\mathbf{x} = \mathbf{b}$ has a solution.*

Proof. We proceed as indicated by the diagram:

$$\begin{array}{ccc} 7 & \rightarrow & 4 \\ \uparrow & & \downarrow \\ 2 & \leftarrow & 3 \end{array}$$

$7 \rightarrow 4$: Let $\mathbf{b} \in R^n$ and $A\mathbf{x} = \mathbf{b}$. Then if $\mathbf{A}_i$ is the ith column of A, $\mathbf{A}_1 x_1 + \cdots + \mathbf{A}_n x_n = \mathbf{b}$, so $\mathbf{b} \in \mathrm{CS}(A)$. Since this is true for all $\mathbf{b} \in R^n$, $\mathrm{CS}(A) = R^n$.

$4 \rightarrow 3$: Since $\dim R^n = n$ and $\mathrm{CS}(A) = R^n$, the n columns of A form a basis for R^n and so are linearly independent.

$3 \rightarrow 2$: Since $\mathrm{CS}(A) \subseteq R^n$ and the n columns of A are linearly independent, they form a basis for $\mathrm{CS}(A)$. Thus $n = \dim[\mathrm{CS}(A)] = r(A)$.

$2 \rightarrow 7$: Since $r(A) = n$, $\dim[\mathrm{CS}(A)] = n$, but $\mathrm{CS}(A) \subseteq R^n$ and $\dim(R^n) = n$, so $\mathrm{CS}(A) = R^n$. Thus if $\mathbf{b} \in R^n$, there are $x_1, \ldots, x_n \in R$ with $\mathbf{b} = \mathbf{A}_1 x_1 + \cdots + \mathbf{A}_n x_n$. If

$$\mathbf{x} = \begin{bmatrix} x_1 \\ \vdots \\ x_n \end{bmatrix}$$

then $A\mathbf{x} = \mathbf{b}$.

Next we show that $1 \rightarrow 7 \rightarrow 1$.

$1 \rightarrow 7$: Since A is invertible, $AA^{-1} = I$. Let $\mathbf{b} \in R^n$; then $\mathbf{b} = I\mathbf{b} = (AA^{-1})\mathbf{b} = A(A^{-1}\mathbf{b})$. Thus $\mathbf{x} = A^{-1}\mathbf{b}$ is a solution of $A\mathbf{x} = \mathbf{b}$.

$7 \rightarrow 1$: Since $\mathbf{e}_1 \in R^n$, there is $\mathbf{x}_1 \in R^n$ with $A\mathbf{x}_1 = \mathbf{e}_1$. In general, for $i = 2, \ldots, n$, there is $\mathbf{x}_i \in R^n$ with $A\mathbf{x}_i = \mathbf{e}_i$. Let X be the $n \times n$ matrix whose ith column is $\mathbf{x}_i$. Then $AX = A[\mathbf{x}_1 | \mathbf{x}_2 | \cdots | \mathbf{x}_n] = [\mathbf{e}_1 | \mathbf{e}_2 | \cdots | \mathbf{e}_n] = I$, so A is invertible.

We leave the proof of

$$\begin{array}{ccc} 2 & \rightarrow & 5 \\ & \nwarrow \swarrow & \\ & 6 & \end{array}$$

as an exercise.

As one application of this theorem, we consider the system of linear equations written in matrix notation

$$A\mathbf{x} = \mathbf{b}$$

where A is a given $n \times n$ square matrix and $\mathbf{x} \in R^n$. If A is nonsingular, then the system $A\mathbf{x} = \mathbf{b}$ has a unique solution $\mathbf{x} = A^{-1}\mathbf{b}$.

Example 4.9.5 Solve the system

$$\begin{bmatrix} 1 & 2 & -4 \\ 2 & 3 & 1 \\ 3 & 5 & -4 \end{bmatrix}\begin{bmatrix} x \\ y \\ z \end{bmatrix} = \begin{bmatrix} -14 \\ 1 \\ -16 \end{bmatrix}$$

by matrix inversion.

Solution The inverse of

$$\begin{bmatrix} 1 & 2 & -4 \\ 2 & 3 & 1 \\ 3 & 5 & -4 \end{bmatrix}$$

is

$$\begin{bmatrix} -17 & -12 & 14 \\ 11 & 8 & -9 \\ 1 & 1 & -1 \end{bmatrix}$$

and so

$$\begin{bmatrix} x \\ y \\ z \end{bmatrix} = \begin{bmatrix} -17 & -12 & 14 \\ 11 & 8 & -9 \\ 1 & 1 & -1 \end{bmatrix}\begin{bmatrix} -14 \\ 1 \\ -16 \end{bmatrix} = \begin{bmatrix} 2 \\ -2 \\ 3 \end{bmatrix}$$ ■

There is no great advantage in using this technique to solve a single equation like $A\mathbf{x} = \mathbf{b}$. However, in numerous applications the coefficient matrix remains the same but different vectors appear on the right. For example, you may want to study $A\mathbf{x} = \mathbf{b}_1$, $A\mathbf{x} = \mathbf{b}_2, \ldots, A\mathbf{x} = \mathbf{b}_{10}$.

If A is the matrix of Example 4.9.5, to solve

$$\begin{bmatrix} 1 & 2 & -4 \\ 2 & 3 & 1 \\ 3 & 5 & -4 \end{bmatrix}\begin{bmatrix} x \\ y \\ z \end{bmatrix} = \begin{bmatrix} 1 \\ 0 \\ 1 \end{bmatrix}$$

we use A^{-1}, which was computed in Example 4.9.5. Thus

$$\begin{bmatrix} x \\ y \\ z \end{bmatrix} = A^{-1}\begin{bmatrix} 1 \\ 0 \\ 1 \end{bmatrix} = \begin{bmatrix} -17 & -12 & 14 \\ 11 & 8 & -9 \\ 1 & 1 & -1 \end{bmatrix}\begin{bmatrix} 1 \\ 0 \\ 1 \end{bmatrix} = \begin{bmatrix} -3 \\ 2 \\ 0 \end{bmatrix}$$

SUMMARY

Theorems to know:

Let A be an $m \times n$ matrix. The following conditions on A are equivalent:

1. A is nonsingular,
2. $r(A) = n$.

3. the columns of A are linearly independent.
4. the columns of A span R^n.
5. the rows of A are linearly independent.
6. the rows of A span R^n.
7. given any $\mathbf{b} \in R^n$, $A\mathbf{x} = \mathbf{b}$ has a solution.

EXERCISE SET 4.9

1. Prove that $(AB)(B^{-1}A^{-1}) = I$.
2. Prove the following conditions equivalent for an $n \times n$ matrix A.
   a. $r(A) = n$
   b. The rows of A span R^n.
   c. The rows of A are linearly independent.
3. Prove that $(A^2)^{-1} = (A^{-1})^2$. (*Hint:* $(AA)^{-1} = A^{-1}A^{-1}$)
4. Prove that $(A^n)^{-1} = (A^{-1})^n$.
5. Prove that $(ABC)^{-1} = C^{-1}B^{-1}A^{-1}$.
6. Find the inverse of

$$\begin{bmatrix} 1 & 0 & 0 \\ 2 & 1 & 0 \\ -4 & 0 & 1 \end{bmatrix}$$

7. A square matrix L is lower triangular if $l_{ij} = 0$ when $i < j$. If L is a lower triangular $n \times n$ matrix with 1s on the diagonal, prove the following.
   a. L is invertible.
   b. L^{-1} is lower triangular.
8. Prove that $AB^{-1} = B^{-1}A$ if and only if $AB = BA$.
9. Prove that if A has a column of all zeros, then A is not invertible.
10. For n a positive integer, show that

$$(ABA^{-1})^n = AB^nA^{-1}$$

11. Show that $(A^{-1}A^T)^T = (A^TA^{-1})^{-1}$.
12. Solve the system

$$\begin{aligned} 4x_1 + 3x_2 &= b_1 \\ 5x_1 + 4x_2 &= b_2 \end{aligned}$$

for each of the following values of $\begin{bmatrix} b_1 \\ b_2 \end{bmatrix}$

a. $\begin{bmatrix} 1 \\ 0 \end{bmatrix}$ b. $\begin{bmatrix} 0 \\ 1 \end{bmatrix}$ c. $\begin{bmatrix} 0 \\ 0 \end{bmatrix}$ d. $\begin{bmatrix} -3 \\ -2 \end{bmatrix}$

13. Solve the system

$$\begin{aligned} x + y - z &= b_1 \\ 2x - y + z &= b_2 \\ 3x - 2y + 4z &= b_3 \end{aligned}$$

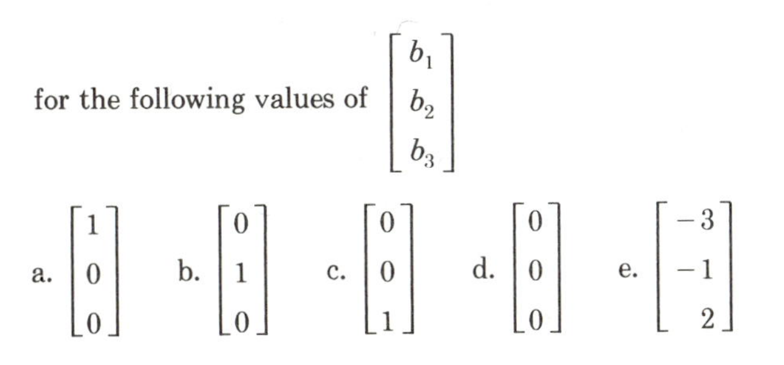

for the following values of $\begin{bmatrix} b_1 \\ b_2 \\ b_3 \end{bmatrix}$

a. $\begin{bmatrix} 1 \\ 0 \\ 0 \end{bmatrix}$ b. $\begin{bmatrix} 0 \\ 1 \\ 0 \end{bmatrix}$ c. $\begin{bmatrix} 0 \\ 0 \\ 1 \end{bmatrix}$ d. $\begin{bmatrix} 0 \\ 0 \\ 0 \end{bmatrix}$ e. $\begin{bmatrix} -3 \\ -1 \\ 2 \end{bmatrix}$

14. Let $A\mathbf{x} = \mathbf{0}$ and A be an $n \times n$ square matrix. Show that the equation always has a solution. When would it have a unique solution?
15. Solve the equation $A\mathbf{X} = \mathbf{B}$, where $\mathbf{X}$ is a 3×3 matrix,

$$A = \begin{bmatrix} 1 & -1 & 1 \\ 2 & 3 & -1 \\ 3 & -1 & 2 \end{bmatrix}$$

and $\mathbf{B}$ is as given.

a. $B = \begin{bmatrix} 1 & 0 & 0 \\ 0 & 1 & 0 \\ 0 & 0 & 1 \end{bmatrix}$ b. $B = \begin{bmatrix} -2 & 1 & 5 \\ 3 & 0 & -4 \\ 4 & -2 & 1 \end{bmatrix}$

16. Let A be a 2×2 matrix. Suppose that $\{\mathbf{v}_1, \mathbf{v}_2\}$ is a basis for R^2 and $A\mathbf{v}_1 = 2\mathbf{v}_1$ and $A\mathbf{v}_2 = -\mathbf{v}_2$. Is A nonsingular? Why?

SECTION 4.10 WHY MULTIPLICATION OF MATRICES IS DEFINED AS WE DEFINED IT (OPTIONAL)

In Chapter 1, we defined the product of two matrices. At that time, we noted that the multiplication of matrices did not seem to proceed in a straightforward manner. We are now in a position to show why we multiply matrices like we do.

Let $f: U \to V$ and $g: V \to W$ be linear transformations. Let $A = \{\mathbf{a}_1, \ldots, \mathbf{a}_n\}$, $B = \{\mathbf{b}_1, \ldots, \mathbf{b}_k\}$, and $C = \{\mathbf{c}_1, \ldots, \mathbf{c}_m\}$ be bases of U, V, and W, respectively. First we require the following proposition.

Proposition 4.10.1
The composite function $g \circ f : V \to W$ defined by $(g \circ f)(\mathbf{u}) = g[f(\mathbf{u})]$ is a linear transformation.

Proof. Let $\mathbf{u}_1, \mathbf{u}_2 \in U$ and $a, b \in R$. Then $f(a\mathbf{u}_1 + b\mathbf{u}_2) = af(\mathbf{u}_1) + bf(\mathbf{u}_2)$, since f is linear. Let $\mathbf{v}_1 = f(\mathbf{u}_1)$ and $\mathbf{v}_2 = f(\mathbf{u}_2)$; then $\mathbf{v}_1, \mathbf{v}_2 \in V$. Since g is linear, $g(a\mathbf{v}_1 + b\mathbf{v}_2) = ag(\mathbf{v}_1) + bg(\mathbf{v}_2)$. Therefore,

$$\begin{aligned}(g \circ f)(a\mathbf{u}_1 + b\mathbf{u}_2) &= g[f(a\mathbf{u}_1 + b\mathbf{u}_2)] = g[af(\mathbf{u}_1) + bf(\mathbf{u}_2)] \\ &= g(a\mathbf{v}_1 + b\mathbf{v}_2) = ag(\mathbf{v}_1) + bg(\mathbf{v}_2) \\ &= ag[f(\mathbf{u}_1)] + bg[f(\mathbf{u}_2)] \\ &= a(g \circ f)(\mathbf{u}_1) + b(g \circ f)(\mathbf{u}_2)\end{aligned}$$

Returning to our definition of matrix multiplication, suppose N is the $k \times m$ matrix representing f with respect to A and B, whereas M is the $n \times k$ matrix representing g with respect to B and C. If P is the matrix representing $g \circ f$ with respect to A and C, then the ith column of P is $[(g \circ f)(\mathbf{a}_i)]_C$.

If

$$[(g \circ f)(\mathbf{a}_i)]_C = \begin{bmatrix} p_{1i} \\ p_{2i} \\ \vdots \\ p_{mi} \end{bmatrix}$$

then

$$(g \circ f)(\mathbf{a}_i) = p_{1i}\mathbf{c}_1 + p_{2i}\mathbf{c}_2 + \cdots + p_{mi}\mathbf{c}_m$$

Since $(g \circ f)(\mathbf{a}_i) = g(f(\mathbf{a}_i))$, we look at $f(\mathbf{a}_i)$. By the definition of N,

$$[f(\mathbf{a}_i)]_B = \begin{bmatrix} n_{1i} \\ n_{2i} \\ \vdots \\ n_{mi} \end{bmatrix}$$

where $f(\mathbf{a}_i) = n_{1i}\mathbf{b}_1 + n_{2i}\mathbf{b}_2 + \cdots + n_{ki}\mathbf{b}_k$. Then

$$\begin{aligned} (g \circ f)(\mathbf{a}_i) &= g(n_{1i}\mathbf{b}_1 + n_{2i}\mathbf{b}_2 + \cdots + n_{ki}\mathbf{b}_k) \\ &= n_{1i}g(\mathbf{b}_1) + n_{2i}g(\mathbf{b}_2) + \cdots + n_{ki}g(\mathbf{b}_k) \end{aligned}$$

By the definition of M, for $j = 1, 2, \ldots, k$,

$$g(\mathbf{b}_j) = m_{1j}\mathbf{c}_1 + m_{2j}\mathbf{c}_2 + \cdots + m_{nj}\mathbf{c}_n$$

Hence

$$\begin{aligned} (g \circ f)(\mathbf{a}_i) &= n_{1i}[m_{11}\mathbf{c}_1 + m_{21}\mathbf{c}_2 + \cdots + m_{n1}\mathbf{c}_n] \\ &\quad + n_{2i}[m_{12}\mathbf{c}_1 + m_{22}\mathbf{c}_2 + \cdots + m_{n2}\mathbf{c}_n] \\ &\quad + \cdots \\ &\quad + n_{ki}[m_{1k}\mathbf{c}_1 + m_{2k}\mathbf{c}_2 + \cdots + m_{nk}\mathbf{c}_n] \\ &= (m_{11}n_{1i} + m_{12}n_{2i} + \cdots + m_{1k}n_{ki})\mathbf{c}_1 \\ &\quad + (m_{21}n_{1i} + m_{22}n_{2i} + \cdots + m_{2k}n_{ki})\mathbf{c}_2 \\ &\quad + \cdots \\ &\quad + (m_{n1}n_{1i} + m_{n2}n_{2i} + \cdots + m_{nk}n_{ki})\mathbf{c}_n \end{aligned}$$

Hence, the entry p_{si} of P is the coefficient of $\mathbf{c}_s$ in $(g \circ f)(\mathbf{a}_i)$, which is

$$m_{s1}n_{1i} + m_{s2}n_{2i} + \cdots + m_{sk}n_{ki}$$

This is also the entry in the sth row and ith column of the product MN as we defined matrix multiplication. Since this is true for all $1 \leqq s \leqq m$ and $1 \leqq i \leqq n$, $MN = P$. Thus if $f: R^m \to R^k$ has matrix N and $g: R^k \to R^n$ has matrix M, then $g \circ f: R^m \to R^n$ has matrix MN.

SUMMARY

Propositions to know:

Let $f: U \to V$ and $g: V \to W$ be linear transformations. Then $g \circ f: U \to W$ defined by $(g \circ f)(\mathbf{u}) = g(f(\mathbf{u}))$ is a linear transformation.

CHAPTER FIVE

DETERMINANTS

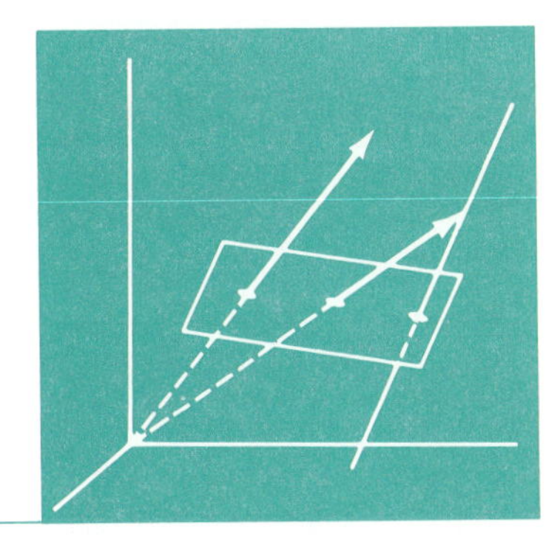

Leibnitz (1646–1716) is given credit for the introduction of determinants in 1693. Nothing was done with this concept until Gabriel Cramer (1704–1752) used determinants in solving systems of linear equations in 1750. Even though credit is given to Leibnitz for the concept and Cramer for its use in solving systems of linear equations, a Japanese mathematician, Seki Kowa (1642–1708), introduced the concept of determinants to solve systems of linear equations at least 10 years earlier than Leibnitz.

Recall that a function from set A to set B is a rule that associates with each element in set A a unique element in set B. In this chapter, set A is the set of square matrices and set B is the set of real numbers.

This function to be developed, det( ), is useful in determining whether a matrix A is invertible. We will see that A is invertible precisely when $\det(A) \neq 0$. Given a square matrix A that is not too large, $\det(A)$ is relatively easy to calculate and so is used to solve the matrix equation $A\mathbf{x} = \mathbf{b}$ for $\mathbf{x}$. This method is called Cramer's rule.

SECTION 5.1 DETERMINANTS

Every square matrix has a real number associated with it, called the **determinant** of the matrix. We are defining a function, det( ), where the elements of the domain are square matrices and the range is the set of all real numbers. The notation we use for the determinant of a square matrix A is $\det(A)$. The vertical bar notation $\det(A) = |A|$, introduced in 1841 by Arthur Cayley (1821–1895), is often used in the literature.

If A is an $n \times n$ matrix, we shall refer to $\det(A)$ as a determinant of the nth order. Determinants for orders 1, 2, and 3 are first defined, and later a definition for determinants of any order n is stated.

Definition 5.1.1

1. The determinant of a 1×1 matrix $A = [a]$ is $\det(A) = a$.
2. The determinant of a 2×2 matrix

$$A = \begin{bmatrix} a_{11} & a_{12} \\ a_{21} & a_{22} \end{bmatrix}$$

is the number $\det(A) = a_{11}a_{22} - a_{12}a_{21}$.

3. The determinant of a 3×3 matrix

$$A = \begin{bmatrix} a_{11} & a_{12} & a_{13} \\ a_{21} & a_{22} & a_{23} \\ a_{31} & a_{32} & a_{33} \end{bmatrix}$$

is the number $\det(A) = a_{11}a_{22}a_{33} + a_{12}a_{23}a_{31} + a_{13}a_{21}a_{32} - a_{13}a_{22}a_{31} - a_{11}a_{23}a_{32} - a_{12}a_{21}a_{33}$.

The definition for the third-order determinant is cumbersome. However, a more useful computational scheme is given later. Until that time, notice that there are six terms, and each term is the product of three factors. Also, each product consists of one element from each row and one element from each column such that no two factors in any one term are elements of the same row or column. Furthermore, all such possible products appear.

Example 5.1.2 Calculate $\det(A)$ for each A.

a. $A = [-2]$ b. $A = \begin{bmatrix} 3 & -4 \\ -5 & 2 \end{bmatrix}$ c. $A = \begin{bmatrix} 2 & -1 & 3 \\ 3 & -2 & 1 \\ 4 & -3 & 2 \end{bmatrix}$

Solution a. $\det(A) = \det([-2]) = -2$

b. $\det(A) = \det\left(\begin{bmatrix} 3 & -4 \\ -5 & 2 \end{bmatrix}\right) = (3)(2) - (-4)(-5) = 6 - 20 = -14$

c.
$$\begin{aligned}\det(A) &= \det\left(\begin{bmatrix} 2 & -1 & 3 \\ 3 & -2 & 1 \\ 4 & -3 & 2 \end{bmatrix}\right) \\ &= (2)(-2)(2) + (-1)(1)(4) \\ &\quad + (3)(3)(-3) - (3)(-2)(4) - (2)(1)(-3) - (-1)(3)(2) \\ &= -8 + (-4) + (-27) - (-24) - (-6) - (-6) \\ &= -39 + 36 = -3\end{aligned}$$

■

Example 5.1.3 Let

$$A = \begin{bmatrix} 2x & 3 \\ -5 & -4 \end{bmatrix}$$

Solve $\det(A) = 7x$ for x.

Solution $\det(A) = \det\left(\begin{bmatrix} 2x & 3 \\ -5 & -4 \end{bmatrix}\right) = (2x)(-4) - (3)(-5) = -8x + 15$

Thus $\det(A) = 7x$ means $-8x + 15 = 7x$, or $x = -1$. ■

Let us now reconsider the square matrix of order 3. Notice that

$$a_{11}\det\left(\begin{bmatrix} a_{22} & a_{23} \\ a_{32} & a_{33} \end{bmatrix}\right) - a_{12}\det\left(\begin{bmatrix} a_{21} & a_{23} \\ a_{31} & a_{33} \end{bmatrix}\right) + a_{13}\det\left(\begin{bmatrix} a_{21} & a_{22} \\ a_{31} & a_{32} \end{bmatrix}\right)$$

$$= a_{11}(a_{22}a_{33} - a_{23}a_{32}) - a_{12}(a_{21}a_{33} - a_{23}a_{31})$$
$$+ a_{13}(a_{21}a_{32} - a_{22}a_{31})$$
$$= a_{11}a_{22}a_{33} - a_{11}a_{23}a_{32} - a_{12}a_{21}a_{33} + a_{12}a_{23}a_{31}$$
$$+ a_{13}a_{21}a_{32} - a_{13}a_{22}a_{31}$$
$$= \det\begin{bmatrix} a_{11} & a_{12} & a_{13} \\ a_{21} & a_{22} & a_{23} \\ a_{31} & a_{32} & a_{33} \end{bmatrix}$$

Thus a 3×3 determinant can be evaluated as a linear combination of three 2×2 determinants. Now notice that for each element in the first row of the 3×3 matrix, there exists a submatrix of order 2, obtained by deleting the row and column in which the element lies. This idea is developed further in evaluating determinants of order n.

Now suppose $A = [a_{ij}]$ is an $n \times n$ matrix. For each pair (i, j) with $1 \leqq i,\ j \leqq n$, let A_{ij} be the matrix obtained from A by deleting the ith row and jth column of A. Thus A_{ij} is an $(n-1) \times (n-1)$ matrix.

Example 5.1.4 a. Let

$$A = \begin{bmatrix} 1 & 2 \\ 3 & 5 \end{bmatrix}$$

Then $A_{11} = [5]$, whereas $A_{21} = [2]$.

b. Let

$$B = \begin{bmatrix} 1 & 0 & 7 \\ 9 & 8 & 2 \\ 4 & 1 & 6 \end{bmatrix}$$

Then

$$B_{23} = \begin{bmatrix} 1 & 0 \\ 4 & 1 \end{bmatrix}$$

c. Let

$$C = \begin{bmatrix} 4 & 2 & 6 & 5 \\ 2 & 0 & 1 & -3 \\ -4 & -1 & 2 & 3 \\ 5 & 7 & -1 & 6 \end{bmatrix}$$

Then

$$C_{31} = \begin{bmatrix} 2 & 6 & 5 \\ 0 & 1 & -3 \\ 7 & -1 & 6 \end{bmatrix}$$

Definition 5.1.5 When $A = [a_{ij}]$ is an $n \times n$ matrix, define

$$\det(A) = \sum_{j=1}^{n} (-1)^{1+j} a_{1j} \det(A_{1j}) \qquad (5.1.1)$$

We refer to (5.1.1) as the **expansion of det(A) by the elements of the first row.**

To illustrate this definition, let

$$A = \begin{bmatrix} a_{11} & a_{12} \\ a_{21} & a_{22} \end{bmatrix}$$

be an arbitrary 2×2 matrix. Then $A_{11} = [a_{22}]$, $A_{12} = [a_{21}]$ and

$$\begin{aligned} \det(A) &= \sum_{j=1}^{2} (-1)^{1+j} a_{1j} \det(A_{1j}) \\ &= (-1)^{1+1} a_{11} \det(A_{11}) + (-1)^{1+2} a_{12} \det(A_{12}) \\ &= (-1)^{2} a_{11} \det([a_{22}]) + (-1)^{3} a_{12} \det([a_{21}]) \\ &= a_{11}a_{22} - a_{12}a_{21} \end{aligned}$$

which agrees with (2) in Definition 5.1.1.

Example 5.1.6 Calculate $\det(A)$, where

$$A = \begin{bmatrix} 1 & 7 & 9 \\ 8 & 6 & 4 \\ 3 & 2 & 5 \end{bmatrix}$$

Solution We first calculate A_{11}, A_{12}, and A_{13}. Using the definition, we obtain

$$A_{11} = \begin{bmatrix} 6 & 4 \\ 2 & 5 \end{bmatrix}, \quad A_{12} = \begin{bmatrix} 8 & 4 \\ 3 & 5 \end{bmatrix}, \quad \text{and} \quad A_{13} = \begin{bmatrix} 8 & 6 \\ 3 & 2 \end{bmatrix}$$

Since

$$\begin{aligned} \det(A) = (-1)^{1+1} a_{11} \det(A_{11}) &+ (-1)^{1+2} a_{12} \det(A_{12}) \\ &+ (-1)^{1+3} a_{13} \det(A_{13}) \end{aligned}$$

we must calculate

$$\det(A_{11}) = \det\left(\begin{bmatrix} 6 & 4 \\ 2 & 5 \end{bmatrix}\right) = (6)(5) - (4)(2) = 30 - 8 = 22$$

$$\det(A_{12}) = \det\left(\begin{bmatrix} 8 & 4 \\ 3 & 5 \end{bmatrix}\right) = (8)(5) - (4)(3) = 40 - 12 = 28$$

and

$$\det(A_{13}) = \det\left(\begin{bmatrix} 8 & 6 \\ 3 & 2 \end{bmatrix}\right) = (8)(2) - (6)(3) = 16 - 18 = -2$$

Thus

$$\begin{aligned}\det(A) &= (-1)^2(1)(22) + (-1)^3(7)(28) + (-1)^4(9)(-2) \\ &= 22 - 196 - 18 = -192\end{aligned}$$

■

Example 5.1.7 Find the value of x if $\det(A) = 0$, where

$$A = \begin{bmatrix} 1 & 2 & 5 \\ 1 & x & 5 \\ 3 & -1 & 2 \end{bmatrix}$$

Solution As in Example 5.1.6, we first calculate A_{11}, A_{12}, and A_{13}, obtaining

$$A_{11} = \begin{bmatrix} x & 5 \\ -1 & 2 \end{bmatrix}, \qquad A_{12} = \begin{bmatrix} 1 & 5 \\ 3 & 2 \end{bmatrix}, \quad \text{and} \quad A_{13} = \begin{bmatrix} 1 & x \\ 3 & -1 \end{bmatrix}$$

Furthermore,

$$\det(A_{11}) = \det\left(\begin{bmatrix} x & 5 \\ -1 & 2 \end{bmatrix}\right) = (x)(2) - (5)(-1) = 2x + 5$$

$$\det(A_{12}) = \det\left(\begin{bmatrix} 1 & 5 \\ 3 & 2 \end{bmatrix}\right) = (1)(2) - (5)(3) = 2 - 15 = -13$$

and

$$\det(A_{13}) = \det\left(\begin{bmatrix} 1 & x \\ 3 & -1 \end{bmatrix}\right) = (1)(-1) - (x)(3) = -1 - 3x$$

Thus

$$\begin{aligned}\det(A) &= (-1)^{1+1}a_{11}\det(A_{11}) + (-1)^{1+2}a_{12}\det(A_{12}) \\ &\qquad + (-1)^{1+3}a_{13}\det(A_{13}) \\ &= (-1)^2(1)(2x+5) + (-1)^3(2)(-13) + (-1)^4(5)(-1-3x) \\ &= 2x + 5 + 26 - 5 - 15x = -13x + 26\end{aligned}$$

and so $\det(A) = 0$ becomes $-13x + 26 = 0$, or $x = 2$. ■

There is nothing sacred about using the first row to evaluate $\det(A)$; that is, in $\det(A) = \sum_{j=1}^{n}(-1)^{1+j}a_{1j}\det(A_{1j})$, we use only the entries a_{1j} that lie in row 1. In the following exercise set for $n = 3$, you will be asked to demonstrate that for the ith row $(i = 1, 2, \ldots, n)$

$$\det(A) = \sum_{j=1}^{n} (-1)^{i+j} a_{ij} \det(A_{ij}) \tag{5.1.2}$$

We refer to (5.1.2) as the **expansion of det(A) by the ith row.** In fact, we can even use the columns. For the jth column $(j = 1, 2, \ldots, n)$,

$$\det(A) = \sum_{i=1}^{n} (-1)^{i+j} a_{ij} \det(A_{ij}) \tag{5.1.3}$$

We refer to (5.1.3) as the **expansion of det(A) by the jth column.** Note the difference between formulas (5.1.2) and (5.1.3).

For a 3×3 matrix

$$A = \begin{bmatrix} a_{11} & a_{12} & a_{13} \\ a_{21} & a_{22} & a_{23} \\ a_{31} & a_{32} & a_{33} \end{bmatrix}$$

using formula (5.1.2) (expansion by the ith row), we have

$$\det(A) = (-1)1^{i+1}a_{i1}\det(A_{i1}) + (-1)^{i+2}a_{i2}\det(A_{i2})$$
$$+(-1)^{i+3}a_{i3}\det(A_{i3})$$

while formula (5.1.3) (expansion by the jth column) yields

$$\det(A) = (-1)^{i+j}a_{1j}\det(A_{1j}) + (-1)^{2+j}a_{2j}\det(A_{2j})$$
$$+(-1)^{3+j}a_{3j}\det(A_{3j})$$

Thus formula (5.1.2) sums on the columns (i.e., the a_{ij} lies in the same row), whereas formula (5.1.3) sums on the rows (i.e., the a_{ij} lies in the same column).

Example 5.1.8 Calculate $\det(A)$, where

$$A = \begin{bmatrix} 1 & 7 & 9 \\ 8 & 6 & 4 \\ 3 & 2 & 5 \end{bmatrix}$$

by expanding by the second row.

Solution

$$\det(A) = (-1)^{2+1}a_{21}\det(A_{21}) + (-1)^{2+2}a_{22}\det(A_{22}) + (-1)^{2+3}a_{23}\det(A_{23})$$

$$= -8\det\left(\begin{bmatrix} 7 & 9 \\ 2 & 5 \end{bmatrix}\right) + 6\det\left(\begin{bmatrix} 1 & 9 \\ 3 & 5 \end{bmatrix}\right) - 4\det\left(\begin{bmatrix} 1 & 7 \\ 3 & 2 \end{bmatrix}\right)$$

$$= -8(35 - 18) + 6(5 - 27) - 4(2 - 21)$$

$$= -8(17) + 6(-22) - 4(-19)$$

$$= -136 - 132 + 76 = -192$$

This agrees with the answer obtained by expansion by the first row as performed in Example 5.1.6. ■

Example 5.1.9 Find $\det(A)$, where

$$A = \begin{bmatrix} 0 & 0 & 1 & 0 \\ 4 & 2 & 1 & 0 \\ 1 & 1 & 3 & 1 \\ 0 & 1 & 0 & 4 \end{bmatrix}$$

Solution We can take advantage of the zeros by expanding by row 1, and then by column 1 obtaining

$$\det(A) = \det\left(\begin{bmatrix} 4 & 2 & 0 \\ 1 & 1 & 1 \\ 0 & 1 & 4 \end{bmatrix}\right)$$

$$= 4\det\left(\begin{bmatrix} 1 & 1 \\ 1 & 4 \end{bmatrix}\right) - 1\det\left(\begin{bmatrix} 2 & 0 \\ 1 & 4 \end{bmatrix}\right)$$

$$= 4(4 - 1) - 1(8 - 0) = 4$$ ■

SUMMARY

Terms to know:

Determinant of an $n \times n$ matrix

Methods to know:

Calculate $\det(A)$ by (1) expansion by rows and (2) expansion by columns.

EXERCISE SET 5.1

Evaluate the determinants of each of the following matrices.

1. $\begin{bmatrix} 1 & 4 \\ 7 & 9 \end{bmatrix}$

2. $\begin{bmatrix} 1 & 3 \\ 2 & 6 \end{bmatrix}$

3. $\begin{bmatrix} 6 & -1 \\ 3 & -2 \end{bmatrix}$

4. $\begin{bmatrix} 5 & 7 \\ 8 & 0 \end{bmatrix}$

5. $\begin{bmatrix} \frac{1}{2} & -\frac{1}{3} \\ \frac{1}{4} & \frac{1}{6} \end{bmatrix}$

6. $\begin{bmatrix} -1.2 & 3.4 \\ 2.7 & 5.3 \end{bmatrix}$

In Exercises 7–9, solve for x.

7. $\det\left(\begin{bmatrix} 7 & -6 \\ x & 9 \end{bmatrix}\right) = 0$

8. $\det\left(\begin{bmatrix} 1 & -1 \\ x & x^2 \end{bmatrix}\right) = 2$

9. $\det\left(\begin{bmatrix} 1-x & 3 \\ 5 & 3-x \end{bmatrix}\right) = 0$

10. For what values of x is

$$\det\left(\begin{bmatrix} x & 8 \\ 32 & x \end{bmatrix}\right) > 0$$

11. First evaluate $\det(A)$ by expanding by the elements of the first row and then by the elements of the second column if

$$A = \begin{bmatrix} 3 & 2 & -6 \\ 1 & 0 & -2 \\ 2 & -1 & -4 \end{bmatrix}$$

In Exercises 12–17, evaluate the determinant, making efficient choices for the row or column by which you expand.

12. $\begin{bmatrix} 1 & 8 & 7 \\ 2 & 0 & 4 \\ 1 & 0 & 3 \end{bmatrix}$

13. $\begin{bmatrix} 1 & 0 & 0 \\ 0 & 1 & 1 \\ 0 & 0 & 2 \end{bmatrix}$

14. $\begin{bmatrix} 8 & 9 & 7 \\ 0 & 4 & 2 \\ 0 & 0 & 9 \end{bmatrix}$

15. $\begin{bmatrix} 1 & 1 & -1 \\ 3 & 0 & 2 \\ -3 & -3 & 3 \end{bmatrix}$

16. $\begin{bmatrix} 3 & 2 & 1 & 1 \\ 1 & 2 & 0 & 0 \\ -1 & 3 & 2 & 1 \\ 4 & 1 & 1 & 4 \end{bmatrix}$

17. $\begin{bmatrix} a & c & b \\ 2a & 2 & 3 \\ c & b & 8a \end{bmatrix}$

18. Find the values of x for which

$$\det\left(\begin{bmatrix} 1 & 2 & -3 \\ 1 & x & -3 \\ 1 & 4 & -x \end{bmatrix}\right) = 0$$

19. Verify the following identity.

$$\det\left(\begin{bmatrix} 1 & a & a^2 \\ 1 & b & b^2 \\ 1 & c & c^2 \end{bmatrix}\right) = (b-a)(c-a)(c-b)$$

20. Let

$$A = \begin{bmatrix} a_{11} & a_{12} & a_{13} \\ a_{21} & a_{22} & a_{23} \\ a_{31} & a_{32} & a_{33} \end{bmatrix}$$

Evaluate $\det(A)$ by expanding by the (a) first row, (b) second row, and (c) third row. Compare your answers with part 3 of Definition 5.1.1.

21. Prove that $\det(A) = 0$ if all the elements of some row or column are zeros.

22. Replace the first column of A in Exercise 20 with ca_{11}, ca_{21}, and ca_{31} to obtain a new matrix B. Then expand $\det(B)$ by the elements of the first column. Verify $\det(B) = c\det(A)$.

23. Replace the second column of A in Exercise 20 with a_{11}, a_{21}, and a_{31} to obtain a matrix B with two identical columns. Expand $\det(B)$ by the elements of the third column and verify that $\det(B) = 0$.

24. Given matrix A in Exercise 20, find A^T and then $\det(A^T)$. Verify $\det(A^T) = \det(A)$.

25. Interchange the first and second rows of A in Exercise 20 to obtain a new matrix B. Verify that $\det(B) = -\det(A)$.

26. Let

$$A = \begin{bmatrix} 3 & -2 \\ -3 & 1 \end{bmatrix} \quad \text{and} \quad B = \begin{bmatrix} 4 & -1 \\ 2 & -3 \end{bmatrix}$$

a. Find AB and verify that $\det(AB) = [\det(A)][\det(B)]$.
b. Find $A + B$ and verify that $\det(A + B) \neq \det(A) + \det(B)$.

27. Replace each element a_{ij} of A in Exercise 20 with ca_{ij} to obtain a new matrix B. What is $\det(B)$? Suppose each element a_{ij} in an $n \times n$ matrix A is replaced by ca_{ij} to obtain a new matrix B. What is $\det(B)$?

28. a. Let

$$A = \begin{bmatrix} 2 & 0 & 0 \\ 0 & 3 & 0 \\ 0 & 0 & 4 \end{bmatrix}$$

Evaluate $\det(A)$.

b. Let

$$A = \begin{bmatrix} a_{11} & 0 & 0 \\ 0 & a_{22} & 0 \\ 0 & 0 & a_{33} \end{bmatrix}$$

Evaluate $\det(A)$.

c. Let A be a diagonal matrix of order n. What (in words) is $\det(A)$?

d. Let A be the identity matrix of order n. What is $\det(A)$?

29. a. Let

$$A = \begin{bmatrix} 2 & -1 & -4 \\ 0 & 3 & 5 \\ 0 & 0 & 4 \end{bmatrix}$$

Evaluate $\det(A)$.

b. Let

$$A = \begin{bmatrix} a_{11} & a_{12} & a_{13} \\ 0 & a_{22} & a_{23} \\ 0 & 0 & a_{33} \end{bmatrix}.$$

Evaluate $\det(A)$.

c. Let A be an upper (or lower) triangular matrix of order n. What (in words) is $\det(A)$?

SECTION 5.2 DETERMINANTS AND ROW OPERATIONS

In Chapter 1, we discussed three operations performed on the rows of a matrix:

1. Multiply a row by a nonzero number.
2. Add a multiple of one row to another.
3. Interchange two rows of a matrix.

We now consider the effects on $\det(A)$ due to the use of row operations on A. First, consider operation (1), and suppose $B = [b_{ij}]$ and $A = [a_{ij}]$ are $n \times n$ matrices where $a_{ij} = b_{ij}$ for $i \neq i_0$, and $b_{i_0j} = ca_{i_0j}$ for all $j = 1, 2, \ldots, n$. In other words, matrix B is obtained from matrix A by multiplying the i_0th row by c. Then, expanding by the i_0th row, we have

$$\begin{aligned} \det(B) &= \sum_{j=1}^{n} (-1)^{i_0+j} b_{i_0j} \det\left(B_{i_0j}\right) \\ &= \sum_{j=1}^{n} (-1)^{i_0+j} ca_{i_0j} \det\left(A_{i_0j}\right) \quad [\text{since } a_{ij} = b_{ij} \text{ for } i \neq i_0] \\ &= c \sum_{j=1}^{n} (-1)^{i_0+j} a_{i_0j} \det\left(A_{i_0j}\right) \\ &= c \det(A) \end{aligned}$$

Thus we have

1′. If a matrix B is obtained from matrix A by multiplying any row of A by the number c, then

$$\det(B) = c \det(A)$$

Corresponding to row operations (2) and (3), we have the following rules.

2′. If a matrix B is obtained from matrix A by adding a multiple of one row to another row (i.e., k row i to row j), then

$$\det(B) = \det(A)$$

3′. If a matrix B is obtained from matrix A by interchanging two rows, then

$$\det(B) = -\det(A)$$

In the case of 2×2 and 3×3 matrices, the proofs of these two rules are easy. You are asked to verify 2′ and 3′ in these cases in the exercises.

Next we observe that in an upper triangular matrix T (i.e., $t_{ij} = 0$ for $i > j$),

$$\det(T) = t_{11}t_{22} \cdots t_{nn}$$

This is easily seen to be true in the 3×3 upper triangular matrix

$$A = \begin{bmatrix} t_{11} & t_{12} & t_{13} \\ 0 & t_{22} & t_{23} \\ 0 & 0 & t_{33} \end{bmatrix}$$

Expanding by the first column twice, we obtain

$$\begin{aligned} \det(T) &= t_{11}\det(T_{11}) - 0\det(T_{12}) + 0\det(T_{13}) \\ &= t_{11}\det\left(\begin{bmatrix} t_{22} & t_{23} \\ 0 & t_{33} \end{bmatrix}\right) \\ &= t_{11}(t_{22}\det([t_{33}])) \\ &= t_{11}t_{22}t_{33} \end{aligned}$$

Let A be an $n \times n$ matrix. Evaluating $\det(A)$ by row or column expansion can be very tedious and prone to error. The amount of calculation can often be reduced by using the following method.

1. Using rules 1′, 2′, and 3′ to reduce A to an upper triangular matrix T. We obtain $\det(A) = k\det(T)$ where the value of the constant k is determined by rules 1′–3′.
2. Since $\det(T)$ is read off as

   $$\det(T) = t_{11}t_{22} \cdots t_{nn}$$

   then

   $$\det(A) = kt_{11}t_{22} \cdots t_{nn}$$

Example 5.2.1 Calculate det(A), where

$$A = \begin{bmatrix} 1 & 3 & 7 & 4 \\ 2 & 9 & 1 & 8 \\ 3 & 1 & 4 & 2 \\ 0 & 1 & 6 & 3 \end{bmatrix}$$

Solution

$$\det(A) = \det\left(\begin{bmatrix} 1 & 3 & 7 & 4 \\ 0 & 3 & -13 & 0 \\ 3 & 1 & 4 & 2 \\ 0 & 1 & 6 & 3 \end{bmatrix}\right) \quad (-2)R1 \text{ to } R2 \quad [\text{rule } 2']$$

$$= \det\left(\begin{bmatrix} 1 & 3 & 7 & 4 \\ 0 & 3 & -13 & 0 \\ 0 & -8 & -17 & -10 \\ 0 & 1 & 6 & 3 \end{bmatrix}\right) \quad (-3)R1 \text{ to } R3 \quad [\text{rule } 2']$$

$$= (-1)^{1+1}\det\left(\begin{bmatrix} 3 & -13 & 0 \\ -8 & -17 & -10 \\ 1 & 6 & 3 \end{bmatrix}\right) \quad \text{expansion by column 1}$$

$$= (-1)\det\left(\begin{bmatrix} 1 & 6 & 3 \\ -8 & -17 & -10 \\ 3 & -13 & 0 \end{bmatrix}\right) \quad \text{interchange } R1 \text{ with } R3 \quad [\text{rule } 3']$$

$$= (-1)\det\left(\begin{bmatrix} 1 & 6 & 3 \\ 0 & 31 & 14 \\ 3 & -13 & 0 \end{bmatrix}\right) \quad (8)R1 \text{ to } R2 \quad [\text{rule } 2']$$

$$= (-1)\det\left(\begin{bmatrix} 1 & 6 & 3 \\ 0 & 31 & 14 \\ 0 & -31 & -9 \end{bmatrix}\right) \quad (-3)R1 \text{ to } R3 \quad [\text{rule } 2']$$

$$= (-1)\det\left(\begin{bmatrix} 1 & 6 & 3 \\ 0 & 31 & 14 \\ 0 & 0 & 5 \end{bmatrix}\right) \quad (1)R2 \text{ to } R3 \quad [\text{rule } 2']$$

$$= (-1)[(1)(31)(5)] = -155$$

■

Example 5.2.2 Calculate det(A), where

$$A = \begin{bmatrix} 1 & 1 & 1 & 1 \\ 2 & 1 & 1 & -2 \\ 1 & -1 & -1 & 1 \\ 2 & -3 & 4 & 1 \end{bmatrix}$$

Solution

$$\det(A) = \det\left(\begin{bmatrix} 1 & 1 & 1 & 1 \\ 0 & -1 & -1 & -4 \\ 1 & -1 & -1 & 1 \\ 2 & -3 & 4 & 1 \end{bmatrix}\right) \quad (-2)R1 \text{ to } R2 \quad [\text{rule } 2']$$

$$= \det\left(\begin{bmatrix} 1 & 1 & 1 & 1 \\ 0 & -1 & -1 & -4 \\ 0 & -2 & -2 & 0 \\ 2 & -3 & 4 & 1 \end{bmatrix}\right) \quad (-1)R1 \text{ to } R3 \quad [\text{rule } 2']$$

$$= \det\left(\begin{bmatrix} 1 & 1 & 1 & 1 \\ 0 & -1 & -1 & -4 \\ 0 & -2 & -2 & 0 \\ 0 & -5 & 2 & -1 \end{bmatrix}\right) \quad (-2)R1 \text{ to } R4 \quad [\text{rule } 2']$$

$$= 1\det\left(\begin{bmatrix} -1 & -1 & -4 \\ -2 & -2 & 0 \\ -5 & 2 & -1 \end{bmatrix}\right) \quad \text{expansion by column 1}$$

$$= \det\left(\begin{bmatrix} -1 & -1 & -4 \\ 0 & 0 & 8 \\ -5 & 2 & -1 \end{bmatrix}\right) \quad (-2)R1 \text{ to } R2 \quad [\text{rule } 2']$$

$$= (-8)\det\left(\begin{bmatrix} -1 & -1 \\ -5 & 2 \end{bmatrix}\right) \quad \text{expansion by Row 2}$$

$$= 8\det\left(\begin{bmatrix} 1 & 1 \\ -5 & 2 \end{bmatrix}\right) \quad (-1)R1 \quad [\text{rule } 1']$$

$$= 8[2 - (-5)] = 56$$

■

The following properties are also very helpful in simplifying the evaluation of determinants:

1. $\det(A^T) = \det(A)$. (Compare Exercise 24, Exercise Set 5.1)

Example 5.2.3 Let

$$A = \begin{bmatrix} 1 & 2 \\ 3 & 4 \end{bmatrix}$$

Show that $\det(A) = \det(A^T)$.

Solution $\det(A) = 1(4) - (2)3 = 4 - 6 = -2$. However,

$$A^T = \begin{bmatrix} 1 & 3 \\ 2 & 4 \end{bmatrix}$$

and $\det(A^T) = 1(4) - (3)2 = 4 - 6 = -2$ ■

2. If two rows of A are identical or proportional, then $\det(A) = 0$. (Compare Exercise 23, Exercise Set 5.1)

Example 5.2.4 Let

$$A = \begin{bmatrix} 1 & 2 \\ 1 & 2 \end{bmatrix}$$

Then $\det(A) = 2 - 2 = 0$ ■

3. If a row of A is identically 0, then $\det(A) = 0$. (Exercise 21, Exercise Set 5.1)
4. If I is the identity matrix, then $\det(I) = 1$.

SUMMARY

Facts to know:

If a matrix B is obtained from a matrix A by multiplying a row of A by the number c, then $\det(B) = c\det(A)$.

If a matrix B is obtained from a matrix A by interchanging two rows of A, then $\det(B) = -\det(A)$.

If a matrix B is obtained from a matrix A by adding a multiple of one row of A to another row of A, then $\det(B) = \det(A)$.

$\det(A^T) = \det(A)$.

If two rows of A are identical or proportional, then $\det(A) = 0$.

If some row of A is identically 0, then $\det(A) = 0$.

If I is the identity matrix, then $\det(I) = 1$.

EXERCISE SET 5.2

1. Explain why $\det(A) = \det(B)$ and then verify this fact by actual evaluation, where

$$A = \begin{bmatrix} 5 & 1 & 8 \\ -5 & 1 & 4 \\ 6 & -1 & 0 \end{bmatrix} \quad \text{and} \quad B = \begin{bmatrix} 5 & -5 & 6 \\ 1 & 1 & -1 \\ 8 & 4 & 0 \end{bmatrix}$$

2. Show by actual expansion that the sign of $\det(A)$ is changed if the second and third row are interchanged, where

$$A = \begin{bmatrix} 9 & 5 & 4 \\ -2 & 9 & 2 \\ 6 & -4 & 8 \end{bmatrix}$$

3. Without evaluating, state why the determinants of the following matrices are zero. Check by evaluating the determinants.

a. $\begin{bmatrix} 2 & 9 & -3 \\ 1 & 2 & 1 \\ 2 & 9 & -3 \end{bmatrix}$ b. $\begin{bmatrix} 4 & 1 & -2 \\ 1 & 2 & 5 \\ -8 & -2 & 4 \end{bmatrix}$

4. Evaluate the det(A), where

$$A = \begin{bmatrix} 1 & 3 & -2 \\ -3 & 6 & 3 \\ 2 & 6 & -4 \end{bmatrix}$$

by factoring out the common terms from any row and then expanding. Check by expanding directly.

5. In the following matrix, multiply each of the elements in the second row by -3, and form a new matrix by adding these results to the corresponding elements in the third row. Show by direct expansion that the value of the original determinant is unchanged.

$$\begin{bmatrix} 1 & 0 & -1 \\ 1 & 3 & -5 \\ 2 & 3 & -3 \end{bmatrix}$$

6. In the matrix of Exercise 5, multiply each of the elements in the first column by 2, and form a new matrix by adding these results to the corresponding elements in the third column. Show by direct expansion that the value of the original determinant is unchanged.

Evaluate the determinants of the following matrices using elementary row operations to simplify the calculations.

7. $\begin{bmatrix} -1 & 0 & 1 \\ 2 & -3 & 2 \\ 1 & -3 & 5 \end{bmatrix}$

8. $\begin{bmatrix} 2 & 1 & 3 \\ 5 & -1 & 3 \\ 7 & 0 & 5 \end{bmatrix}$

9. $\begin{bmatrix} 1 & 3 & -2 & 1 \\ 2 & 0 & -1 & 3 \\ 3 & 0 & 4 & 2 \\ 1 & 2 & -3 & -4 \end{bmatrix}$

10. $\begin{bmatrix} 5 & 2 & 4 & 2 \\ -1 & 0 & 1 & 2 \\ 2 & 2 & -4 & 4 \\ 3 & 1 & -6 & -5 \end{bmatrix}$

11. $\begin{bmatrix} 0 & a & b \\ -a & 0 & -c \\ -b & c & 0 \end{bmatrix}$

12. Show by expanding each determinant and comparing the results that

$$\det\left(\begin{bmatrix} a_{11} & a_{12} \\ b_{11} + c_{11} & b_{12} + c_{12} \end{bmatrix}\right) = \det\left(\begin{bmatrix} a_{11} & a_{12} \\ b_{11} & b_{12} \end{bmatrix}\right) + \det\left(\begin{bmatrix} a_{11} & a_{12} \\ c_{11} & c_{12} \end{bmatrix}\right)$$

13. Determine the roots of the equation det(A) = 0, where

$$A = \begin{bmatrix} 1 & x & x^2 \\ 1 & a & a^2 \\ 1 & b & b^2 \end{bmatrix}$$

Hint: If $x = a$, the first two are equal.

14. Determine the roots of the equation $\det(A) = 0$, where

$$A = \begin{bmatrix} 1 & x & x^2 & x^3 \\ 1 & a & a^2 & a^3 \\ 1 & b & b^2 & b^3 \\ 1 & c & c^2 & c^3 \end{bmatrix}$$

15. Find the roots of the equation $\det(A) = 0$, where

$$A = \begin{bmatrix} 3 & 2x & x^2 \\ 1 & 2 & 3 \\ 9 & 6 & 3 \end{bmatrix}$$

16. Evaluate $\det(A)$, where

$$A = \begin{bmatrix} a_{11} & a_{12} & a_{13} \\ a_{21} & a_{22} & a_{32} \\ ca_{11} & ca_{12} & ca_{13} \end{bmatrix}$$

Notice that the third row is proportional to the first row.

17. Show for a 3×3 matrix: If a matrix B is obtained from matrix A by adding a multiple of one row to another row, then $\det(B) = \det(A)$.

18. Show for a 3×3 matrix: If a matrix B is obtained from matrix A by interchanging two rows, then $\det(B) = -\det(A)$.

In the following sequence of exercises, we verify Rules 2′ and 3′ for general $n \times n$ matrices. These exercises require a knowledge of mathematical induction and are more challenging than the previous exercises.

19. Let A be an $n \times n$ matrix. Let B be the $n \times n$ matrix obtained by interchanging row i and row j of A. Thus if

$$A = \begin{bmatrix} a_{11} & a_{12} & \cdots & a_{1n} \\ \vdots & \vdots & & \vdots \\ a_{i1} & a_{i2} & \cdots & a_{in} \\ \vdots & \vdots & & \vdots \\ a_{j1} & a_{j2} & \cdots & a_{jn} \\ \vdots & \vdots & & \vdots \\ a_{n1} & a_{n2} & \cdots & a_{nn} \end{bmatrix}$$

then

$$B = \begin{bmatrix} a_{11} & a_{12} & \cdots & a_{1n} \\ \vdots & \vdots & & \vdots \\ a_{j1} & a_{j2} & \cdots & a_{jn} \\ \vdots & \vdots & & \vdots \\ a_{i1} & a_{i2} & \cdots & a_{in} \\ \vdots & \vdots & & \vdots \\ a_{n1} & a_{n2} & \cdots & a_{nn} \end{bmatrix}$$

In previous exercises you have verified (3′) for $n = 2$ and $n = 3$. Suppose you have verified (3′) for $n = k$. We will verify (3′) for $n = k + 1$. By mathematical induction, this suffices to prove (3′) for all n.

a. Find $\det(A)$ by expanding by column 1.

b. Find $\det(B)$ by expanding by column 1.

c. Since (3′) is true for $n = k$, each

$$\det(B_{i1}) = -\det(A_{i1})$$

Using this fact, rewrite the equation for $\det(B)$.

d. Noting that $b_{i1} = a_{j1}$, $b_{j1} = a_{i1}$, and $b_{s1} = a_{s1}$ for $s \neq i, j$, substitute for the b_{s1} in the result of part (c).

e. Finally, by factoring out the (-1) in the result of part (d), observe that the result of part (d) is just $-\det(A)$.

20. Let A be an $n \times n$ matrix with row i and row j equal.

a. Let B be the $n \times n$ matrix obtained from A by interchanging row i with row j. Now use (3′) to write $\det(B)$ in terms of $\det(A)$.

b. Noting that $A = B$ and so $\det(A) = \det(B)$, show that $\det(A) = 0$.

Hint: If $x = -x$, what real numbers must x be?

21. Let A be an $n \times n$ matrix and let B be the matrix obtained from A by adding c row i to row j. This is a proof of property 2′.

a. What is b_{j1}?

b. What is b_{j2}?

c. What is b_{jk}?

d. Write out $\det(B)$ expanding by row j.

e. In the formula found in part (d), $A_{jk} = B_{jk}$. Thus $\det(B_{jk}) = \det(A_{jk})$ for $k = 1, 2, \ldots, n$. In the formula in part (d), replace $\det(B_{jk})$ by $\det(A_{jk})$.

f. In part (c), you found that $b_{jk} = ca_{ik} + a_{jk}$. Separate your formula from part (e) into two parts—the first involving only the ca_{ik} and the second involving only the a_{jk}.

g. In part (f), the sum involving only the a_{jk} is $\det(A)$. Replace this sum by $\det(A)$.

h. Let C be the matrix obtained from A by replacing row j by row i. Write $\det(C)$ by expanding by row j. Note that $\det(C) = 0$ since C has two identical rows.

i. Factor out the c the sum found in part (f) involving only the ca_{ik}. Compare this with the formula found for $\det(C)$ found in part (h).

j. From part (i), you should observe that the sum found in part (f) involving only the ca_{ik} is 0. Thus $\det(B)$ equals the sum in part (f), involving only the a_{jk}. In part (g), you found this sum to be $\det(A)$. Thus you have shown that $\det(B) = \det(A)$.

22. In Exercise 19 of section 5.1 you showed that

$$\det\left(\begin{bmatrix} 1 & a & a^2 \\ 1 & b & b^2 \\ 1 & c & c^2 \end{bmatrix}\right) = (b-a)(c-a)(c-b)$$

a. Verify that

$$\det\left(\begin{bmatrix} 1 & a & a^2 & a^3 \\ 1 & b & b^2 & b^3 \\ 1 & c & c^2 & c^3 \\ 1 & d & d^2 & d^3 \end{bmatrix}\right)$$
$$= (b-a)(c-a)(d-a)(c-b)(d-b)(d-c)$$

b. Verify that

$$\det\left(\begin{bmatrix} 1 & a & a^2 & a^3 & a^4 \\ 1 & b & b^2 & b^3 & b^4 \\ 1 & c & c^2 & c^3 & c^4 \\ 1 & d & d^2 & d^3 & d^4 \\ 1 & e & e^2 & e^3 & e^4 \end{bmatrix}\right)$$

$$= (b-a)(c-a)(d-a)(e-a)(c-b)(d-b)(e-b)(d-c)(e-c)(e-d)$$

The following two problems are best solved using the accompanying software package or a hand-held calculator. Compute the determinant of the following matrices.

23. $\begin{bmatrix} 2.1 & 0.9 & 1.1 & 1.2 \\ 1.9 & 4.2 & -1.4 & -1.6 \\ 4.3 & 1.9 & 3.8 & 5.6 \\ -2.7 & 0.9 & 0.9 & 1.1 \end{bmatrix}$

24. $\begin{bmatrix} 4.2 & 2.4 & 5.7 & 5.3 \\ 1.9 & 0 & 1.4 & -3.2 \\ -4.5 & -1.6 & 1.7 & 2.9 \\ 5.2 & 6.7 & -1.3 & 6.4 \end{bmatrix}$

SECTION 5.3 DETERMINANTS AND INVERSES

In this section we will exhibit some uses of determinants. We require the following theorems. The proofs of these theorems are omitted.

Theorem 5.3.1 *For $n \times n$ matrices A and B,* $\det(AB) = [\det(A)][\det(B)]$.

In the case of 2×2 matrices, this theorem can be verified without too much trouble. Let

$$A = \begin{bmatrix} a & b \\ c & d \end{bmatrix} \quad \text{and} \quad B = \begin{bmatrix} r & s \\ p & q \end{bmatrix}$$

Then $\det(A) = ad - bc$ and $\det(B) = rq - sp$. Furthermore,

$$AB = \begin{bmatrix} ar + bp & as + bq \\ cr + dp & cs + dq \end{bmatrix}$$

and

$$\begin{aligned} \det(AB) &= (ar + bp)(cs + dq) - (as + bq)(cr + dp) \\ &= arcs + ardq + bpcs + bpdq - ascr - asdp - bqcr - bqdp \\ &= ardq - crbq - dpas + bpcs \\ &= (ad - bc)(rq - sp) = [\det(A)][\det(B)] \end{aligned}$$

Having verified Theorem 5.3.1 for $n = 2$, we will assume it true for $n > 2$.

Theorem 5.3.2 *If I_n is the $n \times n$ identity matrix, then* $\det(I_n) = 1$.

Theorem 5.3.3 $\det(A^T) = \det(A)$.

This theorem is useful because when we perform a row operation on A to obtain the matrix B, then B^T is obtained from A^T by performing the same operation on the columns of A^T. Thus this theorem says that we can use column operations as well as row operations to simplify our calculation of $\det(A)$.

To be specific, the following rules can be used in calculating $\det(A)$:

1′. If a matrix B is obtained from matrix A by multiplying the ith column of A by the number c, then

$$\det(B) = c[\det(A)]$$

2′. If a matrix B is obtained from matrix A by adding a multiple of one column to another column, then

$$\det(B) = \det(A)$$

3′. If a matrix B is obtained from matrix A by interchanging two columns, then

$$\det(B) = -\det(A)$$

Example 5.3.4 Use column operations to calculate $\det(A)$, where

$$A = \begin{bmatrix} 1 & 2 & 4 \\ 3 & 1 & 2 \\ 2 & 4 & 8 \end{bmatrix}$$

Solution

$$\det\left(\begin{bmatrix} 1 & 2 & 4 \\ 3 & 1 & 2 \\ 2 & 4 & 8 \end{bmatrix}\right) \qquad [\text{add } (-2)C1 \text{ to } C2]$$

$$= \det\left(\begin{bmatrix} 1 & 0 & 4 \\ 3 & -5 & 2 \\ 2 & 0 & 8 \end{bmatrix}\right) \qquad [(-4)C1 \text{ to } C3]$$

$$= \det\left(\begin{bmatrix} 1 & 0 & 0 \\ 3 & -5 & -10 \\ 2 & 0 & 0 \end{bmatrix}\right)$$

$$= 1\det\left(\begin{bmatrix} -5 & -10 \\ 0 & 0 \end{bmatrix}\right) \qquad [\text{expansion by row } 1]$$

$$= 0$$

■

15. Verify that if A and B are square matrices and AB has an inverse, then A and B have inverses.
16. An elementary matrix E is obtained from the identity matrix by the interchange of two rows; by adding to one row a constant multiple of another row; or by multiplying a row by a nonzero constant.

    a. Verify that if E is an elementary matrix and A is any matrix of the same order, then

    $$\det(EA) = [(\det E)][(\det A)]$$

    b. If A is an $n \times n$ nonsingular matrix, then there exist elementary matrices $E_1, E_2, \ldots, E_k$ such that

    $$E_k E_{k-1} \cdots E_2 E_1 A = I$$

    Use this fact to show that if A is nonsingular, then $\det(AB) = [\det(A)][\det(B)]$ for A and B $n \times n$ matrices.

    c. If A is an $n \times n$ singular matrix, then there exist elementary matrices $E_1, E_2, \ldots, E_k$ such that

    $$E_k E_{k-1} \cdots E_2 E_1 A = T$$

    where at least one row of T consists completely of zeros. Use this fact to show for the case where A is singular, then $\det(AB) = [(\det(A)][\det(B)]$ for A and B $n \times n$ matrices.

SECTION 5.4 DETERMINANTS AND SOLUTIONS OF SYSTEMS OF EQUATIONS

Let $A\mathbf{x} = \mathbf{b}$ be a system of equations, where A is an $n \times n$ matrix. If $\mathbf{c}$ is a solution, then $A\mathbf{c} = \mathbf{b}$, or

$$c_1\mathbf{A}_1 + \cdots + c_n\mathbf{A}_n = \mathbf{b}$$

Let $A_{(i)}$ be the matrix obtained from A by replacing the ith column of A by the vector

$$\begin{bmatrix} b_1 \\ b_2 \\ \vdots \\ b_n \end{bmatrix}$$

We can obtain $A_{(i)}$ from A by multiplying the ith column of A by c_i and then adding c_1 (column 1) to column i, then c_2 (column 2) to column i, and so on, until column i becomes $c_1\mathbf{A}_1 + c_2\mathbf{A}_2 + \cdots + c_n\mathbf{A}_n = \mathbf{b}$. Thus the resulting matrix is $A_{(i)}$. In terms of determinants, the only operation changing A to $A_{(i)}$ that makes a difference in the value of $\det(A)$ is the operation of multiplying the ith column by c_i. The other operation adds a multiple of one column to another and so leaves the determinant unchanged, by rule 2′. Thus

$$\det(A_{(i)}) = c_i\det(A)$$

or

$$c_i = \frac{\det(A_{(i)})}{\det(A)} \qquad \text{if } \det(A) \neq 0$$

This method of solving $A\mathbf{x} = \mathbf{b}$ for A an $n \times n$ matrix is called **Cramer's rule**.

Theorem 5.4.1 (Cramer's Rule) *Let A be an $n \times n$ nonsingular matrix and $\mathbf{b}$ be an $n \times 1$ matrix (vector). Then the solution vector $\mathbf{c}$ of $A\mathbf{c} = \mathbf{b}$ is obtained as follows: The ith component of $\mathbf{c}$ is c_i and*

$$c_i = \frac{\det(A_{(i)})}{\det(A)}$$

where $A_{(i)}$ is the matrix obtained from A by replacing the ith column of A by $\mathbf{b}$.

Example 5.4.2 Solve

$$\begin{bmatrix} 3 & 1 \\ 2 & 4 \end{bmatrix}\begin{bmatrix} x_1 \\ x_2 \end{bmatrix} = \begin{bmatrix} 1 \\ 4 \end{bmatrix}$$

using Cramer's rule.

Solution Let

$$A = \begin{bmatrix} 3 & 1 \\ 2 & 4 \end{bmatrix}$$

Then

$$A_{(1)} = \begin{bmatrix} 1 & 1 \\ 4 & 4 \end{bmatrix} \quad \text{and} \quad A_{(2)} = \begin{bmatrix} 3 & 1 \\ 2 & 4 \end{bmatrix}$$

Thus $\det(A_{(1)}) = 0$ and $\det(A_{(2)}) = \det(A) = 10$. So

$$c_1 = \frac{\det(A_{(1)})}{\det(A)} = 0 \quad \text{and} \quad c_2 = \frac{\det(A_{(2)})}{\det(A)} = 1$$

To check,

$$\begin{bmatrix} 3 & 1 \\ 2 & 4 \end{bmatrix}\begin{bmatrix} 0 \\ 1 \end{bmatrix} = \begin{bmatrix} 3 \cdot 0 + 1 \cdot 1 \\ 2 \cdot 0 + 4 \cdot 1 \end{bmatrix} = \begin{bmatrix} 1 \\ 4 \end{bmatrix}$$

■

Example 5.4.3 Using Cramer's rule, solve the system

$$\begin{aligned} x_1 - x_2 + 2x_3 &= 0 \\ 2x_1 + x_2 + 3x_3 &= 1 \\ -2x_1 + 4x_2 - 3x_3 &= -2 \end{aligned}$$

Solution Let

$$A = \begin{bmatrix} 1 & -1 & 2 \\ 2 & 1 & 3 \\ -2 & 4 & -3 \end{bmatrix}$$

Then

$$A_{(1)} = \begin{bmatrix} 0 & -1 & 2 \\ 1 & 1 & 3 \\ -2 & 4 & -3 \end{bmatrix}, \qquad A_{(2)} = \begin{bmatrix} 1 & 0 & 2 \\ 2 & 1 & 3 \\ -2 & -2 & -3 \end{bmatrix},$$

$$A_{(3)} = \begin{bmatrix} 1 & -1 & 0 \\ 2 & 1 & 1 \\ -2 & 4 & -2 \end{bmatrix}$$

Therefore,

$$\det(A) = 5$$
$$\det(A_{(1)}) = 15$$
$$\det(A_{(2)}) = -1$$
$$\det(A_{(3)}) = -8$$

and

$$c_1 = \frac{\det A_{(1)}}{\det A} = \frac{15}{5} = 3$$
$$c_2 = \frac{\det A_{(2)}}{\det A} = \frac{-1}{5}$$
$$c_3 = \frac{\det A_{(3)}}{\det A} = \frac{-8}{5}$$

■

This method is practical only when A is a 2×2 or 3×3 matrix. For $n \geqq 4$, it is usually better to use row operations to reduce partially the augmented matrix to a matrix with many zeros.

SUMMARY

Terms to know:

Cramer's rule

Methods to know:

Determine the solution to a system of linear equations using Cramer's rule.

EXERCISE SET 5.4

Use Cramer's rule to solve each system of equations, if possible.

1. $\begin{bmatrix} 1 & 2 \\ 4 & 7 \end{bmatrix}\begin{bmatrix} x_1 \\ x_2 \end{bmatrix} = \begin{bmatrix} 1 \\ 3 \end{bmatrix}$

2. $\begin{bmatrix} 1 & 8 \\ 9 & 4 \end{bmatrix}\begin{bmatrix} x_1 \\ x_2 \end{bmatrix} = \begin{bmatrix} 7 \\ 9 \end{bmatrix}$

3. $\begin{bmatrix} 1 & 4 \\ 1 & 4 \end{bmatrix}\begin{bmatrix} x_1 \\ x_2 \end{bmatrix} = \begin{bmatrix} 1 \\ 3 \end{bmatrix}$

4. $\begin{bmatrix} 1 & 2 & 3 \\ 7 & 9 & 1 \\ 4 & 2 & 3 \end{bmatrix}\begin{bmatrix} x_1 \\ x_2 \\ x_3 \end{bmatrix} = \begin{bmatrix} 7 \\ 4 \\ 2 \end{bmatrix}$

5. $\begin{bmatrix} 3 & 2 & -3 \\ 4 & -1 & 7 \\ 1 & -3 & 10 \end{bmatrix}\begin{bmatrix} x_1 \\ x_2 \\ x_3 \end{bmatrix} = \begin{bmatrix} 0 \\ 0 \\ 0 \end{bmatrix}$

6. $\begin{bmatrix} 1 & 1 & -1 \\ 2 & 2 & -2 \\ 1 & 1 & -1 \end{bmatrix}\begin{bmatrix} x_1 \\ x_2 \\ x_3 \end{bmatrix} = \begin{bmatrix} 1 \\ 2 \\ 0 \end{bmatrix}$

7. $\begin{aligned} 10x_1 + 11x_2 - x_3 &= 0 \\ -15x_1 + 16x_2 + 12x_3 &= 5 \\ 20x_1 + 22x_2 - 2x_3 &= 0 \end{aligned}$

8. $\begin{aligned} x_1 \qquad - x_3 &= 1 \\ x_1 + x_2 \qquad &= -1 \\ x_2 - x_3 &= -1 \end{aligned}$

9. $\begin{aligned} 3x_1 - x_2 + 5x_3 &= 4 \\ x_1 + 7x_2 - 2x_3 &= -1 \\ 2x_1 - x_2 - x_3 &= 0 \end{aligned}$

10. $\begin{bmatrix} 1 & -1 & 0 \\ 8 & 0 & 14 \\ -3 & 1 & -1 \end{bmatrix}\begin{bmatrix} x_1 \\ x_2 \\ x_3 \end{bmatrix} = \begin{bmatrix} 0 \\ 0 \\ 0 \end{bmatrix}$

11. $\begin{bmatrix} 3 & 1 & -1 \\ 1 & 3 & -1 \\ 1 & 1 & -3 \end{bmatrix}\begin{bmatrix} x_1 \\ x_2 \\ x_3 \end{bmatrix} = \begin{bmatrix} 11 \\ 13 \\ 4 \end{bmatrix}$

12. $\begin{aligned} 2x_1 + x_2 + 3x_3 &= 0 \\ x_2 + 4x_3 &= 2 \\ 6x_1 + x_2 + 8x_3 &= 5 \end{aligned}$

13. $\begin{bmatrix} 1 & 2 & -1 & 0 \\ 0 & 1 & 3 & -1 \\ -1 & 0 & 1 & 4 \\ 5 & -1 & 0 & 1 \end{bmatrix}\begin{bmatrix} x_1 \\ x_2 \\ x_3 \\ x_4 \end{bmatrix} = \begin{bmatrix} 8 \\ 3 \\ -20 \\ 9 \end{bmatrix}$

14. $\begin{bmatrix} 1 & 1 & 1 & 1 \\ 1 & 2 & 3 & 4 \\ 1 & 3 & 6 & 10 \\ 1 & 4 & 10 & 20 \end{bmatrix}\begin{bmatrix} x_1 \\ x_2 \\ x_3 \\ x_4 \end{bmatrix} = \begin{bmatrix} -4 \\ 0 \\ 9 \\ 24 \end{bmatrix}$

15. Using Cramer's rule, verify that if $A\mathbf{x} = \mathbf{0}$ is a system of n equations in n unknowns and if $\det(A) \neq 0$, then the solution must be $\mathbf{x} = \mathbf{0}$.

16. Verify that if a system of n homogeneous linear equations in n variables has a nontrivial solution, then the determinant of the coefficient matrix is zero.

17. Verify that if the determinant of the coefficient matrix is zero in a system of n homogeneous linear equations in n variables, then the system has infinitely many nontrivial solutions.

CHAPTER SIX

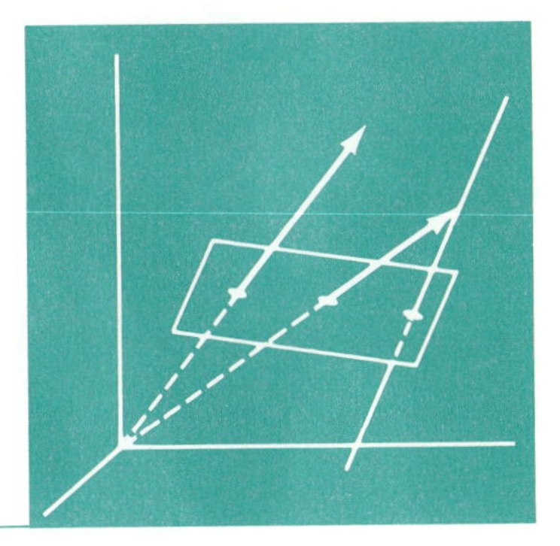

EIGENVALUES AND EIGENVECTORS

In many applications, arbitrarily high powers of a square matrix A must be calculated. In fact, some applications, such as population models, consider the limit of A^k as k goes to $+\infty$, that is, $\lim_{k\to\infty} A^k$. If A is a diagonal matrix,

$$A = \begin{bmatrix} a_1 & 0 & \cdots & 0 \\ 0 & a_2 & \cdots & 0 \\ \vdots & \vdots & & \vdots \\ 0 & 0 & \cdots & a_n \end{bmatrix}$$

then

$$A^k = \begin{bmatrix} a_1^k & 0 & \cdots & 0 \\ 0 & a_2^k & \cdots & 0 \\ \vdots & \vdots & & \vdots \\ 0 & 0 & \cdots & a_n^k \end{bmatrix}$$

is easily calculated. However, if A is not diagonal, A^k may be time-consuming to calculate. In using a computer to calculate A^k, round-off errors may badly affect the accuracy of A^k. In this chapter, we learn how to write certain matrices A as $P^{-1}DP$, where D is diagonal and P is an invertible matrix. Then $A^k = (P^{-1}DP)^k = P^{-1}D^kP$, and the calculation of A^k is again easy.

SECTION 6.1 EIGENVALUES AND EIGENVECTORS

We begin this section with a definition.

Definition 6.1.1 An $n \times n$ matrix A is **diagonalizable** if there exists a nonsingular matrix P such that $P^{-1}AP$ is a diagonal matrix.

Of basic importance to the diagonalizing of A is the concept of an eigenvalue.

Consider the equation $A\mathbf{x} = \lambda\mathbf{x}$, where A is an $n \times n$ matrix, $\mathbf{x} \neq \mathbf{0}$ is in R^n, and λ is a constant. The equation implies that the multiplication of a vector $\mathbf{x}$ by A generates a new vector $\mathbf{b}$, which has the same or opposite (as $\lambda > 0$ or $\lambda < 0$) direction as the original vector $\mathbf{x}$ (that is, $\mathbf{b}$ is proportional to $\mathbf{x}$), or is $\mathbf{0}$ when $\lambda = 0$.

Definition 6.1.2 Let A be an $n \times n$ matrix. A real number λ is an **eigenvalue** of A if there is a nonzero vector $\mathbf{v}$ in R^n satisfying

$$A\mathbf{v} = \lambda\mathbf{v}$$

The vector $\mathbf{v}$ is called an **eigenvector** of A belonging to λ.

REMARK Complex eigenvalues and eigenvectors are considered in more advanced treatments.

Example 6.1.3 For the matrix

$$A = \begin{bmatrix} 1 & 1 & 0 \\ 0 & 2 & 4 \\ 0 & 0 & 1 \end{bmatrix}$$

verify that $A\mathbf{v} = \lambda\mathbf{v}$ for the given λ and $\mathbf{v}$.

Solution Let $\lambda = 1$ and

$$\mathbf{v} = \begin{bmatrix} 1 \\ 0 \\ 0 \end{bmatrix}$$

Then

$$\begin{bmatrix} 1 & 1 & 0 \\ 0 & 2 & 4 \\ 0 & 0 & 1 \end{bmatrix}\begin{bmatrix} 1 \\ 0 \\ 0 \end{bmatrix} = 1\begin{bmatrix} 1 \\ 0 \\ 0 \end{bmatrix}, \quad \text{or} \quad \begin{bmatrix} 1 \\ 0 \\ 0 \end{bmatrix} = \begin{bmatrix} 1 \\ 0 \\ 0 \end{bmatrix}$$

Let $\lambda = 2$ and

$$\mathbf{v} = \begin{bmatrix} 1 \\ 1 \\ 0 \end{bmatrix}$$

Then

$$\begin{bmatrix} 1 & 1 & 0 \\ 0 & 2 & 4 \\ 0 & 0 & 1 \end{bmatrix}\begin{bmatrix} 1 \\ 1 \\ 0 \end{bmatrix} = 2\begin{bmatrix} 1 \\ 1 \\ 0 \end{bmatrix}, \quad \text{or} \quad \begin{bmatrix} 2 \\ 2 \\ 0 \end{bmatrix} = \begin{bmatrix} 2 \\ 2 \\ 0 \end{bmatrix}$$

Actually, for any number k, $k \neq 0$,

$$\begin{bmatrix} k \\ 0 \\ 0 \end{bmatrix}$$

is an eigenvector associated with the eigenvalue $\lambda = 1$ and

$$\begin{bmatrix} k \\ k \\ 0 \end{bmatrix}$$

is an eigenvector associated with the eigenvalue $\lambda = 2$. ■

To find the eigenvalues and associated eigenvectors of A, we first observe that if λ is an eigenvalue of A, there is some $\mathbf{v} \neq \mathbf{0}$ with $A\mathbf{v} = \lambda\mathbf{v}$. Since $I\mathbf{v} = \mathbf{v}$ and λI is a matrix with λ's on the diagonal, we may write $A\mathbf{v} = \lambda I\mathbf{v}$ as $A\mathbf{v} - \lambda I\mathbf{v} = \mathbf{0}$, or $(A - \lambda I)\mathbf{v} = \mathbf{0}$. Since $\mathbf{v} \neq \mathbf{0}$, $A - \lambda I$ must be singular. The set of vectors $\mathbf{v}$ satisfying the homogeneous system $(A - \lambda I)\mathbf{v} = \mathbf{0}$ is the null space of $A - \lambda I$.

Definition 6.1.4 If A is a matrix with eigenvalue λ, the **eigenspace** of A belonging to λ is $NS(A - \lambda I)$.

Example 6.1.5 Given the matrix

$$A = \begin{bmatrix} 3 & 1 & 1 & 3 \\ 2 & 0 & 4 & 2 \\ 1 & 0 & 1 & 2 \\ 3 & 1 & 3 & 1 \end{bmatrix}$$

first write the eigenvalue/eigenvector equation and then rewrite it as a homogeneous system.

Solution Let

$$\mathbf{v} = \begin{bmatrix} x_1 \\ x_2 \\ x_3 \\ x_4 \end{bmatrix}$$

The eigenvalue/eigenvector equation

$$\begin{bmatrix} 3 & 1 & 1 & 3 \\ 2 & 0 & 4 & 2 \\ 1 & 0 & 1 & 2 \\ 3 & 1 & 3 & 1 \end{bmatrix} \begin{bmatrix} x_1 \\ x_2 \\ x_3 \\ x_4 \end{bmatrix} = \lambda \begin{bmatrix} x_1 \\ x_2 \\ x_3 \\ x_4 \end{bmatrix}$$

can be rewritten as

$$\begin{bmatrix} 3 & 1 & 1 & 3 \\ 2 & 0 & 4 & 2 \\ 1 & 0 & 1 & 2 \\ 3 & 1 & 3 & 1 \end{bmatrix} \begin{bmatrix} x_1 \\ x_2 \\ x_3 \\ x_4 \end{bmatrix} = \lambda \begin{bmatrix} 1 & 0 & 0 & 0 \\ 0 & 1 & 0 & 0 \\ 0 & 0 & 1 & 0 \\ 0 & 0 & 0 & 1 \end{bmatrix} \begin{bmatrix} x_1 \\ x_2 \\ x_3 \\ x_4 \end{bmatrix}$$

$$\begin{bmatrix} 3 & 1 & 1 & 3 \\ 2 & 0 & 4 & 2 \\ 1 & 0 & 1 & 2 \\ 3 & 1 & 3 & 1 \end{bmatrix} \begin{bmatrix} x_1 \\ x_2 \\ x_3 \\ x_4 \end{bmatrix} = \begin{bmatrix} \lambda & 0 & 0 & 0 \\ 0 & \lambda & 0 & 0 \\ 0 & 0 & \lambda & 0 \\ 0 & 0 & 0 & \lambda \end{bmatrix} \begin{bmatrix} x_1 \\ x_2 \\ x_3 \\ x_4 \end{bmatrix}$$

$$\begin{bmatrix} 3 & 1 & 1 & 3 \\ 2 & 0 & 4 & 2 \\ 1 & 0 & 1 & 2 \\ 3 & 1 & 3 & 1 \end{bmatrix} \begin{bmatrix} x_1 \\ x_2 \\ x_3 \\ x_4 \end{bmatrix} - \begin{bmatrix} \lambda & 0 & 0 & 0 \\ 0 & \lambda & 0 & 0 \\ 0 & 0 & \lambda & 0 \\ 0 & 0 & 0 & \lambda \end{bmatrix} \begin{bmatrix} x_1 \\ x_2 \\ x_3 \\ x_4 \end{bmatrix} = \begin{bmatrix} 0 \\ 0 \\ 0 \\ 0 \end{bmatrix}$$

$$\begin{bmatrix} 3-\lambda & 1 & 1 & 3 \\ 2 & -\lambda & 4 & 2 \\ 1 & 0 & 1-\lambda & 2 \\ 3 & 1 & 3 & 1-\lambda \end{bmatrix} \begin{bmatrix} x_1 \\ x_2 \\ x_3 \\ x_4 \end{bmatrix} = \begin{bmatrix} 0 \\ 0 \\ 0 \\ 0 \end{bmatrix}$$

■

To find which λ are eigenvalues and to find the eigenspaces of A belonging to λ, we need only to calculate the null space of $A - \lambda I$—that is, obtain solutions to the homogeneous system $(A - \lambda I)\mathbf{v} = \mathbf{0}$. By now, this should be fairly straightforward. However, applying Gauss–Jordan elimination to $A - \lambda I$ is usually complicated and time-consuming because of the λ's involved.

The most commonly used method, at least for small n, is first to determine the λ's through use of the characteristic equation of A and then, for each λ, use the elimination method to determine the associated eigenspace. The following paragraphs detail this method.

For the homogeneous system $(A - \lambda I)\mathbf{v} = \mathbf{0}$ to have a nonzero solution, $A - \lambda I$ must be singular (i.e., the inverse does not exist), which means $\det(A - \lambda I) = 0$. For A an $n \times n$ matrix, $\det(A - \lambda I)$ is a polynomial in λ of degree n called the **characteristic polynomial** of A and $\det(A - \lambda I) = 0$ is the **characteristic equation** of A. In other words, an eigenvalue λ is a zero of the characteristic polynomial.

Example 6.1.6 Determine the eigenvalues and corresponding eigenspaces of

$$A = \begin{bmatrix} 4 & -5 \\ 2 & -3 \end{bmatrix}$$

Solution We need to solve the eigenvalue/eigenvector equation $A\mathbf{v} = \lambda\mathbf{v}$. If

$$\mathbf{v} = \begin{bmatrix} x_1 \\ x_2 \end{bmatrix}$$

then $A\mathbf{v} = \lambda\mathbf{v}$, or $(A - \lambda I)\mathbf{v} = 0$.

$$\begin{bmatrix} 4 & -5 \\ 2 & -3 \end{bmatrix}\begin{bmatrix} x_1 \\ x_2 \end{bmatrix} = \lambda\begin{bmatrix} x_1 \\ x_2 \end{bmatrix}$$

$$\begin{bmatrix} 4 & -5 \\ 2 & -3 \end{bmatrix}\begin{bmatrix} x_1 \\ x_2 \end{bmatrix} = \lambda\begin{bmatrix} 1 & 0 \\ 0 & 1 \end{bmatrix}\begin{bmatrix} x_1 \\ x_2 \end{bmatrix}$$

$$\begin{bmatrix} 4 & -5 \\ 2 & -3 \end{bmatrix}\begin{bmatrix} x_1 \\ x_2 \end{bmatrix} - \lambda\begin{bmatrix} 1 & 0 \\ 0 & 1 \end{bmatrix}\begin{bmatrix} x_1 \\ x_2 \end{bmatrix} = \begin{bmatrix} 0 \\ 0 \end{bmatrix}$$

$$\begin{bmatrix} 4-\lambda & -5 \\ 2 & -3-\lambda \end{bmatrix}\begin{bmatrix} x_1 \\ x_2 \end{bmatrix} = \begin{bmatrix} 0 \\ 0 \end{bmatrix}$$

The characteristic polynomial is

$$\det\left(\begin{bmatrix} 4-\lambda & -5 \\ 2 & -3-\lambda \end{bmatrix}\right) = (4-\lambda)(-3-\lambda) + 10 = \lambda^2 - \lambda - 2$$

and the characteristic equation is

$$\lambda^2 - \lambda - 2 = 0$$

or

$$(\lambda - 2)(\lambda + 1) = 0$$

The eigenvalues are $\lambda_1 = 2$ and $\lambda_2 = -1$.

To calculate the eigenspace of $\lambda_1 = 2$, we must calculate the null space of

$$A - 2I = \begin{bmatrix} 2 & -5 \\ 2 & -5 \end{bmatrix}$$

We row reduce

$$\begin{bmatrix} 2 & -5 \\ 2 & -5 \end{bmatrix}$$

to obtain the equation $2x_1 - 5x_2 = 0$, which has an infinite number of solutions. One solution is $x_1 = 5$ and $x_2 = 2$, that is,

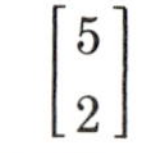

$$\begin{bmatrix} 5 \\ 2 \end{bmatrix}$$

By what we learned in Chapter 3, the eigenspace of A belonging to $\lambda_1 = 2$ has dimension 1 and basis

$$\left\{\begin{bmatrix} 5 \\ 2 \end{bmatrix}\right\}$$

To calculate the eigenspace of $\lambda_2 = -1$, we now calculate the null space of

$$A - (-1)I = \begin{bmatrix} 5 & -5 \\ 2 & -2 \end{bmatrix}$$

Row reducing

$$\begin{bmatrix} 5 & -5 \\ 2 & -2 \end{bmatrix}$$

yields

$$\begin{bmatrix} 1 & -1 \\ 0 & 0 \end{bmatrix}$$

and so $x_1 - x_2 = 0$, which has an infinite number of solutions. One such solution is

$$\begin{bmatrix} 1 \\ 1 \end{bmatrix}$$

The eigenspace belonging to $\lambda_2 = -1$ has dimension 1 and basis

$$\left\{\begin{bmatrix} 1 \\ 1 \end{bmatrix}\right\}$$

Note that

$$\left\{\begin{bmatrix} 5 \\ 2 \end{bmatrix}, \begin{bmatrix} 1 \\ 1 \end{bmatrix}\right\}$$

is a basis of R^2. ■

Example 6.1.7 Determine the eigenvalues and corresponding eigenspaces of

$$A = \begin{bmatrix} 0 & 1 & 0 \\ 0 & 0 & 1 \\ 2 & -2 & 1 \end{bmatrix}$$

Solution To find the eigenvalues of matrix A, we solve $\det(A - \lambda I) = 0$. We first calculate $\det(A - \lambda I)$:

$$\det(A - \lambda I) = \det\left(\begin{bmatrix} -\lambda & 1 & 0 \\ 0 & -\lambda & 1 \\ 2 & -2 & 1-\lambda \end{bmatrix}\right)$$

$$= -\lambda \det\begin{bmatrix} -\lambda & 1 \\ -2 & 1-\lambda \end{bmatrix} - \det\begin{bmatrix} 0 & 1 \\ 2 & 1-\lambda \end{bmatrix}$$

$$= -\lambda(-\lambda + \lambda^2 + 2) - (-2)$$

$$= -\lambda^3 + \lambda^2 - 2\lambda + 2$$

The characteristic equation $\lambda^3 - \lambda^2 + 2\lambda - 2 = 0$ can be factored as $(\lambda - 1)(\lambda^2 + 2) = 0$. The equation $\lambda^2 + 2 = 0$ has complex roots $\lambda = \pm\sqrt{2}\,i$, but we are considering only real eigenvalues, so the only eigenvalue is $\lambda = 1$.

To calculate the eigenspace of $\lambda = 1$, we determine the null space of

$$A - (1)I = \begin{bmatrix} -1 & 1 & 0 \\ 0 & -1 & 1 \\ 2 & -2 & 0 \end{bmatrix}$$

Row reducing this matrix results in

$$\begin{bmatrix} -1 & 1 & 0 \\ 0 & -1 & 1 \\ 0 & 0 & 0 \end{bmatrix}$$

The corresponding system is

$$\begin{aligned} -x_1 + x_2 &= 0 \\ -x_2 + x_3 &= 0 \end{aligned}$$

which has an infinite number of solutions:

$$x_2 \quad \text{arbitrary}$$

$$x_1 = x_2$$

$$x_3 = x_2$$

To find a particular solution, let $x_2 = 1$. Then $x_1 = x_2 = x_3 = 1$, and so

$$\begin{bmatrix} 1 \\ 1 \\ 1 \end{bmatrix}$$

SUMMARY

Terms to know:

Diagonalizable matrix
Eigenvalue
Eigenvector
Eigenspace
Characteristic polynomial
Characteristic equation

Theorems to know:

If U is a triangular matrix, the diagonal entries of U are its eigenvalues.

If D is a diagonal matrix, its eigenvalues are its diagonal entries.

Methods to know:

Determine the eigenvalues of a matrix, and for each eigenvalue λ, find a basis for the eigenspace of λ.

EXERCISE SET 6.1

Find the eigenvalues of the matrices in Exercises 1–9.

1. $\begin{bmatrix} 1 & 3 \\ 2 & 4 \end{bmatrix}$
2. $\begin{bmatrix} 1 & 0 & 1 \\ 0 & 1 & 0 \\ 1 & 0 & -1 \end{bmatrix}$
3. $\begin{bmatrix} 2 & 0 & 1 \\ 0 & 3 & 0 \\ 1 & 0 & -1 \end{bmatrix}$
4. $\begin{bmatrix} 5 & -3 & -9 \\ -3 & 5 & 9 \\ 3 & -3 & -7 \end{bmatrix}$
5. $\begin{bmatrix} 1 & 0 & 1 & 0 \\ 0 & 1 & 0 & 1 \\ 1 & 0 & 1 & 0 \\ 0 & 1 & 0 & 2 \end{bmatrix}$
6. $\begin{bmatrix} 3 & 2 & 2 \\ 1 & 4 & 1 \\ -2 & -4 & -1 \end{bmatrix}$
7. $\begin{bmatrix} -1 & -6 & 3 \\ 3 & 8 & -3 \\ 6 & 12 & -4 \end{bmatrix}$
8. $\begin{bmatrix} -5 & -8 & -12 \\ -6 & -10 & -12 \\ 6 & 10 & 13 \end{bmatrix}$
9. $\begin{bmatrix} 0 & -2 & -2 \\ -2 & -3 & -2 \\ 3 & 6 & 5 \end{bmatrix}$
10. Prove that 0 is an eigenvalue of A if and only if A is singular. What is the corresponding eigenspace?

Compute the eigenvalues and bases for the corresponding eigenspaces for the matrices in Exercises 11–18.

11. $\begin{bmatrix} 1 & 2 & 4 & 3 \\ 0 & 0 & 2 & 1 \\ 0 & 0 & 1 & 1 \\ 0 & 0 & 0 & 2 \end{bmatrix}$ 12. $\begin{bmatrix} 0 & 1 & 2 \\ 0 & 0 & 1 \\ 0 & 0 & 1 \end{bmatrix}$ 13. $\begin{bmatrix} 1 & 1 & 0 \\ 0 & 1 & 1 \\ 0 & 0 & 1 \end{bmatrix}$

14. $\begin{bmatrix} 1 & 0 & 0 \\ 0 & 0 & 0 \\ 0 & 1 & 0 \end{bmatrix}$ 15. $\begin{bmatrix} 1 & 0 & 0 \\ 0 & 1 & 0 \\ 1 & 0 & 2 \end{bmatrix}$ 16. $\begin{bmatrix} 0 & 2 & 0 \\ 0 & 1 & 1 \\ 0 & 0 & 0 \end{bmatrix}$

17. $\begin{bmatrix} 2 & -3 & 4 & 1 \\ 0 & 3 & 1 & 2 \\ 0 & 0 & 1 & 5 \\ 0 & 0 & -2 & 3 \end{bmatrix}$ 18. $\begin{bmatrix} 2 & 4 & 1 \\ 1 & -2 & -1 \\ 0 & 0 & 0 \end{bmatrix}$

19. For what value of k is 3 an eigenvalue of

$$\begin{bmatrix} 3 & 1 & -1 \\ 3 & 5 & -k \\ 3 & k & -1 \end{bmatrix}$$

20. Suppose $\mathbf{v}$ is an eigenvector of A and of B. Show that $\mathbf{v}$ is an eigenvector of $\alpha A + \beta B$, where α and β are any scalars.

21. Prove that if A is nonsingular and has an eigenvalue λ, then A^{-1} has eigenvalue $1/\lambda$.

22. If A is an invertible matrix, prove that the eigenvectors for A and A^{-1} are identical.

    Hint: If $A\mathbf{v} = \lambda\mathbf{v}$, is $\mathbf{v}$ an eigenvector of A^{-1} belonging to $1/\lambda$?

23. Prove Theorem 6.1.10 and Corollary 6.1.11 for a 3×3 matrix.

    Hint: If

$$U = \begin{bmatrix} u_{11} & u_{12} & u_{13} \\ 0 & u_{22} & u_{23} \\ 0 & 0 & u_{33} \end{bmatrix}$$

    what is $\det(U - \lambda I)$?

24. Prove that A and A^T have the same eigenvalues.

    Hint: What is $\det(A - \lambda I)^T$?

25. Prove that if $A = B^{-1}CB$, then A and C have the same eigenvalues.

    Hint: $\det(XY) = \det(X)\det(Y)$.

SECTION 6.2 DIAGONALIZATION

To diagonalize an $n \times n$ matrix once the eigenvectors are known is relatively straightforward, providing that the matrix has n linearly independent eigenvectors. If the matrix does not have n linearly

The set of eigenvectors

$$\left\{\begin{bmatrix} 0 \\ -1 \\ 1 \end{bmatrix}, \begin{bmatrix} 1 \\ 0 \\ 0 \end{bmatrix}, \begin{bmatrix} 0 \\ 1 \\ 1 \end{bmatrix}\right\}$$

is linearly independent by inspection. We now form P, using the eigenvectors as columns:

$$P = \begin{bmatrix} 0 & 1 & 0 \\ -1 & 0 & 1 \\ 1 & 0 & 1 \end{bmatrix}$$

b. To diagonalize A we need to compute P^{-1} and then $P^{-1}AP$. Applying the sequence of row operations $R1 \leftrightarrow R2$, $(1)R1$ to $R3$, $(\frac{1}{2})R3$, $(-1)R1$, and $(1)R3$ to $R1$ to

$$\left[\begin{array}{ccc|ccc} 0 & 1 & 0 & 1 & 0 & 0 \\ -1 & 0 & 1 & 0 & 1 & 0 \\ 1 & 0 & 1 & 0 & 0 & 1 \end{array}\right]$$

results in

$$\left[\begin{array}{ccc|ccc} 1 & 0 & 0 & 0 & -\frac{1}{2} & \frac{1}{2} \\ 0 & 1 & 0 & 1 & 0 & 0 \\ 0 & 0 & 1 & 0 & \frac{1}{2} & \frac{1}{2} \end{array}\right]$$

Thus

$$P^{-1} = \begin{bmatrix} 0 & -\frac{1}{2} & \frac{1}{2} \\ 1 & 0 & 0 \\ 0 & \frac{1}{2} & \frac{1}{2} \end{bmatrix}$$

and

$$P^{-1}AP = \begin{bmatrix} 0 & -\frac{1}{2} & \frac{1}{2} \\ 1 & 0 & 0 \\ 0 & \frac{1}{2} & \frac{1}{2} \end{bmatrix}\begin{bmatrix} 1 & 0 & 0 \\ 0 & 1 & 1 \\ 0 & 1 & 1 \end{bmatrix}\begin{bmatrix} 0 & 1 & 0 \\ -1 & 0 & 1 \\ 1 & 0 & 1 \end{bmatrix}$$

$$= \begin{bmatrix} 0 & 0 & 0 \\ 0 & 1 & 0 \\ 0 & 0 & 2 \end{bmatrix}$$

Observe that the eigenvalues $\lambda_1 = 0$, $\lambda_2 = 1$, and $\lambda_3 = 2$ are the diagonal entries of the diagonal matrix. ■

Example 6.2.3 Find A^{10}, where

$$A = \begin{bmatrix} 1 & 0 & 0 \\ 0 & 1 & 1 \\ 0 & 1 & 1 \end{bmatrix}$$

Solution First, note from Example 6.2.2 that

$$P^{-1}AP = D$$

where

$$D = \begin{bmatrix} 0 & 0 & 0 \\ 0 & 1 & 0 \\ 0 & 0 & 2 \end{bmatrix}$$

so $A = PDP^{-1}$. Second, observe that

$$\begin{aligned} A^{10} &= (PDP^{-1})^{10} \\ &= \underbrace{(PDP^{-1})(PDP^{-1}) \cdots (PDP^{-1})(PDP^{-1})}_{10 \text{ factors}} \\ &= PD^{10}P^{-1} \end{aligned}$$

since $P^{-1}P = I$ nine times. Thus

$$A^{10} = \begin{bmatrix} 0 & 1 & 0 \\ -1 & 0 & 1 \\ 1 & 0 & 1 \end{bmatrix} \begin{bmatrix} 0^{10} & 0 & 0 \\ 0 & 1^{10} & 0 \\ 0 & 0 & 2^{10} \end{bmatrix} \begin{bmatrix} 0 & -\frac{1}{2} & \frac{1}{2} \\ 1 & 0 & 0 \\ 0 & \frac{1}{2} & \frac{1}{2} \end{bmatrix}$$

$$= \begin{bmatrix} 1 & 0 & 0 \\ 0 & 512 & 512 \\ 0 & 512 & 512 \end{bmatrix}$$

■

In general, the procedure in Example 6.2.2 is a lot more economical than multiplying A by itself 10 times. The theory behind this procedure is given in the following two theorems.

Theorem 6.2.4 *Let A be an $n \times n$ matrix. A can be diagonalized if and only if there is a basis $\{\mathbf{v}_1, \mathbf{v}_2, \ldots, \mathbf{v}_n\}$ of R^n consisting of eigenvectors of A.*

Proof. Suppose A can be diagonalized; then there exists a nonsingular matrix P such that

$$P^{-1}AP = \begin{bmatrix} d_1 & 0 & \cdots & 0 \\ 0 & d_2 & \cdots & 0 \\ \vdots & \vdots & & \vdots \\ 0 & 0 & \cdots & d_n \end{bmatrix}$$

Thus

$$AP = P\begin{bmatrix} d_1 & 0 & \cdots & 0 \\ 0 & d_2 & \cdots & 0 \\ \vdots & \vdots & & \vdots \\ 0 & 0 & \cdots & d_n \end{bmatrix}$$

Let $P = [\mathbf{P}_1 \quad \mathbf{P}_2 \quad \cdots \quad \mathbf{P}_n]$, where $\mathbf{P}_i$ is the ith column of P. Then the jth column of AP is $A\mathbf{P}_j$ and the jth column of

$$P\begin{bmatrix} d_1 & 0 & \cdots & 0 \\ 0 & d_2 & \cdots & 0 \\ \vdots & \vdots & & \vdots \\ 0 & 0 & \cdots & d_n \end{bmatrix}$$

is $d_j\mathbf{P}_j$. Thus $A\mathbf{P}_j = d_j\mathbf{P}_j$, so that $\mathbf{P}_j$ is an eigenvector of A. Since P is nonsingular, the columns of P are linearly independent and form a basis of R^n. Hence, there is a basis $\{\mathbf{P}_1, \mathbf{P}_2, \ldots, \mathbf{P}_n\}$ of R^n consisting of eigenvectors of A.

Conversely, let $\{\mathbf{v}_1, \mathbf{v}_2, \ldots, \mathbf{v}_n\}$ be a basis of R^n consisting of eigenvectors of A. Let the eigenvector $\mathbf{v}_i$ belong to the eigenvalue λ_i. Hence $A\mathbf{v}_i = \lambda_i\mathbf{v}_i$. Let P denote the $n \times n$ matrix whose ith column is $\mathbf{v}_i$. Since the $\mathbf{v}_i$ form a basis for R^n, the $\mathbf{v}_i$ are linearly independent and P is nonsingular. Because $A\mathbf{v}_i = \lambda_i\mathbf{v}_i$ for each i,

$$AP = P\begin{bmatrix} \lambda_1 & 0 & \cdots & 0 \\ 0 & \lambda_2 & \cdots & 0 \\ \vdots & \vdots & & \vdots \\ 0 & 0 & \cdots & \lambda_n \end{bmatrix}$$

and

$$P^{-1}AP = \begin{bmatrix} \lambda_1 & 0 & \cdots & 0 \\ 0 & \lambda_2 & \cdots & 0 \\ \vdots & \vdots & & \vdots \\ 0 & 0 & \cdots & \lambda_n \end{bmatrix}$$

which means A is diagonalized.

We may view Theorem 6.2.4 in another light using linear transformations, as in Chapter 3. Let V be an n-dimensional vector space and $f: V \to V$ be a linear transformation. Choose a basis $B = \{\mathbf{v}_1, \ldots, \mathbf{v}_n\}$ of V and represent f with respect to the basis B as the $n \times n$ matrix A. If A has n linearly independent eigenvectors $\{\mathbf{p}_1, \mathbf{p}_2, \ldots, \mathbf{p}_n\}$, then

$P^{-1}AP = D$, where P is the $n \times n$ matrix whose ith column is $\mathbf{p}_i$,

$$D = \begin{bmatrix} \lambda_1 & & & 0 \\ & \lambda_2 & & \\ & & \ddots & \\ 0 & & & \lambda_n \end{bmatrix}$$

and where λ_i is the eigenvalue of A belonging to the eigenvector $\mathbf{p}_i$. Since P may be regarded as the change of basis matrix from the basis $\{\mathbf{p}_1, \mathbf{p}_2, \ldots, \mathbf{p}_n\}$ to the standard basis of R^n, in diagonalizing A we are finding a basis $\{\mathbf{p}_1, \mathbf{p}_2, \ldots, \mathbf{p}_n\}$ of V with respect to which the representation of f is a diagonal matrix.

Theorem 6.2.5 *Let A be an $n \times n$ matrix with distinct eigenvalues $\lambda_1, \lambda_2, \ldots, \lambda_k$. Let $\{\mathbf{v}_1, \mathbf{v}_2, \ldots, \mathbf{v}_k\}$ be a set of eigenvectors with $\mathbf{v}_i$ belonging to λ_i for $1 \leqq i \leqq k$. Then the set $\{\mathbf{v}_1, \mathbf{v}_2, \ldots, \mathbf{v}_k\}$ is linearly independent.*

Proof. Assume $\{\mathbf{v}_1, \mathbf{v}_2, \ldots, \mathbf{v}_k\}$ is linearly dependent. Then $c_1\mathbf{v}_1 + c_2\mathbf{v}_2 + \cdots + c_n\mathbf{v}_k = \mathbf{0}$ with some of the $c_i \neq 0$. Now

$$\begin{aligned} \mathbf{0} = A\mathbf{0} &= A(c_1\mathbf{v}_1 + c_2\mathbf{v}_2 + \cdots + c_k\mathbf{v}_k) \\ &= c_1A\mathbf{v}_1 + c_2A\mathbf{v}_2 + \cdots + c_kA\mathbf{v}_k \\ &= c_1\lambda_1\mathbf{v}_1 + c_2\lambda_2\mathbf{v}_2 + \cdots + c_k\lambda_k\mathbf{v}_k \end{aligned} \tag{6.2.1}$$

Also,

$$\begin{aligned} \mathbf{0} &= \lambda_k(c_1\mathbf{v}_1 + c_2\mathbf{v}_2 + \cdots + c_k\mathbf{v}_k) \\ &= c_1\lambda_k\mathbf{v}_1 + c_2\lambda_k\mathbf{v}_2 + \cdots + c_k\lambda_k\mathbf{v}_k \end{aligned} \tag{6.2.2}$$

Subtracting (6.2.2) from (6.2.1) yields

$$(c_1\lambda_1 - c_1\lambda_k)\mathbf{v}_1 + (c_2\lambda_2 - c_2\lambda_k)\mathbf{v}_2 + \cdots + (c_k\lambda_k - c_k\lambda_k)\mathbf{v}_k = \mathbf{0}$$

Since $\lambda_i - \lambda_k \neq 0$ for $1 \leqq i \leqq k - 1$, then $c_i(\lambda_i - \lambda_k) \neq 0$ for some i, $1 \leqq i \leqq k - 1$, and

$$c_1(\lambda_1 - \lambda_k)\mathbf{v}_1 + c_2(\lambda_2 - \lambda_k)\mathbf{v}_2 + \cdots + c_{k-1}(\lambda_{k-1} - \lambda_k)\mathbf{v}_{k-1} = \mathbf{0}$$

which means the set $\{\mathbf{v}_1, \ldots, \mathbf{v}_{k-1}\}$ is linearly dependent. Continuing in this manner, we arrive at the fact the set $\{\mathbf{v}_1\}$ is linearly dependent. But $\mathbf{v}_1$, being an eigenvector, is not equal to the zero vector and hence $\{\mathbf{v}_1\}$ cannot be linearly dependent. Thus our original assumption that $\{\mathbf{v}_1, \mathbf{v}_2, \ldots, \mathbf{v}_k\}$ is linearly dependent is false.

Before leaving this section, we give an example to illustrate that *not all matrices are diagonalizable.*

Example 6.2.6 Diagonalize the matrix

$$A = \begin{bmatrix} 0 & 1 \\ 0 & 0 \end{bmatrix}$$

Solution The characteristic equation

$$\det(A - \lambda I) = \begin{bmatrix} -\lambda & 1 \\ 0 & -\lambda \end{bmatrix} = \lambda^2 = 0$$

has only one root, $\lambda = 0$. The null space of

$$A - \lambda I = \begin{bmatrix} 0 & 1 \\ 0 & 0 \end{bmatrix}$$

is the solution space of the system

$$x_2 = 0$$

$$x_1 \quad \text{arbitrary}$$

A basis for the eigenspace is

$$\left\{\begin{bmatrix} 1 \\ 0 \end{bmatrix}\right\}$$

and so has dimension 1 and cannot be a basis for R^2. This means A is not diagonalizable.

This proves the following theorem. ■

Theorem 6.2.7 *Not all square matrices are diagonalizable.*

SUMMARY

Theorems to know:

An $n \times n$ matrix A can be diagonalized if and only if there is a basis of R^n consisting of eigenvectors of A.

Let A have distinct eigenvalues $\lambda_1, \lambda_2, \ldots, \lambda_k$ and let $\{\mathbf{v}_1, \ldots, \mathbf{v}_k\}$ be a set of eigenvectors of A with $\mathbf{v}_i$ belonging to λ_i for $i = 1, \ldots, k$. Then the set $\{\mathbf{v}_1, \ldots, \mathbf{v}_k\}$ is linearly independent.

Not all square matrices are diagonalizable.

Methods to know:

Diagonalize a matrix.

EXERCISE SET 6.2

For each matrix A, find (if possible) the diagonal matrix D and nonsingular matrix P for which

$$A = PDP^{-1}$$

1. $\begin{bmatrix} 1 & 2 \\ 2 & 1 \end{bmatrix}$ 2. $\begin{bmatrix} -1 & 2 \\ 0 & 1 \end{bmatrix}$ 3. $\begin{bmatrix} 1 & 0 & 1 \\ 0 & 1 & 0 \\ 1 & 0 & 1 \end{bmatrix}$

4. $\begin{bmatrix} 5 & -3 & -9 \\ -3 & 5 & 9 \\ 3 & -3 & -7 \end{bmatrix}$ 5. $\begin{bmatrix} 3 & 4 & 2 \\ -6 & -7 & -6 \\ 8 & 8 & 9 \end{bmatrix}$ 6. $\begin{bmatrix} 0 & 0 & 1 \\ 0 & 1 & 0 \\ 1 & 0 & 0 \end{bmatrix}$

7. $\begin{bmatrix} 0 & 0 & 0 & 1 \\ 0 & 0 & 1 & 0 \\ 0 & 1 & 0 & 0 \\ 1 & 0 & 0 & 0 \end{bmatrix}$ 8. $\begin{bmatrix} 1 & 0 & 3 \\ 0 & 1 & 2 \\ 0 & 0 & 1 \end{bmatrix}$ 9. $\begin{bmatrix} 1 & 1 & 0 \\ 0 & 1 & 0 \\ 0 & 0 & 1 \end{bmatrix}$

10. If $A = C^{-1}BC$ and $P^{-1}AP = D$, a diagonal matrix, find a matrix Q with $Q^{-1}BQ = D$.

11. If

$$P^{-1}AP = \begin{bmatrix} \lambda_1 & 0 & 0 \\ 0 & \lambda_2 & 0 \\ 0 & 0 & \lambda_3 \end{bmatrix}$$

prove that $\det(A) = \lambda_1\lambda_2\lambda_3$.

12. Verify that

$$\left\{ \begin{bmatrix} 0 \\ -1 \\ 1 \end{bmatrix}, \begin{bmatrix} 1 \\ 0 \\ 0 \end{bmatrix}, \begin{bmatrix} 0 \\ 1 \\ 1 \end{bmatrix} \right\}$$

is a linearly independent set.

13. If $A = C^{-1}BC$, prove that $\det(A) = \det(B)$.

14. A matrix A is *similar* to a matrix B if there is a nonsingular C with $A = C^{-1}BC$. If A is similar to the matrix B, prove that A^n is similar to B^n for n any positive integer.

15. If the matrix A is similar to the matrix B for some matrix C, $A = C^{-1}BC$. Prove that $A^n = C^{-1}B^nC$ for n any positive integer.

SECTION 6.3 ORTHOGONAL MATRICES

In the last section, we showed one use of a basis of R^n different than the standard basis $\{\mathbf{e}_1, \mathbf{e}_2, \ldots, \mathbf{e}_n\}$ in diagonalizing a matrix. However, there are often certain advantages in having a basis of mutually perpendicular unit vectors. For **symmetric matrices** (square matrices having the property $A^T = A$), we will see that a basis of mutually perpendicular unit (length 1) eigenvectors exists that diagonalizes A. Such a basis is called an *orthonormal basis*.

Definition 6.3.1 A basis $\{\mathbf{v}_1, \mathbf{v}_2, \dots, \mathbf{v}_n\}$ of R^n is an **orthonormal basis** if

1. $\mathbf{v}_i \cdot \mathbf{v}_j = 0$ for $i \neq j$,
2. $\|\mathbf{v}_i\| = 1 \qquad$ for $i = 1, 2, \dots, n$

Example 6.3.2 a. The basis

$$\left\{ \begin{bmatrix} 1 \\ 0 \end{bmatrix}, \begin{bmatrix} 0 \\ 1 \end{bmatrix} \right\}$$

of R^2 is orthonormal, since

$$\begin{bmatrix} 1 \\ 0 \end{bmatrix} \cdot \begin{bmatrix} 0 \\ 1 \end{bmatrix} = 0, \qquad \left\| \begin{bmatrix} 1 \\ 0 \end{bmatrix} \right\| = 1, \quad \text{and} \quad \left\| \begin{bmatrix} 0 \\ 1 \end{bmatrix} \right\| = 1$$

b. $\{\mathbf{e}_1, \mathbf{e}_2, \dots, \mathbf{e}_n\}$ is an orthonormal basis of R^n, since $\mathbf{e}_i \cdot \mathbf{e}_j = 0$ for $i \neq j$ and $\|\mathbf{e}_i\| = 1$ for $i = 1, 2, \dots, n$.

c.

$$\mathbf{v}_1 = \begin{bmatrix} \frac{1}{\sqrt{2}} \\ 0 \\ \frac{1}{\sqrt{2}} \end{bmatrix}, \qquad \mathbf{v}_2 = \begin{bmatrix} 0 \\ 1 \\ 0 \end{bmatrix}, \quad \text{and} \quad \mathbf{v}_3 = \begin{bmatrix} \frac{1}{\sqrt{2}} \\ 0 \\ -\frac{1}{\sqrt{2}} \end{bmatrix}$$

form an orthonormal basis for R^3, since $\mathbf{v}_1 \cdot \mathbf{v}_2 = 0$, $\mathbf{v}_1 \cdot \mathbf{v}_3 = 0$, $\mathbf{v}_2 \cdot \mathbf{v}_3 = 0$, and $\|\mathbf{v}_1\| = \|\mathbf{v}_2\| = \|\mathbf{v}_3\| = 1$. The verification that $\{\mathbf{v}_1, \mathbf{v}_2, \mathbf{v}_3\}$ is a basis for R^3 is left as an exercise for the reader. ■

Observe that if $\|\mathbf{v}\| = 1$, then $\mathbf{v} \cdot \mathbf{v} = 1$ since $\|\mathbf{v}\| = \mathbf{v} \cdot \mathbf{v} = 1$. Thus an orthonormal basis of R^n can be described as a basis having the following properties:

1. $\mathbf{v}_i \cdot \mathbf{v}_j = 0$ for $i \neq j$.
2. $\mathbf{v}_i \cdot \mathbf{v}_i = 1$ for $i = 1, 2, \dots, n$.

(Thus $\mathbf{v}_i \mathbf{v}_j = 1$ if $i = j$ and 0 if $i \neq j$.)

Recall from Section 2.2 that two vectors $\mathbf{u}$ and $\mathbf{v}$ in R^n are orthogonal if $\mathbf{u} \cdot \mathbf{v} = 0$. Furthermore, a set of vectors $\{\mathbf{v}_1, \dots, \mathbf{v}_k\}$ in R^n is *orthogonal* provided $\mathbf{v}_i \cdot \mathbf{v}_j = 0$ for $i \neq j$. Also that if $\mathbf{u}$ and $\mathbf{v}$ are nonzero vectors and $\mathbf{u} \cdot \mathbf{v} = \|\mathbf{u}\|\,\|\mathbf{v}\| \cos\theta = 0$, where θ is the angle between $\mathbf{u}$ and $\mathbf{v}$, then $\cos\theta = 0$, or $\theta = \pi/2$, which means $\mathbf{u}$ and $\mathbf{v}$ are perpendicular. Calling a set of vectors orthogonal means the vectors in this set are mutually perpendicular.

A useful property of a set of nonzero orthogonal vectors is that the set is linearly independent. It is often easier to show that a set of vectors is orthogonal than to show the set is linearly independent.

Theorem 6.3.3 *If a set of nonzero vectors $\{\mathbf{v}_1, \mathbf{v}_2, \ldots, \mathbf{v}_k\}$ is orthogonal, then it is linearly independent.*

Proof. Suppose $\{\mathbf{v}_1, \mathbf{v}_2, \ldots, \mathbf{v}_k\}$ is an orthogonal set of vectors and that

$$c_1\mathbf{v}_1 + c_2\mathbf{v}_2 + \cdots + c_k\mathbf{v}_k = \mathbf{0}$$

Then for each i

$$\mathbf{v}_i \cdot (c_1\mathbf{v}_1 + c_2\mathbf{v}_2 + \cdots + c_k\mathbf{v}_k) = \mathbf{v}_i \cdot \mathbf{0}$$

so that

$$c_1\mathbf{v}_i \cdot \mathbf{v}_1 + c_2\mathbf{v}_i \cdot \mathbf{v}_2 + \cdots + c_k\mathbf{v}_i \cdot \mathbf{v}_k = 0 \tag{6.3.1}$$

However, since

$$\mathbf{v}_i \cdot \mathbf{v}_j = 0 \qquad \text{for } i \neq j$$

(6.3.1) becomes

$$c_i\mathbf{v}_i \cdot \mathbf{v}_i = 0$$

or

$$c_i\|\mathbf{v}_i\|^2 = 0$$

and so $c_i = 0$, since $\|\mathbf{v}_i\| \neq 0$.

Doing this for every $i = 1, 2, \ldots, k$ we see that all $c_i = 0$, which means the set $\{\mathbf{v}_1, \mathbf{v}_2, \ldots, \mathbf{v}_k\}$ is linearly independent.

Observe that if $k = n$, then $\{\mathbf{v}_1, \mathbf{v}_2, \ldots, \mathbf{v}_n\}$ would be a basis for R^n.

Example 6.3.4 Show that

$$\left\{\begin{bmatrix}1\\0\\1\end{bmatrix}, \begin{bmatrix}1\\0\\-1\end{bmatrix}, \begin{bmatrix}0\\1\\0\end{bmatrix}\right\}$$

is a basis for R^3.

Solution Since there are three nonzero vectors in the set, we need to show that the set is linearly independent. However,

$$\begin{bmatrix}1\\0\\1\end{bmatrix} \cdot \begin{bmatrix}1\\0\\-1\end{bmatrix} = 1 + 0 - 1 = 0$$

$$\begin{bmatrix}1\\0\\1\end{bmatrix} \cdot \begin{bmatrix}0\\1\\0\end{bmatrix} = 0 + 0 + 0 = 0$$

and

$$\begin{bmatrix} 1 \\ 0 \\ -1 \end{bmatrix} \cdot \begin{bmatrix} 0 \\ 1 \\ 0 \end{bmatrix} = 0 + 0 + 0 = 0$$

This is an orthogonal set and, hence, is linearly independent, which means the set of vectors forms a basis for R^3. ■

Observe that if

$$\mathbf{v}_1 = \begin{bmatrix} 1 \\ 0 \\ 1 \end{bmatrix}, \quad \mathbf{v}_2 = \begin{bmatrix} 1 \\ 0 \\ -1 \end{bmatrix}, \quad \text{and} \quad \mathbf{v}_3 = \begin{bmatrix} 0 \\ 1 \\ 0 \end{bmatrix}$$

then

$$\mathbf{u}_1 = \frac{1}{||\mathbf{v}_1||}\mathbf{v}_1 = \begin{bmatrix} \frac{1}{\sqrt{2}} \\ 0 \\ \frac{1}{\sqrt{2}} \end{bmatrix}, \quad \mathbf{u}_2 = \frac{1}{||\mathbf{v}_2||}\mathbf{v}_2 = \begin{bmatrix} \frac{1}{\sqrt{2}} \\ 0 \\ -\frac{1}{\sqrt{2}} \end{bmatrix}$$

and $\mathbf{u}_3 = \mathbf{v}_3$ form an orthonormal basis of R^3.

We learned in Section 6.2 that when we diagonalize an $n \times n$ matrix A, we find a basis $\{\mathbf{p}_1, \ldots, \mathbf{p}_n\}$ of R^n with respect to which the linear transformation $f: R^n \to R^n$ given by $f(\mathbf{x}) = A\mathbf{x}$ can be represented as a diagonal matrix. We now ask whether the basis $\{\mathbf{p}_1, \ldots, \mathbf{p}_n\}$ can be chosen to be an orthonormal basis. Since the matrix $P = [\mathbf{p}_1, \mathbf{p}_2, \ldots, \mathbf{p}_n]$ is used to diagonalize A, we would expect the orthonormality of the basis $\{\mathbf{p}_1, \ldots, \mathbf{p}_n\}$ to be reflected in P by some matrix property.

Definition 6.3.5 A nonsingular square matrix A such that $A^{-1} = A^T$ is called an **orthogonal matrix**.

REMARK By the uniqueness of inverses, A is orthogonal if and only if either $AA^T = I$ or $A^TA = I$.

An example of an orthogonal matrix is

$$A = \begin{bmatrix} 1 & 0 & 0 \\ 0 & -\frac{\sqrt{3}}{2} & -\frac{1}{2} \\ 0 & -\frac{1}{2} & \frac{\sqrt{3}}{2} \end{bmatrix}$$

since

$$A^T = \begin{bmatrix} 1 & 0 & 0 \\ 0 & -\frac{\sqrt{3}}{2} & -\frac{1}{2} \\ 0 & -\frac{1}{2} & \frac{\sqrt{3}}{2} \end{bmatrix}$$

and

$$AA^T = \begin{bmatrix} 1 & 0 & 0 \\ 0 & -\frac{\sqrt{3}}{2} & -\frac{1}{2} \\ 0 & -\frac{1}{2} & \frac{\sqrt{3}}{2} \end{bmatrix} \begin{bmatrix} 1 & 0 & 0 \\ 0 & -\frac{\sqrt{3}}{2} & -\frac{1}{2} \\ 0 & -\frac{1}{2} & \frac{\sqrt{3}}{2} \end{bmatrix}$$

$$= \begin{bmatrix} 1 & 0 & 0 \\ 0 & 1 & 0 \\ 0 & 0 & 1 \end{bmatrix} = I = A^TA$$

Theorem 6.3.6 *If A is an orthogonal $n \times n$ matrix, its columns form an orthonormal basis for R^n.*

Proof. Since the rows of A^T are the columns of A, then

$$A^TA = \begin{bmatrix} \mathbf{A}_1^T\mathbf{A}_1 & \mathbf{A}_1^T\mathbf{A}_2 & \cdots & \mathbf{A}_1^T\mathbf{A}_n \\ \mathbf{A}_2^T\mathbf{A}_1 & \mathbf{A}_2^T\mathbf{A}_2 & \cdots & \mathbf{A}_2^T\mathbf{A}_n \\ \vdots & \vdots & & \vdots \\ \mathbf{A}_n^T\mathbf{A}_1 & \mathbf{A}_n^T\mathbf{A}_2 & \cdots & \mathbf{A}_n^T\mathbf{A}_n \end{bmatrix} = I$$

where $\mathbf{A}_i$ is the ith column of A. Since A is orthogonal, we have

$$\mathbf{A}_i^T\mathbf{A}_j = 0 \qquad \text{for } i \neq j$$

and

$$\mathbf{A}_i^T\mathbf{A}_i = 1$$

Thus the columns of the orthogonal matrix A form an orthonormal set of vectors; since there are n columns, the columns of A form an orthonormal basis for R^n.

In Exercise 13 of this section you are asked to prove the converse of Theorem 6.3.6.

The orthogonal condition also requires

$$\begin{bmatrix} \frac{1}{\sqrt{6}} \\ \frac{1}{\sqrt{6}} \\ \frac{2}{\sqrt{6}} \end{bmatrix} \cdot \begin{bmatrix} b \\ d \\ f \end{bmatrix} = 0, \quad \text{or} \quad b + d + 2f = 0$$

and

$$\begin{bmatrix} \frac{1}{\sqrt{2}} \\ -\frac{1}{\sqrt{2}} \\ 0 \end{bmatrix} \cdot \begin{bmatrix} b \\ d \\ f \end{bmatrix} = 0, \quad \text{or} \quad b - d = 0$$

The solution of

$$b + d + 2f = 0$$

$$b - d = 0$$

is $b = d$, $f = -d$, and d arbitrarily chosen. Thus

$$\begin{bmatrix} b \\ d \\ f \end{bmatrix} = \begin{bmatrix} d \\ d \\ -d \end{bmatrix} = d \begin{bmatrix} 1 \\ 1 \\ -1 \end{bmatrix}$$

Choosing

$$d = \frac{1}{\sqrt{3}} = \frac{1}{\left\| \begin{bmatrix} 1 \\ 1 \\ -1 \end{bmatrix} \right\|}$$

we have for the third column

$$\begin{bmatrix} \frac{1}{\sqrt{3}} \\ \frac{1}{\sqrt{3}} \\ -\frac{1}{\sqrt{3}} \end{bmatrix}$$

Thus, one such orthogonal matrix is

$$\begin{bmatrix} \frac{1}{\sqrt{6}} & \frac{1}{\sqrt{2}} & \frac{1}{\sqrt{3}} \\ \frac{1}{\sqrt{6}} & -\frac{1}{\sqrt{2}} & \frac{1}{\sqrt{3}} \\ \frac{2}{\sqrt{6}} & 0 & -\frac{1}{\sqrt{3}} \end{bmatrix}$$

■

Definition 6.3.9 A matrix A is **orthogonally diagonalizable** if there is an orthogonal matrix Q such that

$$Q^{-1}AQ = \begin{bmatrix} d_1 & 0 & \cdots & 0 \\ 0 & d_2 & \cdots & 0 \\ \vdots & \vdots & & \vdots \\ 0 & 0 & \cdots & d_n \end{bmatrix}$$

In general, orthogonally diagonalizing A is a more difficult problem than diagonalizing A, since we now require the eigenvectors of A to be orthonormal. However, when A is symmetric ($A^T = A$), the problem is simpler.

Theorem 6.3.10 *Let A be a symmetric $n \times n$ matrix. Then the eigenvectors belonging to different eigenvalues are orthogonal.*

Proof. Let $\mathbf{u}$ be an eigenvector belonging to λ_1 and $\mathbf{v}$ be an eigenvector belonging to λ_2, where $\lambda_1 \neq \lambda_2$. Then

$$\begin{aligned} \lambda_1(\mathbf{u}^T\mathbf{v}) &= (\lambda_1\mathbf{u}^T)\mathbf{v} = (\lambda_1\mathbf{u})^T\mathbf{v} = (A\mathbf{u})^T\mathbf{v} = (\mathbf{u}^TA^T)\mathbf{v} \\ &= \mathbf{u}^T(A^T\mathbf{v}) \\ &= \mathbf{u}^T(A\mathbf{v}) = \mathbf{u}^T(\lambda_2\mathbf{v}) = \lambda_2(\mathbf{u}^T\mathbf{v}) \end{aligned}$$

or $(\lambda_1 - \lambda_2)(\mathbf{u}^T\mathbf{v}) = 0$. Since $\lambda_1 \neq \lambda_2$, then $\mathbf{u}^T\mathbf{v} = 0$. Observe that $\mathbf{u}^T\mathbf{v} = 0$ is equivalent to $\mathbf{u} \cdot \mathbf{v} = 0$, and so $\mathbf{u}$ and $\mathbf{v}$ are orthogonal.

At this point, we know that if (1) A is symmetric and (2) A has n distinct eigenvalues, then A is orthogonally diagonalizable as follows.

1. Let $\mathbf{v}_1, \mathbf{v}_2, \ldots, \mathbf{v}_n$ be eigenvectors belonging to $\lambda_1, \lambda_2, \ldots, \lambda_n$, respectively. Then $\mathbf{v}_i \cdot \mathbf{v}_j = 0$ for $i \neq j$.
2. Let $\mathbf{u}_i = \mathbf{v}_i/\|\mathbf{v}_i\|$ for $i = 1, 2, \ldots, n$.

3. Let Q be the $n \times n$ matrix with $\mathbf{u}_i$ the ith column; now Q has the properties $Q^T = Q^{-1}$ and

$$Q^TAQ = \begin{bmatrix} \lambda_1 & 0 & \cdots & 0 \\ 0 & \lambda_2 & \cdots & 0 \\ \vdots & \vdots & & \vdots \\ 0 & 0 & \cdots & \lambda_n \end{bmatrix}$$

Example 6.3.11 Let

$$A = \begin{bmatrix} 1 & 2 \\ 2 & 1 \end{bmatrix}$$

Find an orthogonal matrix Q for which Q^TAQ is diagonal.

Solution First,

$$\det(A - \lambda I) = \begin{bmatrix} 1-\lambda & 2 \\ 2 & 1-\lambda \end{bmatrix} = \lambda^2 - 2\lambda - 3 = 0$$

when $\lambda = 3$ and -1.

For the eigenvalue $\lambda_1 = 3$,

$$A - \lambda_1 I = \begin{bmatrix} -2 & 2 \\ 2 & -2 \end{bmatrix}$$

and if

$$\begin{bmatrix} -2 & 2 \\ 2 & -2 \end{bmatrix}\begin{bmatrix} x_1 \\ x_2 \end{bmatrix} = 0$$

then $-2x_1 + 2x_2 = 0$, or $x_1 = x_2$, x_2 arbitrary. Thus

$$\mathbf{v}_1 = \begin{bmatrix} 1 \\ 1 \end{bmatrix}$$

is an eigenvector corresponding to $\lambda_1 = 3$.

For the eigenvalue $\lambda_2 = -1$,

$$A - \lambda_2 I = \begin{bmatrix} 2 & 2 \\ 2 & 2 \end{bmatrix}$$

and if

$$\begin{bmatrix} 2 & 2 \\ 2 & 2 \end{bmatrix}\begin{bmatrix} x_1 \\ x_2 \end{bmatrix} = \begin{bmatrix} 0 \\ 0 \end{bmatrix}$$

then $x_1 = -x_2$, x_2 arbitrary. An eigenvector corresponding to $\lambda_2 = -1$ is

$$\mathbf{v}_2 = \begin{bmatrix} 1 \\ -1 \end{bmatrix}$$

Observe

$$\mathbf{v}_1 \cdot \mathbf{v}_2 = \begin{bmatrix} 1 \\ 1 \end{bmatrix} \cdot \begin{bmatrix} 1 \\ -1 \end{bmatrix} = 0$$

so $\mathbf{v}_1$ and $\mathbf{v}_2$ are orthogonal. Let

$$\mathbf{u}_1 = \frac{\mathbf{v}_1}{\|\mathbf{v}_1\|} = \begin{bmatrix} \frac{1}{\sqrt{2}} \\ \frac{1}{\sqrt{2}} \end{bmatrix} \quad \text{and} \quad \mathbf{u}_2 = \frac{\mathbf{v}_2}{\|\mathbf{v}_2\|} = \begin{bmatrix} \frac{1}{\sqrt{2}} \\ -\frac{1}{\sqrt{2}} \end{bmatrix}$$

Then

$$\begin{bmatrix} \frac{1}{\sqrt{2}} & \frac{1}{\sqrt{2}} \\ \frac{1}{\sqrt{2}} & -\frac{1}{\sqrt{2}} \end{bmatrix}^T \begin{bmatrix} 1 & 2 \\ 2 & 1 \end{bmatrix} \begin{bmatrix} \frac{1}{\sqrt{2}} & \frac{1}{\sqrt{2}} \\ \frac{1}{\sqrt{2}} & -\frac{1}{\sqrt{2}} \end{bmatrix} = \begin{bmatrix} 3 & 0 \\ 0 & -1 \end{bmatrix}$$

and A is orthogonally diagonalizable. ■

Must A be of any special type if it is orthogonally diagonalizable? The answer is yes, as shown by the following theorem.

Theorem 6.3.12 *If*

$$Q^TAQ = \begin{bmatrix} \lambda_1 & 0 & \cdots & 0 \\ 0 & \lambda_2 & \cdots & 0 \\ \vdots & \vdots & & \vdots \\ 0 & 0 & \cdots & \lambda_n \end{bmatrix}$$

where Q is orthogonal, then A is symmetric.

Proof. If

$$Q^TAQ = \begin{bmatrix} \lambda_1 & 0 & \cdots & 0 \\ 0 & \lambda_2 & \cdots & 0 \\ \vdots & \vdots & & \vdots \\ 0 & 0 & \cdots & \lambda_n \end{bmatrix}$$

then

$$A = (Q^T)^{-1} \begin{bmatrix} \lambda_1 & 0 & \cdots & 0 \\ 0 & \lambda_2 & \cdots & 0 \\ \vdots & \vdots & & \vdots \\ 0 & 0 & \cdots & \lambda_n \end{bmatrix} Q^{-1}$$

$$= (Q^{-1})^{-1} \begin{bmatrix} \lambda_1 & 0 & \cdots & 0 \\ 0 & \lambda_2 & \cdots & 0 \\ \vdots & \vdots & & \vdots \\ 0 & 0 & \cdots & \lambda_n \end{bmatrix} Q^T$$

[Q is orthogonal; i.e., $Q^T = Q^{-1}$]

$$= Q \begin{bmatrix} \lambda_1 & 0 & \cdots & 0 \\ 0 & \lambda_2 & \cdots & 0 \\ \vdots & \vdots & & \vdots \\ 0 & 0 & \cdots & \lambda_n \end{bmatrix} Q^T$$

Also

$$A^T = \left[Q \begin{bmatrix} \lambda_1 & 0 & \cdots & 0 \\ 0 & \lambda_2 & \cdots & 0 \\ \vdots & \vdots & & \vdots \\ 0 & 0 & \cdots & \lambda_n \end{bmatrix} Q^T \right]^T$$

$$= (Q^T)^T \begin{bmatrix} \lambda_1 & 0 & \cdots & 0 \\ 0 & \lambda_2 & \cdots & 0 \\ \vdots & \vdots & & \vdots \\ 0 & 0 & \cdots & \lambda_n \end{bmatrix}^T Q^T$$

$$= Q \begin{bmatrix} \lambda_1 & 0 & \cdots & 0 \\ 0 & \lambda_2 & \cdots & 0 \\ \vdots & \vdots & & \vdots \\ 0 & 0 & \cdots & \lambda_n \end{bmatrix} Q^T = A$$

and so A is symmetric.

REMARK By Theorem 6.3.12, if A is orthogonally diagonalizable, then A is symmetric. We also know that if A has n distinct eigenvalues, then A is diagonalizable. In Section 6.5, we show that when A is symmetric, even though it may not have n distinct eigenvalues, it can still be orthogonally diagonalized.

SUMMARY

Terms to know:

Orthonormal basis
Orthogonal matrix
Orthogonally diagonalizable matrix

Theorems to know:

An orthogonal set of nonzero vectors is linearly independent.

If A is a symmetric $n \times n$ matrix, eigenvectors belonging to distinct eigenvalues are orthogonal.

An orthogonally diagonalizable matrix is symmetric.

EXERCISE SET 6.3

Enlarge the following matrices so they become square matrices that are orthogonal.

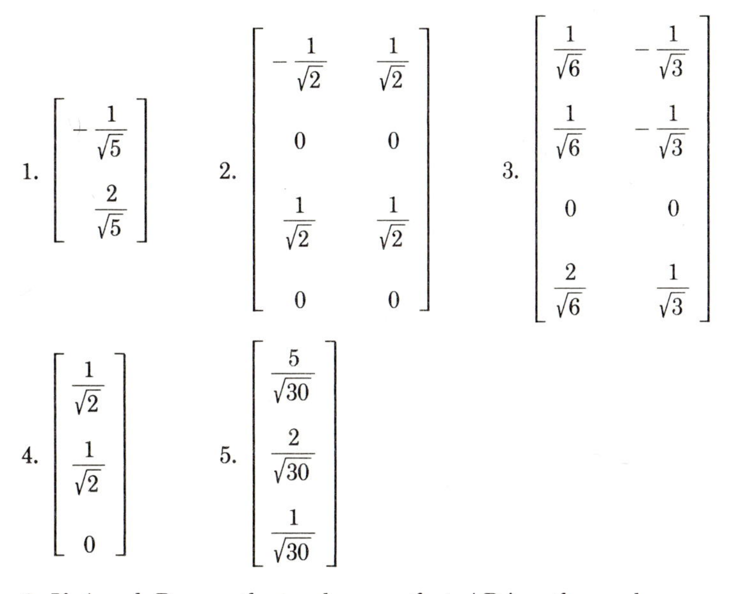

1. $\begin{bmatrix} -\frac{1}{\sqrt{5}} \\ \frac{2}{\sqrt{5}} \end{bmatrix}$

2. $\begin{bmatrix} -\frac{1}{\sqrt{2}} & \frac{1}{\sqrt{2}} \\ 0 & 0 \\ \frac{1}{\sqrt{2}} & \frac{1}{\sqrt{2}} \\ 0 & 0 \end{bmatrix}$

3. $\begin{bmatrix} \frac{1}{\sqrt{6}} & -\frac{1}{\sqrt{3}} \\ \frac{1}{\sqrt{6}} & -\frac{1}{\sqrt{3}} \\ 0 & 0 \\ \frac{2}{\sqrt{6}} & \frac{1}{\sqrt{3}} \end{bmatrix}$

4. $\begin{bmatrix} \frac{1}{\sqrt{2}} \\ \frac{1}{\sqrt{2}} \\ 0 \end{bmatrix}$

5. $\begin{bmatrix} \frac{5}{\sqrt{30}} \\ \frac{2}{\sqrt{30}} \\ \frac{1}{\sqrt{30}} \end{bmatrix}$

6. If A and B are orthogonal, prove that AB is orthogonal.

7. If A and B are orthogonal, is $A + B$ orthogonal?

8. If T is a 3×3 upper triangular matrix and is orthogonal, what must T look like? List all such T.

9. If all $\{\mathbf{v}_1, \mathbf{v}_2, \mathbf{v}_3\}$ is a linearly independent set, must the vectors be orthogonal?

10. Show that

$$\left\{ \begin{bmatrix} \frac{1}{\sqrt{2}} \\ 0 \\ \frac{1}{\sqrt{2}} \end{bmatrix}, \begin{bmatrix} 0 \\ 1 \\ 0 \end{bmatrix}, \begin{bmatrix} \frac{1}{\sqrt{2}} \\ 0 \\ -\frac{1}{\sqrt{2}} \end{bmatrix} \right\}$$

is an orthonormal basis for R^3.

11. Find if possible, an orthogonal matrix Q that diagonalizes A, and then diagonalize A.

a. $A = \begin{bmatrix} 1 & 0 & -2 \\ 0 & 0 & 0 \\ -2 & 0 & 4 \end{bmatrix}$ b. $A = \begin{bmatrix} 3 & 1 & 1 \\ 1 & 1 & 1 \\ -1 & 0 & 1 \end{bmatrix}$

12. Find an orthogonal matrix A that is symmetric and whose first column is

$$\begin{bmatrix} \frac{3}{5} \\ 0 \\ \frac{4}{5} \end{bmatrix}$$

13. Let A be an $n \times n$ matrix whose columns form an orthonormal basis of R^n. Prove that A is an orthogonal matrix.

SECTION 6.4 PROJECTIONS AND THE GRAM-SCHMIDT PROCESS

In high school, we learned to project a point $P = (a, b)$ onto a line L by dropping a perpendicular from P to L. The projection Q is the point where L and the perpendicular meet (see Figure 6.1). We also learned that the length of $\overline{PQ}$ is the *shortest distance* from P to L.

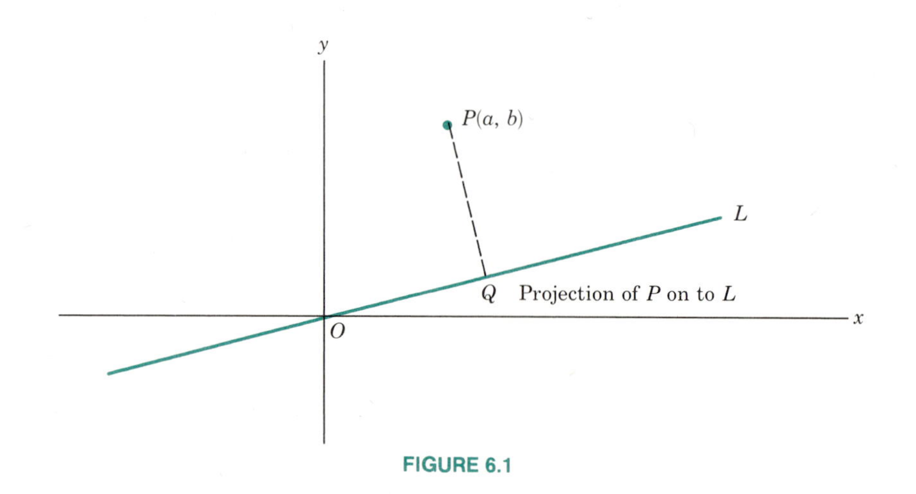

FIGURE 6.1

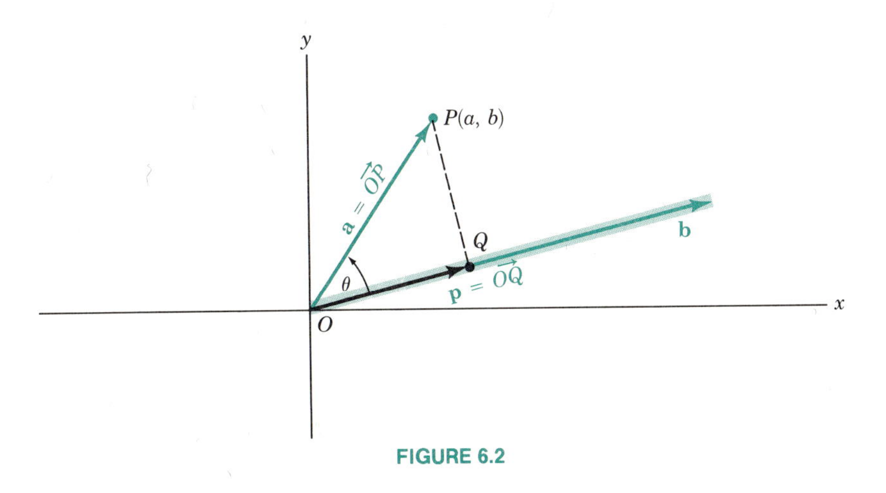

FIGURE 6.2

Furthermore, if the equation of line L is $Ax + By + C = 0$, then the distance from P to L is

$$\frac{|Aa + Bb + C|}{\sqrt{A^2 + B^2}}$$

We note that if P is on line L, then the distance from P to L is 0.

We generalize this notation slightly by considering the position vector

$$\mathbf{a} = \overrightarrow{OP} = \begin{bmatrix} a \\ b \end{bmatrix}$$

(instead of point P) and a vector $\mathbf{b}$ with tail at the origin O and lying on the line L (instead of line L). See Figure 6.2. Let $\mathbf{p} = \overrightarrow{OQ}$ be the projection of $\mathbf{a}$ on $\mathbf{b}$. By right triangle trigonometry, $\|\mathbf{p}\| = \|\mathbf{a}\|\cos\theta$. But $\mathbf{a} \cdot \mathbf{b} = \|\mathbf{a}\|\,\|\mathbf{b}\|\cos\theta$, and so

$$\|\mathbf{p}\| = \|\mathbf{a}\|\cos\theta = \frac{\|\mathbf{a}\|\,\|\mathbf{b}\|\cos\theta}{\|\mathbf{b}\|} = \frac{\mathbf{a} \cdot \mathbf{b}}{\|\mathbf{b}\|}$$

The unit vector in the same direction as $\mathbf{b}$ is $\mathbf{b}/\|\mathbf{b}\|$. Thus

$$\mathbf{p} = \left(\frac{\mathbf{a} \cdot \mathbf{b}}{\|\mathbf{b}\|}\right)\frac{\mathbf{b}}{\|\mathbf{b}\|} = \left(\frac{\mathbf{a} \cdot \mathbf{b}}{\|\mathbf{b}\|^2}\right)\mathbf{b} = \frac{\mathbf{a} \cdot \mathbf{b}}{\mathbf{b} \cdot \mathbf{b}}\mathbf{b}$$

and a vector perpendicular to $\mathbf{b}$ is $\overrightarrow{QP} = \overrightarrow{OP} - \overrightarrow{OQ} = \mathbf{a} - \mathbf{p}$, where

$$\mathbf{a} - \mathbf{p} = \mathbf{a} - \frac{\mathbf{a} \cdot \mathbf{b}}{\mathbf{b} \cdot \mathbf{b}}\mathbf{b}$$

Note that if $\mathbf{a} = k\mathbf{b}$, then $\mathbf{a} \cdot \mathbf{b} = k\mathbf{b} \cdot \mathbf{b}$ and $\mathbf{a} - \mathbf{p} = k\mathbf{b} - k\mathbf{b} = \mathbf{0}$ so that $\mathbf{p} = \mathbf{a}$.

The projection of the vector $\mathbf{a}$ on $\mathbf{b}$ is denoted by $\text{proj}_{\mathbf{b}}\mathbf{a}$.

Example 6.4.1 Find the projection $\mathbf{p}$ of the vector $\mathbf{a}$ on $\mathbf{b}$, where

$$\mathbf{a} = \begin{bmatrix} 2 \\ 0 \end{bmatrix} \quad \text{and} \quad \mathbf{b} = \begin{bmatrix} \frac{3}{2} \\ \frac{\sqrt{3}}{2} \end{bmatrix}$$

and find a vector $\mathbf{u}$ perpendicular to $\mathbf{b}$, such that $\mathbf{a} = \mathbf{p} + \mathbf{u}$.

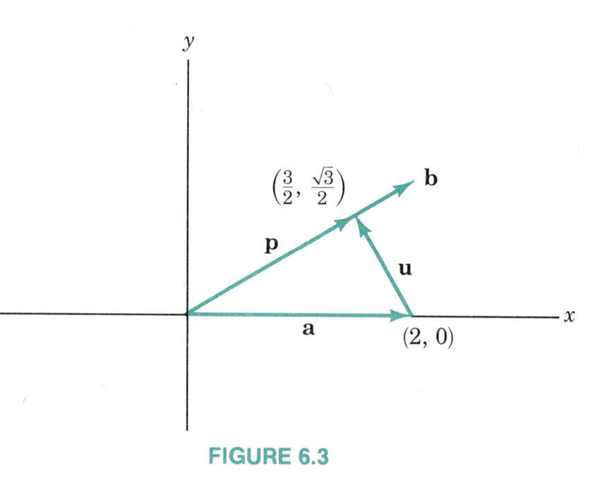

FIGURE 6.3

Solution See Figure 6.3.

First,

$$\mathbf{a} \cdot \mathbf{b} = \begin{bmatrix} 2 \\ 0 \end{bmatrix} \cdot \begin{bmatrix} \frac{3}{2} \\ \frac{\sqrt{3}}{2} \end{bmatrix} = 3 \quad \text{and} \quad \mathbf{b} \cdot \mathbf{b} = \begin{bmatrix} \frac{3}{2} \\ \frac{\sqrt{3}}{2} \end{bmatrix} \cdot \begin{bmatrix} \frac{3}{2} \\ \frac{\sqrt{3}}{2} \end{bmatrix} = 3$$

Thus

$$\mathbf{p} = \frac{\mathbf{a} \cdot \mathbf{b}}{\mathbf{b} \cdot \mathbf{b}} \mathbf{b} = \frac{3}{3} \begin{bmatrix} \frac{3}{2} \\ \frac{\sqrt{3}}{2} \end{bmatrix} = \begin{bmatrix} \frac{3}{2} \\ \frac{\sqrt{3}}{2} \end{bmatrix}$$

and

$$\mathbf{u} = \mathbf{a} - \mathbf{p} = \begin{bmatrix} 2 \\ 0 \end{bmatrix} - \begin{bmatrix} \frac{3}{2} \\ \frac{\sqrt{3}}{2} \end{bmatrix} = \begin{bmatrix} \frac{1}{2} \\ -\frac{\sqrt{3}}{2} \end{bmatrix}$$

Observe that

$$\mathbf{u} \cdot \mathbf{b} = \begin{bmatrix} \frac{1}{2} \\ -\frac{\sqrt{3}}{2} \end{bmatrix} \cdot \begin{bmatrix} \frac{3}{2} \\ \frac{\sqrt{3}}{2} \end{bmatrix} = \frac{3}{4} - \frac{3}{4} = 0$$

and that $\mathbf{a} = \mathbf{p} + \mathbf{u}$. The *distance* from the point A to the line L is

$$\|\mathbf{u}\| = \left[\left(\frac{1}{2}\right)^2 + \left(-\frac{\sqrt{3}}{2}\right)^2\right]^{1/2} = (1)^{1/2} = 1 \qquad \blacksquare$$

Next, in R^n we consider the projection of a vector onto a plane generated by two linearly independent vectors $\mathbf{b}_1$ and $\mathbf{b}_2$. The plane consists of all linear combinations of $\mathbf{b}_1$ and $\mathbf{b}_2$, that is, $\langle \mathbf{b}_1, \mathbf{b}_2 \rangle$. We alternately regard the plane as the column space of the matrix B whose columns are $\mathbf{b}_1$ and $\mathbf{b}_2$. Thus

$$B = [\mathbf{b}_1 \quad \mathbf{b}_2]$$

Let $\mathbf{a}$ be a vector not lying in this plane. We wish to find a vector

$$\mathbf{x} = B\begin{bmatrix} x_1 \\ x_2 \end{bmatrix}$$

in the plane for which $\mathbf{a} - \mathbf{x}$ is perpendicular to the plane. (In R^3, Figure 6.4 shows the situation we are discussing.)

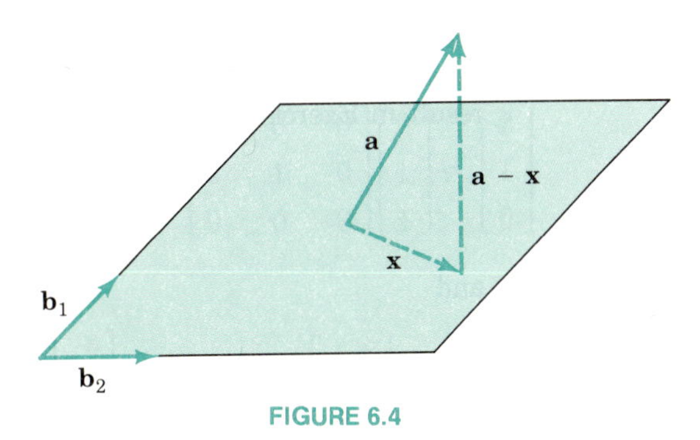

FIGURE 6.4

Since any vector in the plane is in the column space of B, any vector in the plane can be denoted by

$$B\begin{bmatrix} y_1 \\ y_2 \end{bmatrix}$$

If

$$\mathbf{a} - B\begin{bmatrix} x_1 \\ x_2 \end{bmatrix}$$

the projection of $\mathbf{b}$ onto the column space of B—that is, onto the space spanned by $\{\mathbf{c}_1, \ldots, \mathbf{c}_i\}$.

Finally, if $\mathbf{c}_{i+1} = \mathbf{b} - \{$projection of $\mathbf{b}$ onto the column space of $B\}$ or

$$\mathbf{c}_{i+1} = \mathbf{b} - \left\{\frac{\mathbf{c}_1 \cdot \mathbf{b}}{\mathbf{c}_1 \cdot \mathbf{c}_1}\mathbf{c}_1 + \frac{\mathbf{c}_2 \cdot \mathbf{b}}{\mathbf{c}_2 \cdot \mathbf{c}_2}\mathbf{c}_2 + \cdots + \frac{\mathbf{c}_i \cdot \mathbf{b}}{\mathbf{c}_i \cdot \mathbf{c}_i}\mathbf{c}_i\right\}$$

then as before $\{\mathbf{c}_1, \ldots, \mathbf{c}_{i+1}\}$ is an orthogonal set.

Example 6.4.3 Given the basis

$$\left\{\begin{bmatrix} 1 \\ 0 \\ 1 \end{bmatrix}, \begin{bmatrix} 1 \\ 1 \\ 0 \end{bmatrix}, \begin{bmatrix} 0 \\ 1 \\ 1 \end{bmatrix}\right\}$$

use the Gram-Schmidt process to obtain a related orthonormal basis of R^3.

Solution First, we obtain an orthogonal basis $\{\mathbf{c}_1, \mathbf{c}_2, \mathbf{c}_3\}$. Let

$$\mathbf{b}_1 = \begin{bmatrix} 1 \\ 0 \\ 1 \end{bmatrix}, \quad \mathbf{b}_2 = \begin{bmatrix} 1 \\ 1 \\ 0 \end{bmatrix}, \quad \text{and} \quad \mathbf{b}_3 = \begin{bmatrix} 0 \\ 1 \\ 1 \end{bmatrix}$$

Then

$$\mathbf{c}_1 = \mathbf{b}_1 = \begin{bmatrix} 1 \\ 0 \\ 1 \end{bmatrix}$$

and $\mathbf{c}_1 \cdot \mathbf{c}_1 = 2$.

To determine $\mathbf{c}_2$, first we calculate

$$\mathbf{c}_1 \cdot \mathbf{b}_2 = \begin{bmatrix} 1 \\ 0 \\ 1 \end{bmatrix} \cdot \begin{bmatrix} 1 \\ 1 \\ 0 \end{bmatrix} = 1$$

Then

$$\mathbf{c}_2 = \mathbf{b}_2 - \frac{\mathbf{b}_2 \cdot \mathbf{c}_1}{\mathbf{c}_1 \cdot \mathbf{c}_1}\mathbf{c}_1 = \begin{bmatrix} 1 \\ 1 \\ 0 \end{bmatrix} - \frac{1}{2}\begin{bmatrix} 1 \\ 0 \\ 1 \end{bmatrix} = \begin{bmatrix} \frac{1}{2} \\ 1 \\ -\frac{1}{2} \end{bmatrix}$$

Furthermore, $\mathbf{c}_2 \cdot \mathbf{c}_2 = \frac{3}{2}$.

Now,

$$\mathbf{c}_3 = \mathbf{b}_3 - \frac{\mathbf{b}_3 \cdot \mathbf{c}_1}{\mathbf{c}_1 \cdot \mathbf{c}_1}\mathbf{c}_1 - \frac{\mathbf{b}_3 \cdot \mathbf{c}_2}{\mathbf{c}_2 \cdot \mathbf{c}_2}\mathbf{c}_2$$

where

$$\mathbf{b}_3 \cdot \mathbf{c}_1 = \begin{bmatrix} 0 \\ 1 \\ 1 \end{bmatrix} \cdot \begin{bmatrix} 1 \\ 0 \\ 1 \end{bmatrix} = 1$$

and

$$\mathbf{b}_3 \cdot \mathbf{c}_2 = \begin{bmatrix} 0 \\ 1 \\ 1 \end{bmatrix} \cdot \begin{bmatrix} \frac{1}{2} \\ 1 \\ -\frac{1}{2} \end{bmatrix} = \frac{1}{2}$$

Thus,

$$\mathbf{c}_3 = \begin{bmatrix} 0 \\ 1 \\ 1 \end{bmatrix} - \frac{1}{2}\begin{bmatrix} 1 \\ 0 \\ 1 \end{bmatrix} - \frac{\frac{1}{2}}{\frac{3}{2}}\begin{bmatrix} \frac{1}{2} \\ 1 \\ -\frac{1}{2} \end{bmatrix} = \begin{bmatrix} -\frac{2}{3} \\ \frac{2}{3} \\ \frac{2}{3} \end{bmatrix}$$

and

$$\mathbf{c}_3 \cdot \mathbf{c}_3 = \frac{4}{3}$$

Next we normalize the basis $\{\mathbf{c}_1, \mathbf{c}_2, \mathbf{c}_3\}$, obtaining

$$\mathbf{q}_1 = \frac{\mathbf{c}_1}{\|\mathbf{c}_1\|} = \frac{1}{\sqrt{2}}\begin{bmatrix} 1 \\ 0 \\ 1 \end{bmatrix} = \begin{bmatrix} \frac{1}{\sqrt{2}} \\ 0 \\ \frac{1}{\sqrt{2}} \end{bmatrix}$$

$$\mathbf{q}_2 = \frac{\mathbf{c}_2}{\|\mathbf{c}_2\|} = \frac{\sqrt{2}}{\sqrt{3}}\begin{bmatrix} \frac{1}{2} \\ 1 \\ -\frac{1}{2} \end{bmatrix} = \begin{bmatrix} \frac{1}{\sqrt{6}} \\ \frac{\sqrt{2}}{\sqrt{3}} \\ -\frac{1}{\sqrt{6}} \end{bmatrix}$$

and

$$\mathbf{q}_3 = \frac{\mathbf{c}_3}{\|\mathbf{c}_3\|} = \frac{\sqrt{3}}{\sqrt{4}}\begin{bmatrix} -\frac{2}{3} \\ \frac{2}{3} \\ \frac{2}{3} \end{bmatrix} = \begin{bmatrix} -\frac{\sqrt{3}}{3} \\ \frac{\sqrt{3}}{3} \\ \frac{\sqrt{3}}{3} \end{bmatrix}$$

The set $\{\mathbf{q}_1, \mathbf{q}_2, \mathbf{q}_3\}$ is an orthonormal basis of R^3. ■

we have

$$A^T = \begin{bmatrix} 1 & 3 & 1 \\ -1 & -2 & 1 \end{bmatrix}, \qquad A^TA = \begin{bmatrix} 11 & -6 \\ -6 & 6 \end{bmatrix}, \quad \text{and}$$

$$(A^TA)^{-1} = \begin{bmatrix} \frac{1}{5} & \frac{1}{5} \\ \frac{1}{5} & \frac{11}{30} \end{bmatrix}$$

The best solution is

$$\mathbf{x}_0 = (A^TA)^{-1}A^T\mathbf{b} = \begin{bmatrix} \frac{1}{5} & \frac{1}{5} \\ \frac{1}{5} & \frac{11}{30} \end{bmatrix} \begin{bmatrix} 1 & 3 & 1 \\ -1 & -2 & 1 \end{bmatrix} \begin{bmatrix} 2 \\ 1 \\ 1 \end{bmatrix}$$

$$= \begin{bmatrix} \frac{3}{5} \\ \frac{1}{10} \end{bmatrix}$$

■

Example 6.5.3 Find the best solution to the system of equations

$$x_1 + 2x_2 + x_3 = 3$$

$$x_1 + 3x_2 + 2x_3 = 5$$

$$x_1 + 4x_2 + 4x_3 = 10$$

Solution The augmented matrix

$$\left[\begin{array}{ccc|c} 1 & 2 & 1 & 3 \\ 1 & 3 & 2 & 5 \\ 1 & 4 & 4 & 10 \end{array}\right]$$

has the row echelon form

$$\left[\begin{array}{ccc|c} 1 & 2 & 1 & 3 \\ 0 & 1 & 1 & 2 \\ 0 & 0 & 1 & 3 \end{array}\right]$$

Even though the system is inconsistent, we proceed as before to form the coefficient matrix

$$A = \begin{bmatrix} 1 & 2 & 1 \\ 1 & 3 & 2 \\ 1 & 4 & 4 \end{bmatrix}$$

Then

$$A^T = \begin{bmatrix} 1 & 1 & 1 \\ 2 & 3 & 4 \\ 1 & 2 & 4 \end{bmatrix} \quad \text{and} \quad (A^TA)^{-1} = \begin{bmatrix} 3 & 9 & 7 \\ 9 & 29 & 24 \\ 7 & 24 & 21 \end{bmatrix}^{-1}$$

$$= \begin{bmatrix} 33 & -21 & 13 \\ -21 & 14 & -9 \\ 13 & -9 & 6 \end{bmatrix}$$

and so

$$A^+ = (A^TA)^{-1}A^T = \begin{bmatrix} 4 & -4 & 1 \\ -2 & 3 & -1 \\ 1 & -2 & 1 \end{bmatrix}$$

Hence

$$\mathbf{x}_0 = A^+\mathbf{b} = \begin{bmatrix} 4 & -4 & 1 \\ -2 & 3 & -1 \\ 1 & -2 & 1 \end{bmatrix} \begin{bmatrix} 3 \\ 5 \\ 10 \end{bmatrix} = \begin{bmatrix} 2 \\ -1 \\ 3 \end{bmatrix}$$ ■

REMARK In Example 6.5.3, the coefficient matrix

$$\begin{bmatrix} 1 & 2 & 1 \\ 1 & 3 & 2 \\ 1 & 4 & 4 \end{bmatrix}$$

is invertible and so the system has a unique solution $x_1 = 2$, $x_2 = -1$, $x_3 = 3$. In this case, $A^+ = A^{-1}$.

SUMMARY

Methods to know:

Find the "best" solution to an inconsistent system of equations

EXERCISE SET 6.5

Find the generalized inverse matrix A^+ and the best solution to the following systems of equations.

1. $$\begin{aligned} x_1 + x_2 &= 9 \\ 4x_1 + 6x_2 &= 46 \\ 3x_1 + 4x_2 &= 30 \end{aligned}$$

2. $$\begin{aligned} 6x_1 &= 46 \\ -x_1 - 3x_2 &= -1 \\ 5x_1 + 4x_2 &= 42 \end{aligned}$$

3. $$\begin{aligned} x_1 + 5x_2 + 6x_3 &= 17 \\ x_1 + 2x_2 + 3x_3 &= 7 \\ x_1 + x_2 + 2x_3 &= 3 \end{aligned}$$

4. $$\begin{aligned} x_1 + 6x_2 &= 25 \\ 2x_1 - 13x_2 &= 46 \\ x_1 + 6x_2 &= 21 \end{aligned}$$

5. $x_1 + 5x_2 = 13$
$x_1 + 2x_2 = 8$
$x_1 + x_2 = 5$

6. $x_1 + 5x_2 = 21$
$x_1 - 2x_2 = 5$
$x_1 + x_2 = 8$

7. $x_1 + x_2 + 2x_3 = 9$
$4x_1 + 6x_2 + 8x_3 = 46$
$3x_1 + 4x_2 + 6x_3 = 30$

8. $3x_1 = 5$
$4x_1 = 5$

9. $x_1 + x_2 = 5$
$2x_1 + 4x_2 = 5$
$x_1 + 3x_2 = 6$

10. $x_1 + x_2 = 2$
$2x_1 + 5x_2 = 5$
$x_1 + 4x_2 = 5$

11. $x_1 + 2x_2 = 1$
$3x_1 + 2x_2 = 5$
$3x_1 + 8x_2 = 7$

12. $x_1 = 1$
$5x_1 + x_2 = 6$
$7x_1 + 4x_2 = 12$

13. Suppose $A\mathbf{x} = \mathbf{b}$ has a solution $\mathbf{x}_0$ and suppose A is nonsingular. Show that $A^+ = A^{-1}$.

14. If the columns of A are orthonormal, what is A^+?

SECTION 6.6 DIAGONALIZATION REVISITED

Recall that given an $n \times n$ matrix A, we wish to find an orthogonal matrix Q for which $Q^{-1}AQ$ is diagonal. From earlier results, we know

1. A must be symmetric.
2. The columns of Q must be eigenvectors of A.
3. Eigenvectors belonging to different eigenvalues are orthogonal.

Theorem 6.6.1 *A matrix A is orthogonally diagonalizable if and only if A is symmetric.*

In Section 6.3 we proved that if A is an orthogonally diagonalizable matrix, then A must be symmetric. The converse of this theorem is not proved in this text.

Theorem 6.6.1 says that for any symmetric matrix, we can find an orthogonal Q (which means $Q^{-1} = Q^T$) such that Q^TAQ is a diagonal matrix. To do this, we use the following procedure.

1. Calculate the eigenvalues for A.
2. Find a basis for each eigenspace of A.
3. For each eigenspace of dimension 2 or more, apply the Gram-Schmidt process to the basis found in Step 2 to obtain an orthogonal basis.
4. Normalize each basis vector.
5. Put the normalized basis vectors as columns of Q.

We apply this process in Example 6.6.2.

Example 6.6.2 For

$$A = \begin{bmatrix} 7 & -2 & 1 \\ -2 & 10 & -2 \\ 1 & -2 & 7 \end{bmatrix}$$

find Q such that Q is orthogonal and Q^TAQ is a diagonal matrix.

Solution *STEP 1.* We compute the eigenvalues for A. Since

$$A - \lambda I = \begin{bmatrix} 7-\lambda & -2 & 1 \\ -2 & 10-\lambda & -2 \\ 1 & -2 & 7-\lambda \end{bmatrix}$$

and $\det(A - \lambda I) = (\lambda - 12)(\lambda - 6)^2 = 0$ for $\lambda = 12$ and 6, there are two distinct eigenvalues, $\lambda_1 = 12$ and $\lambda_2 = 6$.

STEP 2. We now find bases for the eigenspaces of A corresponding to $\lambda_1 = 12$ and $\lambda_2 = 6$.

For $\lambda_1 = 12$,

$$A - 12I = \begin{bmatrix} -5 & -2 & 1 \\ -2 & -2 & -2 \\ 1 & -2 & -5 \end{bmatrix}$$

and an eigenvector $\mathbf{v}_1$ for $\lambda_1 = 12$ is

$$\mathbf{v}_1 = \begin{bmatrix} 1 \\ -2 \\ 1 \end{bmatrix}$$

The eigenspace is the set of all vectors

$$\alpha \begin{bmatrix} 1 \\ -2 \\ 1 \end{bmatrix}$$

for α any scalar, and the dimension is 1.

For $\lambda_2 = 6$,

$$A - 6I = \begin{bmatrix} 1 & -2 & 1 \\ -2 & 4 & -2 \\ 1 & -2 & 1 \end{bmatrix}$$

and for

$$\mathbf{x} = \begin{bmatrix} x_1 \\ x_2 \\ x_3 \end{bmatrix}$$

$$\begin{bmatrix} 1 & -2 & 1 \\ -2 & 4 & -2 \\ 1 & -2 & 1 \end{bmatrix} \begin{bmatrix} x_1 \\ x_2 \\ x_3 \end{bmatrix} = \begin{bmatrix} 0 \\ 0 \\ 0 \end{bmatrix}$$

when

$$\mathbf{x} = \begin{bmatrix} 2x_2 - x_3 \\ x_2 \\ x_3 \end{bmatrix}$$

for x_2 and x_3 arbitrarily chosen. Now,

$$\mathbf{x} = \begin{bmatrix} 2x_2 - x_3 \\ x_2 \\ x_3 \end{bmatrix} = \begin{bmatrix} 2x_2 \\ x_2 \\ 0 \end{bmatrix} + \begin{bmatrix} -x_3 \\ 0 \\ x_3 \end{bmatrix} = x_2 \begin{bmatrix} 2 \\ 1 \\ 0 \end{bmatrix} + x_3 \begin{bmatrix} -1 \\ 0 \\ 1 \end{bmatrix}$$

which means the solution space is spanned by the set of linearly independent vectors

$$\left\{ \begin{bmatrix} 2 \\ 1 \\ 0 \end{bmatrix}, \begin{bmatrix} -1 \\ 0 \\ 1 \end{bmatrix} \right\}$$

Thus

$$\mathbf{v}_2 = \begin{bmatrix} 2 \\ 1 \\ 0 \end{bmatrix} \quad \text{and} \quad \mathbf{v}_3 = \begin{bmatrix} -1 \\ 0 \\ 1 \end{bmatrix}$$

form a basis for the eigenspace of A belonging to $\lambda_2 = 6$. The dimension of this eigenspace is 2.

STEP 3. Next we orthogonalize the basis consisting of

$$\mathbf{v}_2 = \begin{bmatrix} 2 \\ 1 \\ 0 \end{bmatrix} \quad \text{and} \quad \mathbf{v}_3 = \begin{bmatrix} -1 \\ 0 \\ 1 \end{bmatrix}$$

using the Gram-Schmidt process as follows:

$$\mathbf{c}_2 = \mathbf{v}_2 = \begin{bmatrix} 2 \\ 1 \\ 0 \end{bmatrix}$$

$$\mathbf{c}_3 = \mathbf{v}_3 - \frac{\mathbf{v}_3 \cdot \mathbf{c}_2}{\mathbf{c}_2 \cdot \mathbf{c}_2}\mathbf{c}_2$$

$$= \begin{bmatrix} -1 \\ 0 \\ 1 \end{bmatrix} - \left(\frac{-2}{5}\right)\begin{bmatrix} 2 \\ 1 \\ 0 \end{bmatrix} = \begin{bmatrix} -\frac{1}{5} \\ \frac{2}{5} \\ 1 \end{bmatrix}$$

STEP 4. The orthogonal vectors

$$\mathbf{v}_1 = \begin{bmatrix} 1 \\ -2 \\ 1 \end{bmatrix}, \quad \mathbf{c}_2 = \begin{bmatrix} 2 \\ 1 \\ 0 \end{bmatrix}, \quad \text{and} \quad \mathbf{c}_3 = \begin{bmatrix} -\frac{1}{5} \\ \frac{2}{5} \\ 1 \end{bmatrix}$$

are normalized, obtaining

$$\frac{\mathbf{v}_1}{\|\mathbf{v}_1\|} = \begin{bmatrix} \frac{1}{\sqrt{6}} \\ -\frac{2}{\sqrt{6}} \\ \frac{1}{\sqrt{6}} \end{bmatrix}, \quad \frac{\mathbf{c}_2}{\|\mathbf{c}_2\|} = \begin{bmatrix} \frac{2}{\sqrt{5}} \\ \frac{1}{\sqrt{5}} \\ 0 \end{bmatrix}, \quad \text{and} \quad \frac{\mathbf{c}_3}{\|\mathbf{c}_3\|} = \begin{bmatrix} -\frac{1}{\sqrt{30}} \\ \frac{2}{\sqrt{30}} \\ \frac{5}{\sqrt{30}} \end{bmatrix}$$

STEP 5. Finally,

$$Q = \begin{bmatrix} \frac{1}{\sqrt{6}} & \frac{2}{\sqrt{5}} & -\frac{1}{\sqrt{30}} \\ -\frac{2}{\sqrt{6}} & \frac{1}{\sqrt{5}} & \frac{2}{\sqrt{30}} \\ \frac{1}{\sqrt{6}} & 0 & \frac{5}{\sqrt{30}} \end{bmatrix}$$

Check:

$$Q^TAQ = \begin{bmatrix} 12 & 0 & 0 \\ 0 & 6 & 0 \\ 0 & 0 & 6 \end{bmatrix}$$ ■

SUMMARY

Theorems to know:

A matrix A is orthogonally diagonalizable if and only if A is symmetric.

Methods to know:

Orthogonally diagonalize a matrix.

EXERCISE SET 6.6

For each matrix A, find an orthogonal matrix Q and a diagonal matrix D such that $Q^TAQ = D$.

1. $A = \begin{bmatrix} \frac{5}{4} & -\frac{\sqrt{3}}{4} \\ -\frac{\sqrt{3}}{4} & \frac{7}{4} \end{bmatrix}$

2. $A = \begin{bmatrix} 0 & -1 \\ -1 & 0 \end{bmatrix}$

3. $A = \begin{bmatrix} 3 & -\sqrt{3} \\ -\sqrt{3} & 1 \end{bmatrix}$

4. $A = \begin{bmatrix} \frac{3}{2} & -\frac{1}{2} & 0 \\ -\frac{1}{2} & \frac{3}{2} & 0 \\ 0 & 0 & 1 \end{bmatrix}$

5. $A = \begin{bmatrix} 2 & -1 & -1 \\ -1 & \frac{1}{2} & \frac{1}{2} \\ -1 & \frac{1}{2} & \frac{1}{2} \end{bmatrix}$

6. $A = \frac{1}{30}\begin{bmatrix} 31 & -5 & 2 \\ -5 & 55 & -10 \\ 2 & -10 & 34 \end{bmatrix}$

7. $A = \begin{bmatrix} \sqrt{2} & 0 & \sqrt{2} \\ 0 & 0 & 0 \\ \sqrt{2} & 0 & \sqrt{2} \end{bmatrix}$

8. $A = \begin{bmatrix} 4 & 0 & 1 \\ 0 & 3 & 0 \\ 1 & 0 & 4 \end{bmatrix}$

9. $A = \frac{1}{6}\begin{bmatrix} 1 & -1 & -2 \\ -1 & 1 & 2 \\ -2 & 2 & 4 \end{bmatrix}$

10. $A = \frac{1}{9}\begin{bmatrix} -35 & -28 & -14 \\ -28 & 35 & -14 \\ -14 & -14 & 56 \end{bmatrix}$

11. $A = \frac{1}{30}\begin{bmatrix} -27 & 15 & 6 \\ 15 & 45 & 30 \\ 6 & 30 & -18 \end{bmatrix}$

12. $A = \frac{1}{6}\begin{bmatrix} 9 & 3 & 0 \\ 3 & 9 & 0 \\ 0 & 0 & 6 \end{bmatrix}$

SECTION 6.7 PLANE ANALYTIC GEOMETRY: AN APPLICATION (OPTIONAL)

There are several important standard equations.

The standard equation of an **ellipse** is

$$\frac{x^2}{a^2} + \frac{y^2}{b^2} = 1$$

and its graph is shown in Figure 6.5.

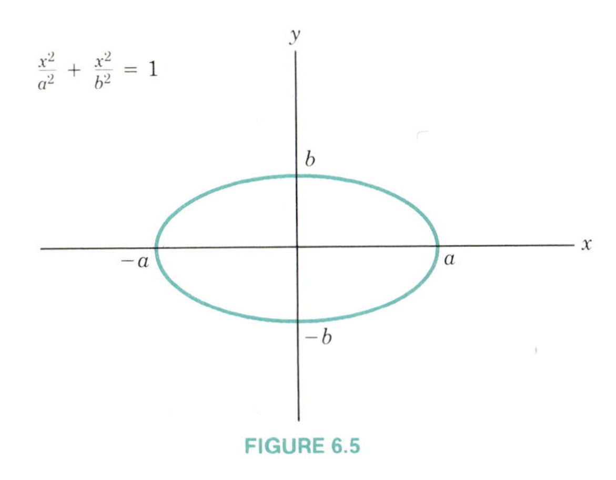

FIGURE 6.5

The standard equations of a **parabola** are $y = ax^2$ or $x = by^2$, and their graphs are shown in Figure 6.6.

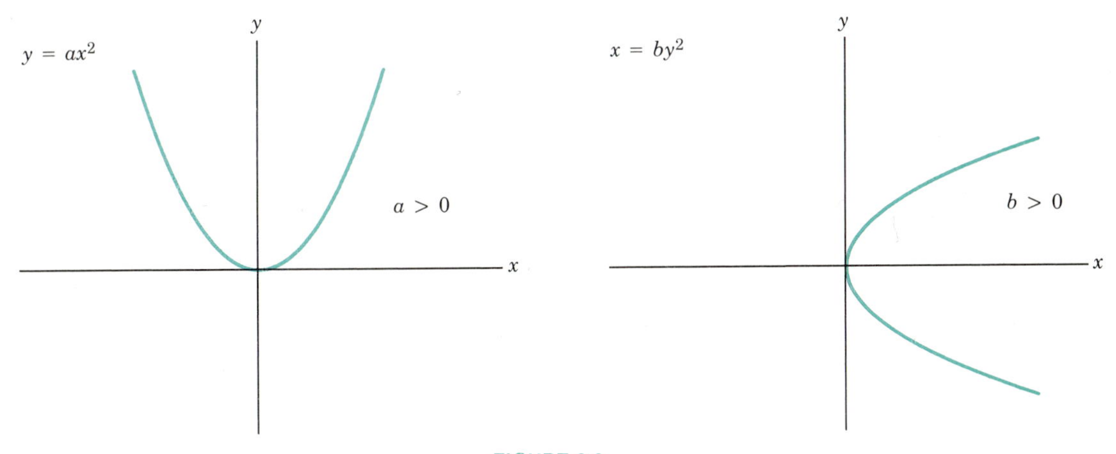

FIGURE 6.6

The standard equations of a **hyperbola** are

$$\frac{x^2}{a^2} - \frac{y^2}{b^2} = 1 \quad \text{or} \quad \frac{y^2}{b^2} - \frac{x^2}{a^2} = 1$$

and their graphs are shown in Figure 6.7.

The **general form of a quadratic equation** is

$$ax^2 + 2bxy + cy^2 + dx + ey + f = 0$$

where a, b, c, d, e, and f are real numbers and at least one of a, b, or c is nonzero.

The main purpose in this section is to show how to recognize a general quadratic equation as either a parabola, hyperbola, ellipse, or a degenerate case, such as $x^2 + 2xy + y^2 - 5 = 0$.

Associated with any quadratic equation is its **quadratic form**

$$ax^2 + 2bxy + cy^2$$

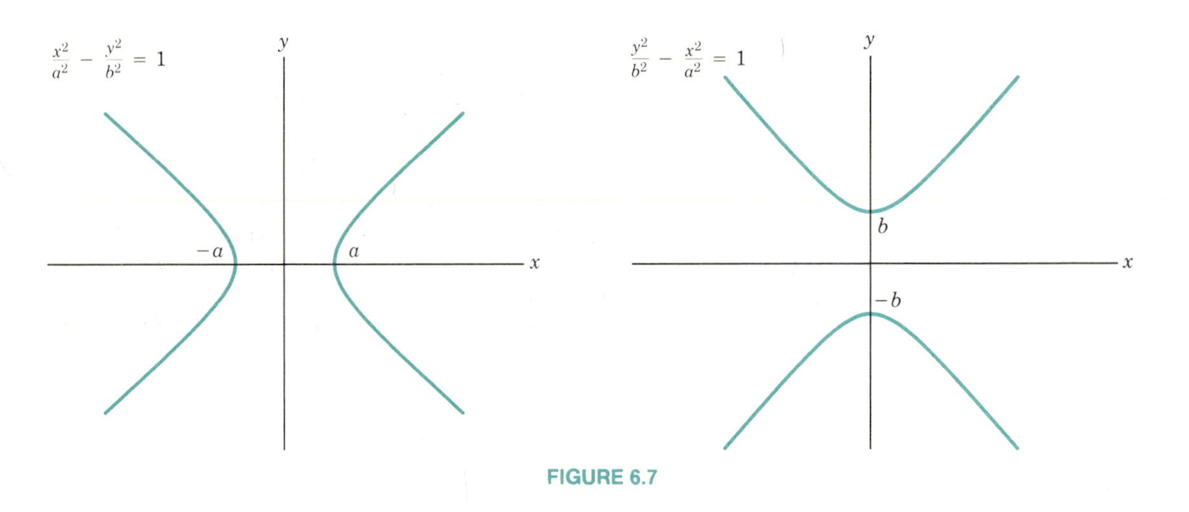

FIGURE 6.7

Example 6.7.1 a. $6x^2 + 4xy$ is the quadratic form associated with $6x^2 + 4xy + 3x + 2y + 1 = 0$.

b. $2x^2 + 3y^2$ is the quadratic form associated with $2x^2 + 3y^2 + 7x = 0$.

c. $2x^2 - 8xy - 3y^2$ is the quadratic form associated with $2x^2 - 8xy - 3y^2 + 7x = 0$. ■

Definition 6.7.2 If

$$\mathbf{x} = \begin{bmatrix} x \\ y \end{bmatrix} \quad \text{and} \quad Q = \begin{bmatrix} a & b \\ b & c \end{bmatrix}$$

then $\mathbf{x}^T Q\mathbf{x} = ax^2 + 2bxy + cy^2$. Q is a symmetric matrix and is called the **matrix of the quadratic form $\mathbf{x}^T Q\mathbf{x}$.**

Example 6.7.3 a. The matrix of the quadratic form of $6x^2 + 4xy$ is

$$Q = \begin{bmatrix} 6 & 2 \\ 2 & 0 \end{bmatrix}$$

b. The matrix of the quadratic form of $2x^2 + 3y^2$ is

$$Q = \begin{bmatrix} 2 & 0 \\ 0 & 3 \end{bmatrix}$$

c. The matrix of the quadratic form $2x^2 - 8xy - 3y^2$ is

$$\begin{bmatrix} 2 & -4 \\ -4 & -3 \end{bmatrix}.$$

■

In general, we can rewrite

$$ax^2 + 2bxy + cy^2 + dx + ey + f = 0$$

as

$$\mathbf{x}^T \begin{bmatrix} a & b \\ b & c \end{bmatrix} \mathbf{x} + \begin{bmatrix} d \\ e \end{bmatrix}^T \mathbf{x} + f = 0$$

or $\mathbf{x}^T Q\mathbf{x} + K\mathbf{x} + f = 0$, where $K = \begin{bmatrix} d \\ e \end{bmatrix}^T$.

A linear transformation $T\colon R^2 \to R^2$ of special interest is a rotation. A rotation about the origin in a counterclockwise direction leaves the distance from a point to the origin unchanged.

Let $\mathbf{v}$ be a vector in R^2 that is in standard position. A counterclockwise rotation of

$$\mathbf{v} = \begin{bmatrix} x \\ y \end{bmatrix}$$

through the angle α results in

$$\mathbf{u} = \begin{bmatrix} x_1 \\ y_1 \end{bmatrix}$$

See Figure 6.8.

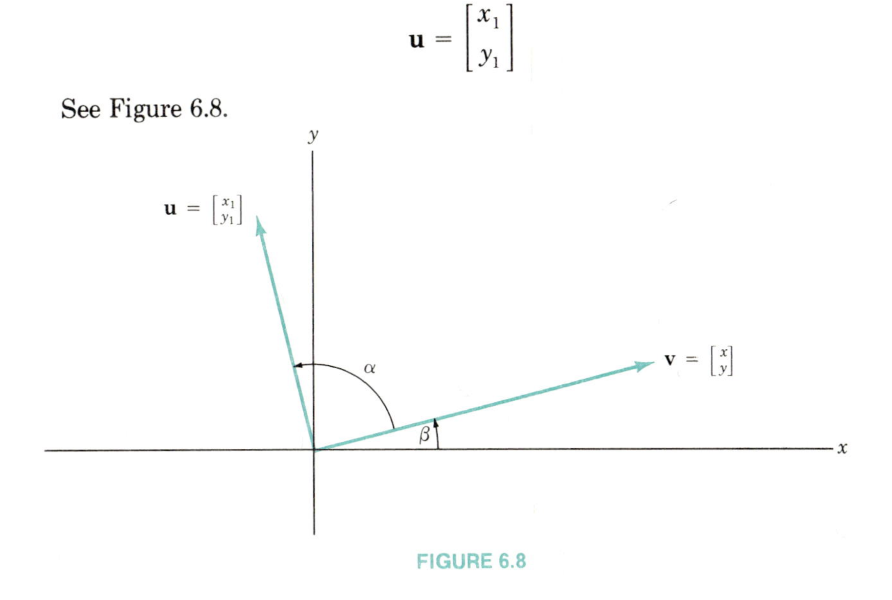

FIGURE 6.8

Theorem 6.7.4 *If*

$$\mathbf{v} = \begin{bmatrix} x \\ y \end{bmatrix}$$

then the rotated vector $\mathbf{u}$ *is*

$$\mathbf{u} = \begin{bmatrix} x_1 \\ y_1 \end{bmatrix} = \begin{bmatrix} x\cos\alpha - y\sin\alpha \\ x\sin\alpha + y\cos\alpha \end{bmatrix}$$

Proof. Let $\|\mathbf{v}\| = L$. Then $\|\mathbf{u}\| = L$. Let β be the angle between $\mathbf{v}$ and the positive x-axis. Then

$$x = L\cos\beta$$

$$y = L\sin\beta$$

The facts that (1) M is orthogonal and (2) $\det(M) = 1$ are left as exercises. Going the other way—that is, given a matrix

$$N = \begin{bmatrix} a & -b \\ b & a \end{bmatrix}$$

that is orthogonal such that $\det N = 1$—then N represents a rotation transformation. For

$$\det N = a^2 + b^2 = 1$$

implies

$$0 \leqq |a| \leqq 1, \quad (\text{or} \quad -1 \leqq a \leqq 1)$$

and

$$0 \leqq |b| \leqq 1, \quad (\text{or} \quad -1 \leqq b \leqq 1)$$

Because $-1 \leqq a \leqq 1$ and $-1 \leqq b \leqq 1$, there is an angle α, $0 \leqq \alpha \leqq 2\pi$, such that $a = \cos\alpha$ and $b = \sin\alpha$. Thus

$$N = \begin{bmatrix} a & -b \\ b & a \end{bmatrix} = \begin{bmatrix} \cos\alpha & -\sin\alpha \\ \sin\alpha & \cos\alpha \end{bmatrix}$$

represents the rotation transformation $T_\alpha: R^2 \to R^2$.

We summarize this result as Theorem 6.7.6.

Theorem 6.7.6 *A* 2×2 *matrix*

$$N = \begin{bmatrix} a & -b \\ b & a \end{bmatrix}$$

is a matrix representation of a rotation transformation if and only if $\det N = 1$.

We now develop a procedure to relate rotations to quadratic forms and conic sections. We begin with a quadratic equation $\mathbf{x}^T Q \mathbf{x} + K\mathbf{x} + f = 0$.

1. We find the eigenvalues α and β of Q and, using the Gram-Schmidt process, find an orthonormal basis $\{\mathbf{v}_1, \mathbf{v}_2\}$ of R^2. In Figure 6.9, we see that $\mathbf{v}_2$ is orthogonal to $\mathbf{v}_1$ and is of length 1, so if

$$\mathbf{v}_1 = \begin{bmatrix} a \\ b \end{bmatrix}$$

then

$$\mathbf{v}_2 = \begin{bmatrix} -b \\ a \end{bmatrix} \quad \text{or} \quad \mathbf{v}_2 = \begin{bmatrix} b \\ -a \end{bmatrix}$$

These are the *only vectors of unit length orthogonal to* $\mathbf{v}_1$.

If $P = [\mathbf{v}_1 \quad \mathbf{v}_2]$, then P is orthogonal, and $P^T A P$ is a diagonal matrix. We have seen that P represents a rotation if and only if $\det(P) = 1$; thus if

$$\mathbf{v}_1 = \begin{bmatrix} a \\ b \end{bmatrix}$$

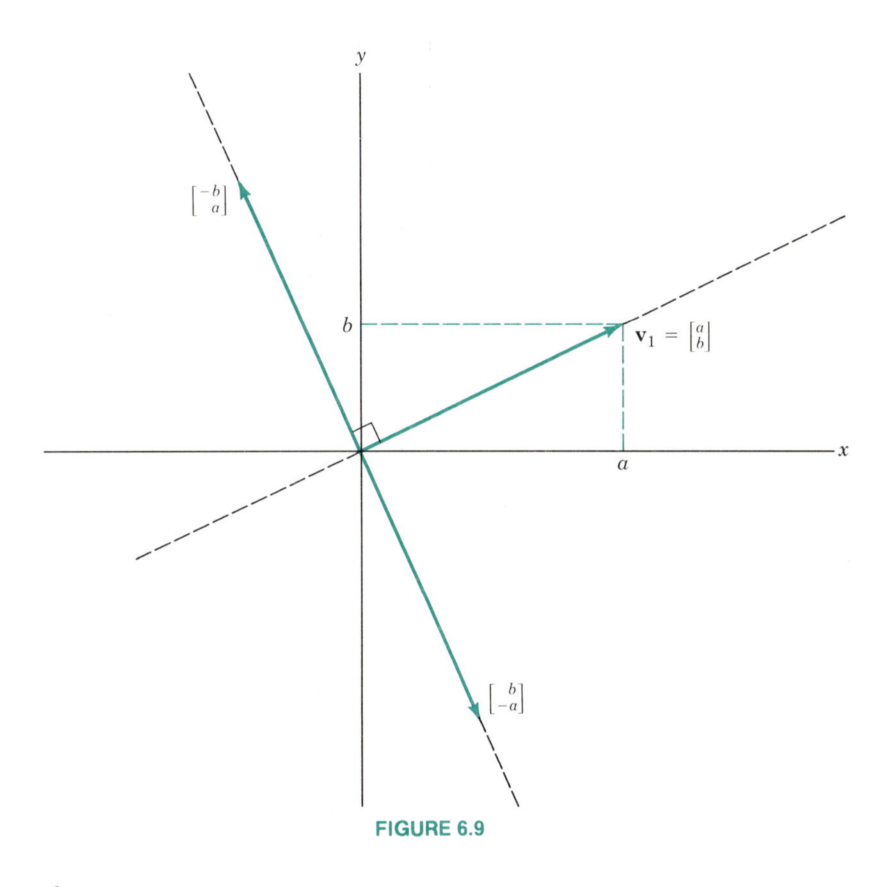

FIGURE 6.9

then

$$P = \begin{bmatrix} a & -b \\ b & a \end{bmatrix}$$

2. Let $\mathbf{x} = P\mathbf{x}'$. Substituting this into the quadratic equation yields

$$(\mathbf{x}')^T P^T Q P(\mathbf{x}') + (KP)\mathbf{x}' + f = 0$$

Denoting KP by

$$\begin{bmatrix} \mathbf{d}' \\ \mathbf{e}' \end{bmatrix}^T$$

we rewrite this equation as

$$\alpha(x')^2 + \beta(y')^2 + d'x' + e'y' + f = 0$$

3. Next, we complete the square, if necessary, on $\alpha(x')^2 + d'x'$ and $\beta(y')^2 + e'y'$ and rewrite the equation in standard form.

Example 6.7.7 Is $x^2 + \sqrt{3}\,xy + 2y^2 + \sqrt{3}\,x - y - \frac{1}{2} = 0$ an equation of a parabola, ellipse, or hyperbola?

Solution The associated quadratic form is

$$x^2 + \sqrt{3}\,xy + 2y^2$$

whose matrix is

$$Q = \begin{bmatrix} 1 & \frac{\sqrt{3}}{2} \\ \frac{\sqrt{3}}{2} & 2 \end{bmatrix}$$

The characteristic equation of Q is

$$\det(Q - \lambda I) = \begin{bmatrix} 1 - \lambda & \frac{\sqrt{3}}{2} \\ \frac{\sqrt{3}}{2} & 2 - \lambda \end{bmatrix} = \lambda^2 - 3\lambda + \frac{5}{4} = 0$$

The solutions determined by using the quadratic formula are

$$\lambda = \frac{3 \pm \sqrt{4}}{2} = \frac{5}{2} \quad \text{and} \quad \frac{1}{2}$$

For $\lambda_1 = \frac{5}{2}$,

$$Q - \left(\frac{5}{2}\right)I = \begin{bmatrix} -\frac{3}{2} & \frac{\sqrt{3}}{2} \\ \frac{\sqrt{3}}{2} & -\frac{1}{2} \end{bmatrix}$$

and an associated eigenvector is

$$\begin{bmatrix} 1 \\ \sqrt{3} \end{bmatrix} = \mathbf{v}_1$$

For $\lambda_2 = \frac{1}{2}$,

$$Q - \left(\frac{1}{2}\right)I = \begin{bmatrix} \frac{1}{2} & \frac{\sqrt{3}}{2} \\ \frac{\sqrt{3}}{2} & \frac{3}{2} \end{bmatrix}$$

and an associated eigenvector is

$$\begin{bmatrix} \sqrt{3} \\ -1 \end{bmatrix} = \mathbf{v}_2$$

We normalize these eigenvectors to obtain the orthogonal matrix

$$P = \begin{bmatrix} \frac{1}{2} & \frac{\sqrt{3}}{2} \\ \frac{\sqrt{3}}{2} & -\frac{1}{2} \end{bmatrix}$$

However, $\det(P) = -1$ and so we interchange the columns of P to obtain

$$P_1 = \begin{bmatrix} \frac{\sqrt{3}}{2} & \frac{1}{2} \\ -\frac{1}{2} & \frac{\sqrt{3}}{2} \end{bmatrix},$$

so that $\det(P_1) = 1$ and P_1 is the matrix of a rotation. Then

$$P_1^T Q P_1 = \begin{bmatrix} \frac{1}{2} & 0 \\ 0 & \frac{5}{2} \end{bmatrix}, \qquad K = \begin{bmatrix} \sqrt{3} \\ -1 \end{bmatrix}^T, \quad \text{and} \quad KP_1 = \begin{bmatrix} 2 \\ 0 \end{bmatrix}$$

This rotation of

$$x^2 + \sqrt{3}\,xy + 2y^2 + \sqrt{3}\,x - y - \frac{1}{2} = 0$$

gives

$$\frac{1}{2}(x')^2 + \frac{5}{2}(y')^2 + 2x' - \frac{1}{2} = 0$$

Completing the square on $\frac{1}{2}(x')^2 + 2x'$, we have

$$\frac{1}{2}\left[(x')^2 + 4x'\right] + \frac{5}{2}(y')^2 = \frac{1}{2}$$

$$\left[(x')^2 + 4x' + 4\right] + 5(y')^2 = 1 + 4$$

$$(x' + 2)^2 + 5(y')^2 = 5$$

$$\frac{(x' + 2)^2}{5} + \frac{(y')^2}{1} = 1$$

which is an equation of an ellipse centered at $(-2, 0)$. ■

Example 6.7.8 Determine whether

$$x^2 + 2xy + y^2 - 8x - 8y = 0$$

is the equation of an ellipse, hyperbola, or parabola.

Solution The associated quadratic form is

$$x^2 + 2xy + y^2$$

whose matrix is

$$Q = \begin{bmatrix} 1 & 1 \\ 1 & 1 \end{bmatrix}$$

The characteristic equation of Q is

$$\det(Q - \lambda I) = \begin{bmatrix} 1-\lambda & 1 \\ 1 & 1-\lambda \end{bmatrix} = \lambda^2 - 2\lambda = 0$$

and its roots are $\lambda_1 = 0$ and $\lambda_2 = 2$.

For $\lambda_1 = 0$,

$$Q - (0)I = \begin{bmatrix} 1 & 1 \\ 1 & 1 \end{bmatrix}$$

and a corresponding eigenvector is

$$\begin{bmatrix} 1 \\ -1 \end{bmatrix}$$

For $\lambda_2 = 2$,

$$Q - 2I = \begin{bmatrix} -1 & 1 \\ 1 & -1 \end{bmatrix}$$

and a corresponding eigenvector is

$$\begin{bmatrix} 1 \\ 1 \end{bmatrix}$$

Normalizing the eigenvectors, we have

$$\mathbf{v}_1 = \begin{bmatrix} \frac{1}{\sqrt{2}} \\ -\frac{1}{\sqrt{2}} \end{bmatrix} \quad \text{and} \quad \mathbf{v}_2 = \begin{bmatrix} \frac{1}{\sqrt{2}} \\ \frac{1}{\sqrt{2}} \end{bmatrix}$$

If

$$P = \begin{bmatrix} \frac{1}{\sqrt{2}} & \frac{1}{\sqrt{2}} \\ -\frac{1}{\sqrt{2}} & \frac{1}{\sqrt{2}} \end{bmatrix}$$

then $\det(P) = 1$.

Substituting

$$\begin{bmatrix} x \\ y \end{bmatrix} = \begin{bmatrix} \frac{1}{\sqrt{2}} & \frac{1}{\sqrt{2}} \\ -\frac{1}{\sqrt{2}} & \frac{1}{\sqrt{2}} \end{bmatrix} \begin{bmatrix} x' \\ y' \end{bmatrix}$$

into our original equation and noting that

$$P^TQP = \begin{bmatrix} 0 & 0 \\ 0 & 2 \end{bmatrix}$$

the equation becomes

$$2(y')^2 - \frac{16}{\sqrt{2}}(y') = 0$$

where

$$\begin{bmatrix} x' \\ y' \end{bmatrix}^T P^TQP \begin{bmatrix} x' \\ y' \end{bmatrix} = 2(y')^2$$

and

$$KP = \begin{bmatrix} -8 \\ -8 \end{bmatrix}^T \begin{bmatrix} \frac{1}{\sqrt{2}} & \frac{1}{\sqrt{2}} \\ -\frac{1}{\sqrt{2}} & \frac{1}{\sqrt{2}} \end{bmatrix} = \begin{bmatrix} 0 \\ -\frac{16}{\sqrt{2}} \end{bmatrix}$$

We now complete the square:

$$\begin{aligned} 2(y')^2 - \frac{16}{\sqrt{2}}y' &= 2\left((y')^2 - \frac{8}{\sqrt{2}}y'\right) \\ &= 2\left((y')^2 - \frac{8}{\sqrt{2}}y' + 8\right) - 16 \\ &= 2\left(y' - \frac{4}{\sqrt{2}}\right)^2 - 16 \end{aligned}$$

and so

$$2\left(y' - \frac{4}{\sqrt{2}}\right)^2 - 16 = 0$$

$$\left(y' - \frac{4}{\sqrt{2}}\right)^2 = 8$$

and

$$y' = \pm\sqrt{8} + \frac{4}{\sqrt{2}} = \pm 2\sqrt{2} + 2\sqrt{2} = 0 \quad \text{or} \quad 4\sqrt{2}$$ ■

This is a *degenerate case* of the two straight lines $y' = 0$ and $y' = 4\sqrt{2}$.

REMARK The matrix

$$P = \begin{bmatrix} \frac{1}{\sqrt{2}} & \frac{1}{\sqrt{2}} \\ -\frac{1}{\sqrt{2}} & \frac{1}{\sqrt{2}} \end{bmatrix}$$

is the matrix of the rotation through an angle of $-\pi/4$ radians.

SUMMARY

Terms to know:

Standard equations of an ellipse, parabola, and hyperbola
Quadratic form
Matrix of a quadratic form

Methods to know:

Given a quadratic equation, determine whether it represents an ellipse, parabola, or hyperbola or degenerate form by diagonalizing the matrix of its associated quadratic form.

EXERCISE SET 6.7

1. Consider a counterclockwise rotation of a vector through α radians about the origin. Show that the rotation is a linear transformation.
2. Let M be a 2×2 matrix representing a counterclockwise rotation. Prove each of the following.
   a. $\det(M) = 1$.
   b. M is orthogonal.
3. Determine the matrix of the linear transformation rotating a vector through the origin α radians counterclockwise for each α.
   a. $\alpha = 0, \pi/2, \pi$, and $3\pi/2$
   b. $\alpha = \pi/4, \pi/3, \pi/6$
   c. $\alpha = -\pi/3, 11\pi/3, -3\pi/4, 5\pi/4$
4. The line $y = x$ is rotated about the origin through $\pi/2$ radians. Find an equation of the new line.
5. The circle $x^2 + y^2 = 1$ is rotated counterclockwise about the origin through $\pi/4$ radians. Find an equation of the new circle.
6. The ellipse

$$\frac{x^2}{9} + \frac{y^2}{4} = 1$$

   is rotated counterclockwise about the origin through $\pi/4$ radians. Find an equation of the new ellipse.
7. The hyperbola $x^2 - y^2 = 1$ is rotated counterclockwise about the origin through $\pi/4$ radians. Find an equation of the new hyperbola.

8. A triangle with vertices $(0,0)$, $(1,0)$, and $(0,1)$ is rotated counterclockwise about the origin through $\pi/4$ radians. Prove that the area of the rotated triangle equals the area of the original triangle.

9. Let

$$T = \begin{bmatrix} a & b \\ c & d \end{bmatrix}$$

be the matrix of a counterclockwise rotation about the origin. We know that if

$$T\begin{bmatrix} x \\ y \end{bmatrix} = \begin{bmatrix} x_1 \\ y_1 \end{bmatrix}$$

then $x^2 + y^2 = x_1^2 + y_1^2$ for all

$$\begin{bmatrix} x \\ y \end{bmatrix}$$

a. Let $a = \frac{4}{5}$. Find all possible values for b, c, and d.

b. Compute $a^2 + b^2$, $b^2 + d^2$, and $ab + cd$

c. If T and

$$U = \begin{bmatrix} \alpha & \beta \\ \gamma & \delta \end{bmatrix}$$

are rotations about the origin, is TU a rotation?

In Exercises 10–20, classify each quadratic equation as the equation of an ellipse, a hyperbola, a parabola, or a degenerate case.

10. $\frac{5}{4}x^2 - \frac{\sqrt{3}}{2}xy + \frac{7}{4}y^2 - 6 = 0$
11. $\frac{7}{4}x^2 + \frac{\sqrt{3}}{2}xy + \frac{5}{4}y^2 - 7 = 0$
12. $-\frac{7}{2}x^2 - \sqrt{3}\,xy - \frac{5}{2}y^2 + 6 = 0$
13. $-6xy + 2\sqrt{3}\,y^2 - 8 = 0$
14. $-2xy - 6 = 0$
15. $2x^2 - 2xy + 2y^2 - 17 = 0$
16. $x^2 + 4xy + y^2 - 13 = 0$
17. $\sqrt{3}\,x^2 + \sqrt{3}\,y^2 - 6 = 0$
18. $x^2 + 6xy + y^2 - 13 = 0$
19. $2\sqrt{3}\,xy + 2y^2 - 37 = 0$
20. $3x^2 - 2\sqrt{3}\,xy + y^2 - 15 = 0$
21. $x^2 + 2xy + y^2 + 1 = 0$
22. $x^2 - 4xy + 4y^2 - 15 = 0$
23. $x^2 + 6xy + 9y^2 - 22 = 0$
24. $9x^2 - 6xy + y^2 - 31 = 0$
25. $x^2 + 8xy + 16y^2 - 1 = 0$

SECTION 6.8 SIMPLE MODELS OF POPULATION DYNAMICS (OPTIONAL)

The investigation of the growth and decline of a population (e.g., a population of humans) is one of the oldest branches of mathematical ecology. To understand how changes in the size and composition of a population occurs, it is imperative to study the factors that determine the changes and the interrelationships between them. Oftentimes it is difficult or impossible to investigate these relationships experimentally. In this case, a **population model** helps us to investigate these relationships mathematically.

Definition 6.8.1 A **population model** specifies a population and rules for bringing the population forward in time through the occurrence of births, deaths, marriages, and migrations into and out of the population.

There are a variety of assumptions that can be made, depending on the specific aims of the investigators. While complex assumptions may yield better agreement with data obtained from a real population, simple assumptions often add more understanding about the nature of the underlying principles that affect the population. The following example may shed light on what a population model actually is.

Example 6.8.2 Consider a human population on an isolated island, where no one is permitted to enter or leave the island. We are interested in $P(t)$, the population of the island as a function of time. When we first measure the population (say on January 1), we arbitrarily set the time $t = 0$. We again measure the size of the population on January 1 of each succeeding year. So, we let $P(t)$ be the population after t years. Thus

$$P(t+1) = P(t) + \begin{matrix}\text{gain or loss of population in between} \\ \text{year } t \text{ and year } t+1.\end{matrix}$$

For the purposes of this example, we assume that the population increases each year at a fixed rate proportional to the population present the previous year. Thus the gain or loss from year t to year $t + 1$ is $kP(t)$ for some constant k.

If $k = 1$, the population doubles each year, since $P(t + 1) = P(t) + 1 \cdot P(t) = 2P(t)$. If $k = -\frac{1}{2}$, the population is halved each year, since $P(t + 1) = P(t) + (-\frac{1}{2})P(t) = \frac{1}{2}P(t)$.

In general, we have

$$P(t+1) = P(t) + kP(t) = (1 + k)P(t)$$

Hence, at $t = 1$,

$$P(1) = P(0 + 1) = (1 + k)P(0)$$

at $t = 2$,

$$P(2) = P(1 + 1) = (1 + k)P(1) = (1 + k)^2 P(0)$$

and at $t = n - 1$,

$$\begin{aligned} P(n) &= P(n-1) + kP(n-1) \\ &= (1+k)P(n-1) \\ &= (1+k)^n P(0) \end{aligned}$$

■

We would like to know what happens to the population as t gets very large. The symbol $t \to \infty$ means t is getting very large without bound.

Definition 6.8.3 Given a population model and $P(t)$, the **asymptotic behavior** of $P(t)$ is

$$\lim_{t \to +\infty} P(t)$$

In Example 6.8.2, if $k = -\frac{1}{2}$, then $P(n) = (1 - \frac{1}{2})^n P(0) = (\frac{1}{2})^n P(0)$. As $n \to +\infty$, $(\frac{1}{2})^n = \dfrac{1}{2^n} \to 0$; asymptotically, the population $P(t) \to 0$. Thus the population faces extinction.

In Example 6.8.2, if $k = 0$, then $P(n) = (1 + 0)^n P(0) = 1^n P(0) = P(0)$. As $n \to +\infty$, $P(n) = P(0)$ and the population remains constant.

In Example 6.8.2, if $k = 1$, then $P(n) = (1 + 1)^n P(0) = 2^n P(0)$. As $n \to +\infty$, $2^n \to +\infty$, so asymptotically the population $P(t) \to +\infty$ and the island becomes overpopulated.

In general, for $k > 1$, $P(t) \to +\infty$. The larger the value of k, the faster the population grows.

We next consider a population situated on two nearby islands.

Example 6.8.4 Consider a population situated on two islands, island 1 and island 2. People are allowed to move from one island to the other, but no contact with the outside world is allowed. Again, we are interested in the population on each island. Let $P_1(t)$ be the population on island 1 after t years and let $P_2(t)$ be the population at that time on island 2. $P_1(0)$ and $P_2(0)$ are the initial populations of the islands. At January 1 of year $t + 1$,

$$P_1(t + 1) = P_1(t) + g_1(t) - m_{12}(t) + m_{21}(t)$$

where $g_1(t)$ is the net population gain of island 1 due to births and deaths between year t and year $t + 1$, $m_{12}(t)$ is the number of people moving from island 1 to island 2, and $m_{21}(t)$ is the number of people moving to island 1 from island 2. Likewise, for island 2

$$P_2(t+1) = P_2(t) + g_2(t) + m_{12}(t) - m_{21}(t)$$

As is usual in this type of model, we assume that

$$g_1(t) = b_1 P_1(t)$$

$$g_2(t) = b_2 P_2(t)$$

$$m_{12}(t) = k_1 P_1(t)$$

$$m_{21}(t) = k_2 P_2(t) \qquad \text{for constants } b_1, b_2, k_1, k_2 > 0.$$

We let

$$\mathbf{P}(t) = \begin{bmatrix} P_1(t) \\ P_2(t) \end{bmatrix}$$

a vector. Then

$$\mathbf{P}(t+1) = G\mathbf{P}(t)$$

where

$$G = \begin{bmatrix} 1 + b_1 - k_1 & k_2 \\ k_1 & 1 + b_2 - k_2 \end{bmatrix}$$

Then, as in Example 6.7.2,

$$\mathbf{P}(1) = G\mathbf{P}(0)$$

$$\mathbf{P}(2) = G\mathbf{P}(1) = G^2\mathbf{P}(0)$$

$$\mathbf{P}(3) = G\mathbf{P}(2) = G^3\mathbf{P}(0)$$

and, in general,

$$\mathbf{P}(t) = G^t\mathbf{P}(0)$$

■

To study the asymptotic behavior of $\mathbf{P}(t)$, we must examine $\lim_{t\to+\infty} G^t$. Before we can study this asymptotic behavior, we must be able to determine the matrix G, and we would like to have an easy way to compute G^t for all $t > 0$.

Example 6.8.5 Suppose initially that

$$\mathbf{P}(0) = \begin{bmatrix} 1000 \\ 3000 \end{bmatrix}$$

Further, we suppose

$$\mathbf{P}(1) = \begin{bmatrix} 900 \\ 2700 \end{bmatrix} \quad \text{and} \quad \mathbf{P}(2) = \begin{bmatrix} 810 \\ 2430 \end{bmatrix}$$

Calculate G.

Solution Let

$$G = \begin{bmatrix} g_{11} & g_{12} \\ g_{21} & g_{22} \end{bmatrix}$$

From $\mathbf{P}(1) = G\mathbf{P}(0)$, we obtain

$$900 = 1000g_{11} + 3000g_{12}$$

$$2700 = 1000g_{21} + 3000g_{22}$$

From $\mathbf{P}(2) = G\mathbf{P}(1)$, we obtain

$$810 = 900g_{11} + 2700g_{12}$$

$$2430 = 900g_{21} + 2700g_{22}$$

These four equations yield a linear system of four equations in four unknowns. The solution of this system is

$$g_{11} = -0.6$$

$$g_{12} = 0.5$$

$$g_{21} = -3$$

$$g_{22} = 1.9$$

thus

$$G = \begin{bmatrix} -0.6 & 0.5 \\ -3 & 1.9 \end{bmatrix}$$

■

Having found G, we wish to determine the asymptotic behavior of $\mathbf{P}(t)$. To do this, we determine $\lim_{t\to\infty} G^t$. If we can diagonalize G as $G = Q^{-1}DQ$, then $G^t = Q^{-1}D^tQ$, whereas if

$$D = \begin{bmatrix} d_{11} & 0 \\ 0 & d_{22} \end{bmatrix}$$

then

$$D^t = \begin{bmatrix} d_{11}^t & 0 \\ 0 & d_{22}^t \end{bmatrix}$$

We illustrate this process for G.

Example 6.8.6 Let

$$G = \begin{bmatrix} -0.6 & 0.5 \\ -3 & 1.9 \end{bmatrix}$$

and $\mathbf{P}(t) = G^t\mathbf{P}(0)$. Determine the asymptotic behavior of $\mathbf{P}(t)$.

Solution Since $\det(G - \lambda I) = \lambda^2 - 1.3\lambda + 0.36 = 0$ we have $\lambda = 0.9$ and $\lambda = 0.4$. For $\lambda_1 = 0.9$ an eigenvector is

$$\mathbf{v}_1 = \begin{bmatrix} 1 \\ 3 \end{bmatrix}$$

whereas for $\lambda_2 = 0.4$ an eigenvector is

$$\mathbf{v}_2 = \begin{bmatrix} 1 \\ 2 \end{bmatrix}$$

Thus for

$$Q = \begin{bmatrix} 1 & 1 \\ 3 & 2 \end{bmatrix}, \qquad Q^{-1} = \begin{bmatrix} -2 & 1 \\ 3 & -1 \end{bmatrix}$$

and $G = Q^{-1}DQ$, where

$$D = \begin{bmatrix} 0.9 & 0 \\ 0 & 0.4 \end{bmatrix}$$

Thus

$$G^t = Q^{-1}D^tQ = Q^{-1}\begin{bmatrix} (0.9)^t & 0 \\ 0 & (0.4)^t \end{bmatrix}Q$$

and as $t \to +\infty$, $(0.9)^t \to 0$ and $(0.4)^t \to 0$, so

$$D^t \to \begin{bmatrix} 0 & 0 \\ 0 & 0 \end{bmatrix}$$

and $\mathbf{P}(t) \to 0$. Thus the populations on both islands die out. Note in this example that the value of $\mathbf{P}(0)$ is immaterial to the asymptotic behavior of $\mathbf{P}(t)$, since the population dies out. ■

Example 6.8.7 Determine the asymptotic behavior of $\mathbf{P}(t) = G^t\mathbf{P}(0)$ with

$$G = \begin{bmatrix} 1 & \frac{\sqrt{3}}{2} \\ \frac{\sqrt{3}}{2} & 2 \end{bmatrix} \quad \text{if} \quad \mathbf{P}(0) = \begin{bmatrix} 2 \\ 2 \end{bmatrix}$$

Solution G has eigenvalues $\lambda_1 = \frac{5}{2}$ and $\lambda_2 = \frac{1}{2}$, and

$$G = Q^{-1}\begin{bmatrix} \frac{5}{2} & 0 \\ 0 & \frac{1}{2} \end{bmatrix} Q \quad \text{for} \quad Q = \begin{bmatrix} \frac{1}{2} & \frac{\sqrt{3}}{2} \\ \frac{\sqrt{3}}{2} & -\frac{1}{2} \end{bmatrix}$$

Furthermore,

$$G^t = Q^{-1}\begin{bmatrix} \left(\frac{5}{2}\right)^t & 0 \\ 0 & \left(\frac{1}{2}\right)^t \end{bmatrix} Q$$

(Note that

$$\lim_{t \to +\infty} \begin{bmatrix} \left(\frac{5}{2}\right)^t & 0 \\ 0 & \left(\frac{1}{2}\right)^t \end{bmatrix} = \begin{bmatrix} +\infty & 0 \\ 0 & 0 \end{bmatrix}$$

so we do not know, at this point, that the population dies out.) Hence we examine

$$\begin{aligned} \lim_{t \to +\infty} \mathbf{P}(t) &= \lim_{t \to \infty} Q^{-1} D^t Q \mathbf{P}(0) \\ &= \lim_{t \to +\infty} \begin{bmatrix} \frac{1}{2} & \frac{\sqrt{3}}{2} \\ \frac{\sqrt{3}}{2} & -\frac{1}{2} \end{bmatrix} \begin{bmatrix} \left(\frac{5}{2}\right)^t & 0 \\ 0 & \left(\frac{1}{2}\right)^t \end{bmatrix} \begin{bmatrix} \frac{1}{2} & \frac{\sqrt{3}}{2} \\ \frac{\sqrt{3}}{2} & -\frac{1}{2} \end{bmatrix} \begin{bmatrix} 2 \\ 2 \end{bmatrix} \\ &= \lim_{t \to +\infty} \frac{1}{4} \begin{bmatrix} \left(\frac{5}{2}\right)^t + 3\left(\frac{1}{2}\right)^t & \sqrt{3}\left[\left(\frac{5}{2}\right)^t - \left(\frac{1}{2}\right)^t\right] \\ \sqrt{3}\left[\left(\frac{5}{2}\right)^t - \left(\frac{1}{2}\right)^t\right] & 3\left(\frac{5}{2}\right)^t + \left(\frac{1}{2}\right)^t \end{bmatrix} \begin{bmatrix} 2 \\ 2 \end{bmatrix} \end{aligned}$$

but as $t \to +\infty$, $(\frac{5}{2})^t \to +\infty$, $(\frac{1}{2})^t \to 0$, and so

$$\lim_{t \to +\infty} \mathbf{P}(t) = \begin{bmatrix} +\infty \\ +\infty \end{bmatrix}$$

Both islands become overpopulated. ■

SUMMARY

Terms to know:

Population model

Asymptotic behavior

Methods to know:

Determine the asymptotic behavior of a population given by the function $\mathbf{P}(t)$.

EXERCISE SET 6.8

1. Develop the matrix equation for the population of three islands with migration between islands allowed but with no contact with the outside world.

2. Let

$$\mathbf{P}(0) = \begin{bmatrix} 1600 \\ 1100 \\ 900 \end{bmatrix} \quad \text{and} \quad G = \begin{bmatrix} \frac{1}{3} & \frac{1}{4} & \frac{1}{3} \\ \frac{1}{3} & \frac{1}{2} & \frac{1}{3} \\ \frac{1}{3} & \frac{1}{4} & \frac{1}{3} \end{bmatrix}$$

in Exercise 1. Find $\mathbf{P}(2)$.

3. In the population of two islands, let

$$\mathbf{P}(0) = \begin{bmatrix} 100 \\ 200 \end{bmatrix}, \quad \mathbf{P}(1) = \begin{bmatrix} 100 \\ 100 \end{bmatrix}, \quad \text{and} \quad \mathbf{P}(2) = \begin{bmatrix} 100 \\ 50 \end{bmatrix}$$

a. Determine the matrix G with $\mathbf{P}(t) = G^t\mathbf{P}(0)$.
b. Determine the asymptotic behavior of $\mathbf{P}(t)$.

4. In the population of two islands, let

$$\mathbf{P}(0) = \begin{bmatrix} 100 \\ 100 \end{bmatrix}, \quad \mathbf{P}(1) = \begin{bmatrix} 75 \\ 50 \end{bmatrix}, \quad \text{and} \quad \mathbf{P}(2) = \begin{bmatrix} 50 \\ 18.75 \end{bmatrix}$$

a. Give a reasonable interpretation of the 18.75 in $\mathbf{P}(2)$.
b. Determine the matrix G satisfying $\mathbf{P}(t) = G^t\mathbf{P}(0)$.
c. Determine the asymptotic behavior of $\mathbf{P}(t)$.

5. In the population of two islands, let

$$\mathbf{P}(0) = \begin{bmatrix} 200 \\ 100 \end{bmatrix}, \quad \mathbf{P}(1) = \begin{bmatrix} 300 \\ 500 \end{bmatrix}, \quad \text{and} \quad \mathbf{P}(2) = \begin{bmatrix} 1300 \\ 1100 \end{bmatrix}$$

a. Determine the matrix G satisfying $\mathbf{P}(t) = G^t\mathbf{P}(0)$.
b. Determine the asymptotic behavior of $\mathbf{P}(t)$.

6. In the population of two islands, let

$$\mathbf{P}(0) = \begin{bmatrix} 100 \\ 200 \end{bmatrix}, \quad \mathbf{P}(1) = \begin{bmatrix} 300 \\ 200 \end{bmatrix}, \quad \text{and} \quad \mathbf{P}(2) = \begin{bmatrix} 100 \\ 200 \end{bmatrix}$$

a. Determine the matrix G satisfying $\mathbf{P}(t) = G^t\mathbf{P}(0)$.
b. Determine the asymptotic behavior of $\mathbf{P}(t)$.

ANSWERS TO ODD NUMBERED EXERCISES

CHAPTER ONE

Section 1.1

1. Augmented matrix $\left[\begin{array}{ccc|c} 3 & 4 & 7 & 4 \\ 2 & 6 & 0 & 3 \end{array}\right]$, nonhomogeneous

3. Augmented matrix $\left[\begin{array}{cc|c} 2 & 4 & 7 \\ 8 & 3 & 4 \end{array}\right]$, nonhomogeneous

5. Augmented matrix $\left[\begin{array}{ccc|c} 7 & 9 & 4 & 3 \\ 7 & 0 & 4 & 6 \\ 0 & 2 & 7 & 5 \\ 8 & 3 & 0 & 2 \end{array}\right]$, nonhomogeneous

7. Augmented matrix $\left[\begin{array}{cc|c} 6 & 4 & 0 \\ 7 & 2 & 6 \end{array}\right]$, nonhomogeneous

9. Augmented matrix $\left[\begin{array}{cc|c} 3 & 4 & 0 \\ 2 & 7 & 0 \end{array}\right]$, homogeneous

11. Augmented matrix $\left[\begin{array}{ccc|c} 3 & 4 & 6 & 7 \\ 1 & -1 & -1 & 8 \end{array}\right]$, nonhomogeneous

13. Augmented matrix $\left[\begin{array}{cccc|c} 2 & 3 & 0 & -1 & 2 \\ 1 & 0 & -1 & 1 & -1 \\ 0 & 1 & 2 & -3 & 1 \end{array}\right]$, nonhomogeneous

15. $\begin{aligned} 3x_1 + 2x_2 + x_3 &= 2 \\ x_1 + 2x_2 - x_3 &= 0 \\ 3x_1 + 2x_2 + x_3 &= 2 \end{aligned}$

17. $\begin{aligned} -2x_1 + 4x_2 + 2x_3 &= 0 \\ -x_1 + 2x_2 + x_3 &= 0 \end{aligned}$

19. $\begin{aligned} 2x_1 + 3x_2 - x_3 + 4x_4 + 2x_5 + 5x_6 &= 7 \\ 6x_1 - 5x_2 + 3x_3 \qquad + 2x_5 \qquad &= 6 \\ x_1 + 3x_2 + 5x_3 + 7x_4 + 9x_5 + 7x_6 &= 2 \\ 4x_1 + 2x_2 + 6x_3 + 8x_4 + 2x_5 + 4x_6 &= 1 \end{aligned}$

21. $3x_1 - 4x_2 + 6x_3 + 2x_4 = 7$

23. We list them in the order a_{11}, a_{21}, and a_{32}.

1. 3, 2	2. $\pi, 4, \pi$	3. 2, 8	4. 3
5. 7, 7, 2	6. 1, 0, 0	7. 6, 7	8. 1, 0, 0
9. 3, 2	10. 1, 1, 1	11. 3, 1	12. 1, 1, −1
13. 2, 1, 1	14. $2 - t, 2, -1$		

25. Augmented matrix $\left[\begin{array}{ccc|c} x_1 & y_1 & 1 & z_1 \\ x_2 & y_2 & 1 & z_2 \\ x_3 & y_3 & 1 & z_3 \end{array}\right]$

27. $A + 2B + C = -5$
$3A + 4B + C = -25$
$A + 4B + C = -17$

Section 1.2

1. $x = \frac{1}{2},\ y = 2$
3. $x = 1,\ y = \frac{1}{6},\ z = \frac{5}{6}$
5. $x = -\frac{1}{2},\ y = \frac{5}{2},\ z = \frac{1}{2},\ w = 0$
7. $x = y = z = 0$
9. $x = y = 0$
11. $x = \frac{3}{2}y + 4$, y arbitrary
13. $x_1 = -1,\ x_2 = -2,\ x_3 = 3,\ x_4 = -4$
15. $x_1 = x_2 = x_3 = 0$
17. $x_1 = 3x_3,\ x_2 = -2x_3$, x_3 arbitrary
19. $x_1 = -2x_3,\ x_2 = -x_3$, x_3 arbitrary
21. $m = -\frac{7}{4},\ b = -\frac{1}{4}$
23. $a = \frac{17}{30},\ b = -\frac{41}{30},\ c = 0$
25. $x = \pm\sqrt{7}/2,\ y = \pm\sqrt{z^2 - \frac{5}{4}}$, and z arbitrarily chosen with respect to the conditions $z \leqq -\sqrt{5}/2$ or $z \geqq \sqrt{5}/2$.
27. $\lambda = \pm 2$
29. $\lambda = 0$
31. $x = -10.861,\ y = -13.462,\ z = -31.832,\ w = 6.657$
33. $x_1 = -1.812x_3 + 0.087x_4 - 1.228$
$x_2 = -0.876x_4 + 1.605x_5 - 2.649$
x_3, x_4, x_5 arbitrary
35. $A = -5.5,\ B = -1.681,\ C = 0.698$

Section 1.3

3. The trivial solution—namely all $x_i = 0$
5. All your solutions must satisfy $x_1 = x_4 = 0$; x_2, x_3 arbitrary
7. All your solutions must satisfy $x_1 = -\frac{3}{2}x_3 - \frac{1}{2}x_4$, where x_2, x_3, and x_4 are arbitrarily chosen real numbers.

Section 1.4

1. $x = 12\frac{2}{3}$ yds^3, $y = 8$ yds^3, $z = 7$ yds^3
3. \$129,444 at 5%, \$53,472.20 at 8%, and \$27,083.30 at 12%.
5. $x = 150$ acres of soybeans, $y = 850$ acres of corn

7. a. at most $6m + 12n$ microseconds;
   b. at most $20m + 36n$ microseconds

9. $x_1 = 60$ units of food, $x_2 = 5$ units of clothing, and $x_3 = 4$ units of shelter

11. \$126,551 at 5.1%, \$56,047.30 at 8.2%, and \$28,512.40 at 12.1%

13. \$3750 at 7.25%, \$1149.43 at 8.75%, and \$3473.68 at 9.5%

Section 1.5

1. $I_1 = \frac{1}{4}$, $I_2 = \frac{1}{4}$
3. $I_1 = \frac{20}{7}$, $I_2 = \frac{9}{7}$
5. $I_1 = \frac{13}{30}$, $I_2 = \frac{3}{10}$, $I_3 = \frac{23}{30}$
7. $I_1 = \frac{8}{5}$, $I_2 = \frac{4}{5}$, $I_3 = \frac{8}{5}$
9. $I_1 = \frac{14}{13}$, $I_2 = \frac{8}{13}$, $I_3 = \frac{12}{13}$
11. $I_1 = \frac{94}{93}$, $I_2 = \frac{32}{31}$, $I_3 = \frac{16}{93}$
13. $t_1 = \frac{300}{7}$, $t_2 = \frac{360}{7}$, $t_3 = \frac{300}{7}$
15. $t_1 = 30$, $t_2 = 30$, $t_3 = 30$
17. $t_1 = 67.643$, $t_2 = 81.929$, $t_3 = 67.643$, $t_4 = 67.643$, $t_5 = 81.929$, $t_6 = 67.643$

Section 1.6

1. b. the higher the GPA, the higher the salary
3. b. the lower the price per loaf, the more bread that is sold
5. \$1075.67
7. \$106,880
9. 203 loaves sold
11. a. $y = 138.550e^{-0.0961x}$
    b. $y = 52.88e^{0.2421x}$

Section 1.7

1. $\begin{bmatrix} 2 & 2 & 4 \\ 9 & 2 & 10 \\ 6 & 2 & 2 \end{bmatrix}$ 3. $\begin{bmatrix} 9 & 11 & 11 \\ 11 & 8 & 6 \end{bmatrix}$ 5. $\begin{bmatrix} 0 & 0 & 0 \\ 0 & 0 & 0 \\ 0 & 0 & 0 \end{bmatrix}$

7. Cannot be added 9. $\begin{bmatrix} 6 & -10 & 1 \\ 6 & -3 & 5 \end{bmatrix}$ 11. $\begin{bmatrix} 6 & -20 & 1 \\ 6 & -3 & 5 \end{bmatrix}$

13. $\begin{bmatrix} 6 & -8 & -3 \\ 14 & -9 & 1 \end{bmatrix}$ 15. $\begin{bmatrix} 3 & -6 & 0 \\ 12 & -6 & 3 \end{bmatrix}$ 17. $x = -2$, $y = \frac{2}{3}$

19. $X = \begin{bmatrix} 4 & -2 & -2 \\ 1 & 3 & -3 \\ -7 & 2 & 6 \end{bmatrix}$

Section 1.8

1. $\begin{bmatrix} 15 & 30 & 22 & 9 \\ 17 & 41 & 29 & 11 \end{bmatrix}$ 3. $\begin{bmatrix} 1 & 0 \\ 0 & 1 \end{bmatrix}$ 5. $\begin{bmatrix} 38 & 28 \\ 2 & 1 \end{bmatrix}$

7. $\begin{bmatrix} 13 \\ 15 \end{bmatrix}$ 9. $\begin{bmatrix} 184 \\ 594 \end{bmatrix}$

11. a. $A = \begin{bmatrix} 1 & 2 \\ 2 & 4 \end{bmatrix}$, $B = \begin{bmatrix} 1 & 1 \\ 1 & 1 \end{bmatrix}$ b. $A = \begin{bmatrix} 1 & 0 \\ 0 & 1 \end{bmatrix}$, $B = \begin{bmatrix} 5 & 6 \\ -3 & 2 \end{bmatrix}$

13. $A = \begin{bmatrix} 2 & 1 \\ 4 & 2 \end{bmatrix}$, $B = \begin{bmatrix} -1 & 2 \\ 2 & -4 \end{bmatrix}$

15. a. $B = \begin{bmatrix} \frac{1}{13} & \frac{3}{13} \\ \frac{5}{13} & \frac{2}{13} \end{bmatrix}$, b. $B = \begin{bmatrix} 1 & -\frac{2}{3} & -\frac{11}{6} \\ 0 & \frac{1}{3} & \frac{2}{3} \\ 0 & 0 & -\frac{1}{2} \end{bmatrix}$

19. $A = \begin{bmatrix} 1 & \frac{1}{5} & \frac{3}{5} \\ 0 & -\frac{2}{5} & -\frac{6}{5} \\ 0 & \frac{3}{5} & -\frac{6}{5} \end{bmatrix}$

21. a. $A^2 = \begin{bmatrix} 2 & 0 & 1 & 0 \\ 0 & 1 & 0 & 1 \\ 1 & 0 & 1 & 0 \\ 0 & 1 & 0 & 2 \end{bmatrix}$ b. $A^3 = \begin{bmatrix} 0 & 2 & 0 & 3 \\ 2 & 0 & 1 & 0 \\ 0 & 1 & 0 & 2 \\ 3 & 0 & 2 & 0 \end{bmatrix}$

c. $A + A^2 = \begin{bmatrix} 2 & 1 & 1 & 1 \\ 1 & 1 & 0 & 1 \\ 1 & 0 & 1 & 1 \\ 1 & 1 & 1 & 2 \end{bmatrix}$ d. $A + A^2 + A^3 = \begin{bmatrix} 2 & 3 & 1 & 4 \\ 3 & 1 & 1 & 1 \\ 1 & 1 & 1 & 3 \\ 4 & 1 & 3 & 2 \end{bmatrix}$

27. a. Multiply a $1 \times n$ matrix by a $n \times 1$ matrix, $n \geqq 1$.
b. Multiply an $m \times p$ matrix by a $p \times 1$ matrix, $m, p \geqq 1$.
c. Multiply a $1 \times p$ matrix by a $p \times n$ matrix, $p, n \geqq 1$.
d. Multiply an $n \times p$ matrix by a $p \times n$ matrix, $p, n \geqq 1$.

Section 1.9

1. $\begin{bmatrix} \frac{1}{4} & -\frac{3}{4} \\ -\frac{1}{12} & \frac{7}{12} \end{bmatrix}$ 3. $\begin{bmatrix} 0 & 1 \\ 1 & 0 \end{bmatrix}$ 5. $\begin{bmatrix} 1 & -3 & 1 \\ 0 & 1 & -1 \\ 0 & 0 & 1 \end{bmatrix}$

7. $\begin{bmatrix} \frac{1}{2} & \frac{3}{2} & -\frac{1}{2} & 0 \\ 0 & 1 & -1 & 1 \\ 0 & 0 & -\frac{1}{2} & \frac{2}{3} \\ 0 & 0 & 0 & \frac{1}{3} \end{bmatrix}$

9. The matrix is not a square matrix so the inverse does not exist.

11. Any λ different from -7 and 9.

13. a. $A = \begin{bmatrix} 2 & -1 \\ 5 & -3 \end{bmatrix}, \quad B = \begin{bmatrix} -2 & 1 \\ -5 & 3 \end{bmatrix}$

b. Let $B = -A$ where A is any invertible matrix.

21. If $AB = 0$, then $0 = A^{-1}0 = A^{-1}(AB) = (A^{-1}A)B = IB = B$

23. $(A_1A_2A_3)^{-1} = ((A_1A_2)A_3)^{-1} = A_3^{-1}(A_1A_2)^{-1} = A_3^{-1}(A_2^{-1}A_1^{-1})$
$= A_3^{-1}A_2^{-1}A_1^{-1}$

27. $C = CI = C(AD) = (CA)D = ID = D$

29. $\begin{bmatrix} 0.988 & 0.623 & -0.609 & 0.317 \\ 0.840 & 1.282 & -0.302 & -0.591 \\ 0.741 & 0.290 & 0.411 & -0.377 \\ -0.203 & 0.187 & 0.317 & 0.595 \end{bmatrix}$

31. $\begin{bmatrix} 0.532 & -0.543 & 0.543 \\ 0.487 & 0.562 & -0.562 \\ -0.522 & 0.480 & 0.525 \end{bmatrix}$

Section 1.10

1. $x_1 = 3,\ x_2 = -4,\ x_3 = 2$
3. $x_1 = -25,\ x_2 = -38,\ x_3 = -18$
5. $x_1 = 2,\ x_2 = 1,\ x_3 = 3$

Section 1.11

1. $\begin{bmatrix} 1 & -2 & -3 \\ 0 & 1 & -\frac{3}{2} \\ 0 & 0 & \frac{1}{2} \end{bmatrix}$ 3. $\begin{bmatrix} \frac{1}{7} & 0 & 0 \\ 0 & \frac{1}{5} & 0 \\ 0 & 0 & -\frac{1}{2} \end{bmatrix}$ 5. $\begin{bmatrix} \frac{1}{7} & 0 & 0 \\ -\frac{1}{14} & -\frac{1}{2} & 0 \\ \frac{1}{42} & \frac{1}{6} & \frac{1}{3} \end{bmatrix}$

7. i. $\begin{bmatrix} 1 & 7 \\ 0 & 2 \end{bmatrix}$, ii. $\begin{bmatrix} 1 & 4 & 2 \\ 0 & 2 & 2 \\ 0 & 0 & 0 \end{bmatrix}$ 9. $\begin{bmatrix} 1 & -7 \\ 3 & 3 \\ 4 & 4 \end{bmatrix}$

11. $\begin{bmatrix} 0 & 2 & 5 & 0 \\ 1 & 4 & -3 & 2 \end{bmatrix}$ 13. $A^T = (BB^T)^T = (B^T)^TB^T = BB^T = A$

17. $S = \begin{bmatrix} -1 & 2 & \frac{3}{2} \\ 2 & 1 & \frac{3}{2} \\ \frac{3}{2} & \frac{3}{2} & 5 \end{bmatrix}, \quad T = \begin{bmatrix} 0 & -2 & \frac{3}{2} \\ 2 & 0 & \frac{1}{2} \\ -\frac{3}{2} & -\frac{1}{2} & 0 \end{bmatrix}$

19. $S = \begin{bmatrix} -6 & \frac{3}{2} & \frac{5}{2} \\ \frac{3}{2} & 4 & \frac{3}{2} \\ \frac{5}{2} & \frac{3}{2} & 0 \end{bmatrix}, \quad T = \begin{bmatrix} 0 & \frac{1}{2} & \frac{5}{2} \\ -\frac{1}{2} & 0 & \frac{3}{2} \\ -\frac{5}{2} & -\frac{3}{2} & 0 \end{bmatrix}$

7. c. The line passing through $\begin{bmatrix} 0 \\ 0 \end{bmatrix}$ in the direction $\begin{bmatrix} 2 \\ -3 \end{bmatrix}$

d. the plane $3x + y - 3z$

Section 3.3

1. Yes 3. Yes 5. No

7. a. $\langle \mathbf{x}_1, \mathbf{x}_2 \rangle = \left\{ \begin{bmatrix} 3c_1 + c_2 \\ c_1 + 3c_2 \end{bmatrix} : c_1, c_2 \text{ in } R \right\} = R^2$

b. $\begin{bmatrix} 7 \\ -5 \end{bmatrix} = \frac{13}{4}\begin{bmatrix} 3 \\ 1 \end{bmatrix} + \left(\frac{-11}{4}\right)\begin{bmatrix} 1 \\ 3 \end{bmatrix}$

15. a. $\left\{ \begin{bmatrix} \frac{5}{3} \\ \frac{2}{3} \\ 1 \\ 0 \end{bmatrix}, \begin{bmatrix} 0 \\ 1 \\ 0 \\ 1 \end{bmatrix} \right\}$ b. $\left\{ \begin{bmatrix} -\frac{13}{6} \\ \frac{3}{2} \\ -\frac{1}{3} \\ -\frac{2}{3} \\ 1 \end{bmatrix} \right\}$

Section 3.4

1. linearly independent
3. linearly dependent
5. linearly independent
7. linearly dependent
9. linearly dependent
11. linearly dependent
13. linearly dependent
15. a. linearly independent
    b. linearly independent
    c. linearly dependent
    d. linearly dependent
17. linearly independent
23. a. linearly independent
    b. linearly dependent

Section 3.5

1. Is a basis
3. Is a basis
5. linearly dependent; not a basis
7. linearly dependent; not a basis

9. a. Add $\begin{bmatrix} 1 \\ 0 \\ 0 \end{bmatrix}$ to the set b. Add x^3 to the set

$$2 - 3x + x^2 + 4x^3 = -5(1) + 3(2 - x) + 1(x^2 + 1) + 4(x^3).$$

11. $\left\{ \begin{bmatrix} -\frac{1}{2} \\ 1 \\ 0 \end{bmatrix}, \begin{bmatrix} -\frac{3}{2} \\ 0 \\ 1 \end{bmatrix} \right\}$

15. b. Add $\begin{bmatrix} 1 & 0 \\ 0 & 0 \end{bmatrix}$ to the set

17. Yes

19. Yes

21. $\{1 + 3x, x^2, x^3\}$

23. a. $\left\{\begin{bmatrix} 1 \\ -2 \\ 1 \end{bmatrix}\right\}$ b. $\left\{\begin{bmatrix} -2 \\ 1 \\ -1 \\ 0 \\ 1 \end{bmatrix}\right\}$

c. $\left\{\begin{bmatrix} 0 \\ 0 \\ 1 \\ 0 \\ 0 \\ 0 \\ 0 \end{bmatrix}, \begin{bmatrix} 0 \\ -6 \\ 0 \\ 1 \\ 0 \\ 0 \\ 0 \end{bmatrix}, \begin{bmatrix} -\frac{1}{2} \\ 0 \\ 0 \\ 0 \\ 1 \\ 0 \\ 0 \end{bmatrix}, \begin{bmatrix} -\frac{1}{2} \\ 0 \\ 0 \\ 0 \\ 0 \\ 1 \\ 0 \end{bmatrix}, \begin{bmatrix} \frac{1}{2} \\ -3 \\ 0 \\ 0 \\ 0 \\ 0 \\ 1 \end{bmatrix}\right\}$

Section 3.6

1. a. $\begin{bmatrix} 1 \\ 0 \\ 1 \end{bmatrix} = 1\begin{bmatrix} 1 \\ 0 \\ 1 \end{bmatrix} + 0\begin{bmatrix} -1 \\ 1 \\ 0 \end{bmatrix} + 0\begin{bmatrix} 0 \\ -1 \\ 1 \end{bmatrix}$

$\begin{bmatrix} 2 \\ 1 \\ 4 \end{bmatrix} = \frac{7}{2}\begin{bmatrix} 1 \\ 0 \\ 1 \end{bmatrix} + \frac{3}{2}\begin{bmatrix} -1 \\ 1 \\ 0 \end{bmatrix} + \frac{1}{2}\begin{bmatrix} 0 \\ -1 \\ 1 \end{bmatrix}$

$\begin{bmatrix} -2 \\ 1 \\ 3 \end{bmatrix} = 1\begin{bmatrix} 1 \\ 0 \\ 1 \end{bmatrix} + 3\begin{bmatrix} -1 \\ 1 \\ 0 \end{bmatrix} + 2\begin{bmatrix} 0 \\ -1 \\ 1 \end{bmatrix}$

$\begin{bmatrix} 4 \\ 0 \\ -1 \end{bmatrix} = \frac{3}{2}\begin{bmatrix} 1 \\ 0 \\ 1 \end{bmatrix} - \frac{5}{2}\begin{bmatrix} -1 \\ 1 \\ 0 \end{bmatrix} - \frac{5}{2}\begin{bmatrix} 0 \\ -1 \\ 1 \end{bmatrix}$

b. $A^T = \begin{bmatrix} 1 & \frac{7}{2} & 1 & \frac{3}{2} \\ 0 & \frac{3}{2} & 3 & -\frac{5}{2} \\ 0 & \frac{1}{2} & 2 & -\frac{5}{2} \end{bmatrix}$

5. a. $\left\{\begin{bmatrix} 1 \\ 2 \\ 3 \end{bmatrix}, \begin{bmatrix} 1 \\ -1 \\ 4 \end{bmatrix}\right\}$ b. $\left\{\begin{bmatrix} 1 \\ -1 \\ 1 \end{bmatrix}, \begin{bmatrix} 2 \\ -1 \\ 1 \end{bmatrix}, \begin{bmatrix} 0 \\ -3 \\ 1 \end{bmatrix}\right\}$

c. $\left\{\begin{bmatrix} 1 \\ 0 \\ 0 \\ 0 \end{bmatrix}, \begin{bmatrix} 0 \\ 1 \\ 0 \\ 0 \end{bmatrix}, \begin{bmatrix} 0 \\ 0 \\ 1 \\ 0 \end{bmatrix}, \begin{bmatrix} 0 \\ 0 \\ 0 \\ 1 \end{bmatrix}\right\}$ d. $\left\{\begin{bmatrix} 1 \\ 0 \end{bmatrix}, \begin{bmatrix} 0 \\ 1 \end{bmatrix}\right\}$

e. $\left\{\begin{bmatrix} 1 \\ -1 \\ 1 \\ 0 \end{bmatrix}, \begin{bmatrix} -1 \\ 1 \\ 0 \\ 1 \end{bmatrix}, \begin{bmatrix} 1 \\ 0 \\ 0 \\ 1 \end{bmatrix}, \begin{bmatrix} 0 \\ 1 \\ 1 \\ 0 \end{bmatrix}\right\}$

CHAPTER FOUR

Section 4.1

1. Yes 3. Yes 5. No 7. Yes
9. No 11. Yes 13. No 15. Yes
17. No 19. Yes 21. No

27. S interchanges the x- and y-axes while T projects $\mathbf{x}$ onto the x-axis

$$(S+T)\begin{bmatrix} x \\ y \end{bmatrix} = \begin{bmatrix} x+y \\ x \end{bmatrix}, \quad TS\begin{bmatrix} x \\ y \end{bmatrix} = \begin{bmatrix} y \\ 0 \end{bmatrix}, \quad ST\begin{bmatrix} x \\ y \end{bmatrix} = \begin{bmatrix} 0 \\ x \end{bmatrix}$$

$$S^2\begin{bmatrix} x \\ y \end{bmatrix} = \begin{bmatrix} x \\ y \end{bmatrix}, \quad T^2\begin{bmatrix} x \\ y \end{bmatrix} = \begin{bmatrix} x \\ 0 \end{bmatrix}$$

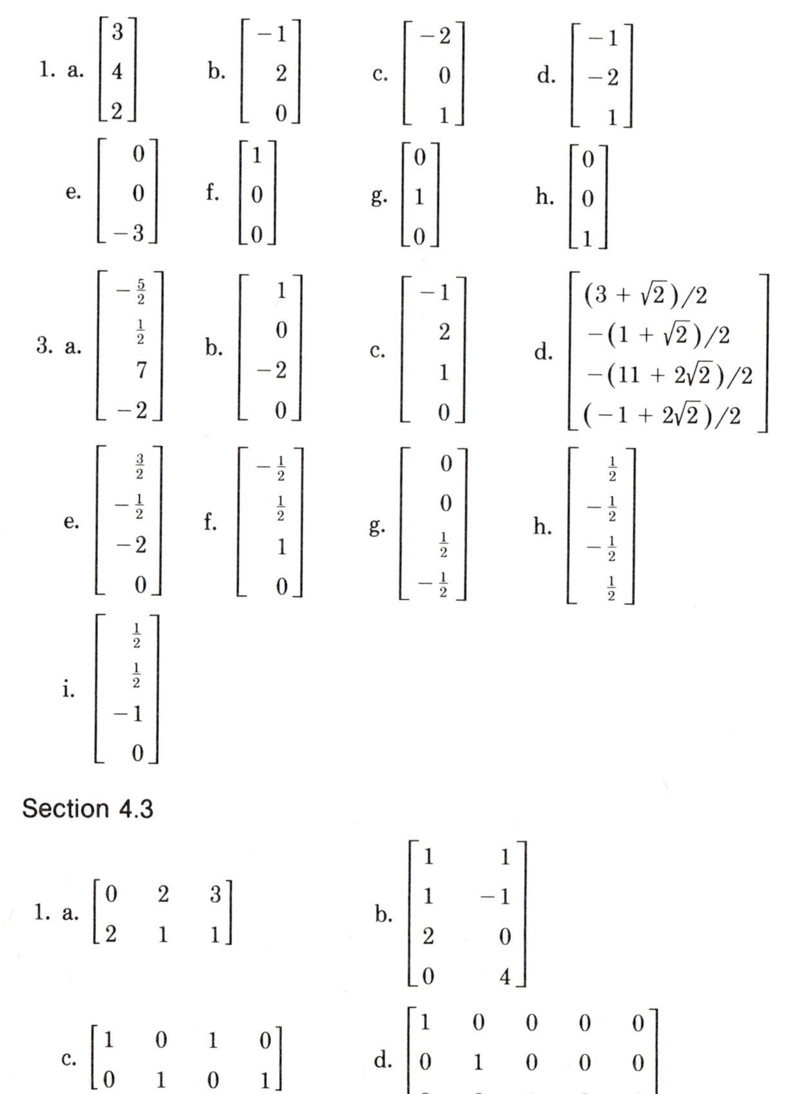

Section 4.2

1. a. $\begin{bmatrix} 3 \\ 4 \\ 2 \end{bmatrix}$ b. $\begin{bmatrix} -1 \\ 2 \\ 0 \end{bmatrix}$ c. $\begin{bmatrix} -2 \\ 0 \\ 1 \end{bmatrix}$ d. $\begin{bmatrix} -1 \\ -2 \\ 1 \end{bmatrix}$

e. $\begin{bmatrix} 0 \\ 0 \\ -3 \end{bmatrix}$ f. $\begin{bmatrix} 1 \\ 0 \\ 0 \end{bmatrix}$ g. $\begin{bmatrix} 0 \\ 1 \\ 0 \end{bmatrix}$ h. $\begin{bmatrix} 0 \\ 0 \\ 1 \end{bmatrix}$

3. a. $\begin{bmatrix} -\frac{5}{2} \\ \frac{1}{2} \\ 7 \\ -2 \end{bmatrix}$ b. $\begin{bmatrix} 1 \\ 0 \\ -2 \\ 0 \end{bmatrix}$ c. $\begin{bmatrix} -1 \\ 2 \\ 1 \\ 0 \end{bmatrix}$ d. $\begin{bmatrix} (3+\sqrt{2})/2 \\ -(1+\sqrt{2})/2 \\ -(11+2\sqrt{2})/2 \\ (-1+2\sqrt{2})/2 \end{bmatrix}$

e. $\begin{bmatrix} \frac{3}{2} \\ -\frac{1}{2} \\ -2 \\ 0 \end{bmatrix}$ f. $\begin{bmatrix} -\frac{1}{2} \\ \frac{1}{2} \\ 1 \\ 0 \end{bmatrix}$ g. $\begin{bmatrix} 0 \\ 0 \\ \frac{1}{2} \\ -\frac{1}{2} \end{bmatrix}$ h. $\begin{bmatrix} \frac{1}{2} \\ -\frac{1}{2} \\ -\frac{1}{2} \\ \frac{1}{2} \end{bmatrix}$

i. $\begin{bmatrix} \frac{1}{2} \\ \frac{1}{2} \\ -1 \\ 0 \end{bmatrix}$

Section 4.3

1. a. $\begin{bmatrix} 0 & 2 & 3 \\ 2 & 1 & 1 \end{bmatrix}$ b. $\begin{bmatrix} 1 & 1 \\ 1 & -1 \\ 2 & 0 \\ 0 & 4 \end{bmatrix}$

c. $\begin{bmatrix} 1 & 0 & 1 & 0 \\ 0 & 1 & 0 & 1 \end{bmatrix}$ d. $\begin{bmatrix} 1 & 0 & 0 & 0 & 0 \\ 0 & 1 & 0 & 0 & 0 \\ 0 & 0 & 1 & 0 & 0 \end{bmatrix}$

3. a. $\begin{bmatrix} \frac{9}{2} & -\frac{11}{2} \\ -\frac{1}{2} & \frac{3}{2} \end{bmatrix}$ b. $\begin{bmatrix} \frac{17}{3} & -\frac{11}{3} \\ \frac{4}{3} & -\frac{4}{3} \end{bmatrix}$

5. a. $\begin{bmatrix} 1 & -1 & 0 \\ 0 & 1 & -1 \\ 1 & -2 & 0 \\ -1 & 1 & 1 \end{bmatrix}$ b. $\begin{bmatrix} 0 & 1 & -2 \\ 1 & -1 & 0 \\ -1 & 3 & -4 \\ 0 & -1 & 3 \end{bmatrix}$

c. $\begin{bmatrix} 1 & -1 & 0 \\ 0 & 1 & -1 \\ 1 & -2 & 2 \\ -1 & 1 & 0 \end{bmatrix}$ d. $\begin{bmatrix} 1 & -1 & 0 \\ 0 & 0 & 0 \\ 1 & -1 & -1 \\ -1 & 1 & 1 \end{bmatrix}$

e. $\begin{bmatrix} 0 & 0 & 0 \\ 0 & 0 & 0 \\ 0 & 0 & 1 \\ 0 & 0 & 0 \end{bmatrix}$

Section 4.4

1. $P = \begin{bmatrix} 1 & 1 & \frac{1}{2} \\ 0 & 0 & \frac{1}{2} \\ -2 & -1 & \frac{3}{2} \end{bmatrix}$ 3. $P = \begin{bmatrix} 2 & 0 & 0 \\ -1 & 0 & 1 \\ 0 & 1 & -1 \end{bmatrix}$

5. $P = \begin{bmatrix} -\frac{1}{4} & \frac{5}{4} & -\frac{3}{2} \\ \frac{1}{4} & -\frac{1}{4} & -\frac{1}{2} \\ \frac{1}{2} & \frac{3}{2} & 0 \end{bmatrix}$ 7. $P = \begin{bmatrix} -\frac{9}{2} & -\frac{5}{2} & \frac{7}{2} \\ \frac{9}{2} & \frac{9}{2} & \frac{5}{2} \\ -6 & 0 & 12 \end{bmatrix}$

9. $P = \begin{bmatrix} 1 & \frac{3}{2} & \frac{7}{2} \\ -2 & -\frac{3}{2} & -\frac{1}{2} \\ 0 & 2 & 9 \end{bmatrix}$

15. a. $N = \begin{bmatrix} -1 & -\frac{1}{3} & \frac{5}{3} \\ 0 & \frac{2}{3} & \frac{8}{3} \\ 1 & \frac{2}{3} & -\frac{1}{3} \\ 1 & \frac{2}{3} & \frac{2}{3} \end{bmatrix}$ b. $P = \begin{bmatrix} -3 & -1 & 3 \\ 5 & 1 & -4 \\ -1 & 0 & 1 \end{bmatrix}$

c. $Q = \begin{bmatrix} \frac{2}{3} & \frac{2}{3} & 0 & \frac{1}{3} \\ -\frac{1}{3} & -\frac{1}{3} & 1 & -\frac{2}{3} \\ \frac{2}{3} & -\frac{1}{3} & 0 & \frac{1}{3} \\ -\frac{1}{3} & \frac{2}{3} & 0 & \frac{1}{3} \end{bmatrix}$

Section 4.5

1. $\left\{\begin{bmatrix} 1 \\ 0 \\ -\frac{1}{2} \end{bmatrix}, \begin{bmatrix} 0 \\ 0 \\ -\frac{1}{2} \end{bmatrix}\right\}$ 3. $\left\{\begin{bmatrix} 1 \\ 1 \\ 2 \\ 0 \end{bmatrix}, \begin{bmatrix} 0 \\ -2 \\ -2 \\ 3 \end{bmatrix}\right\}$

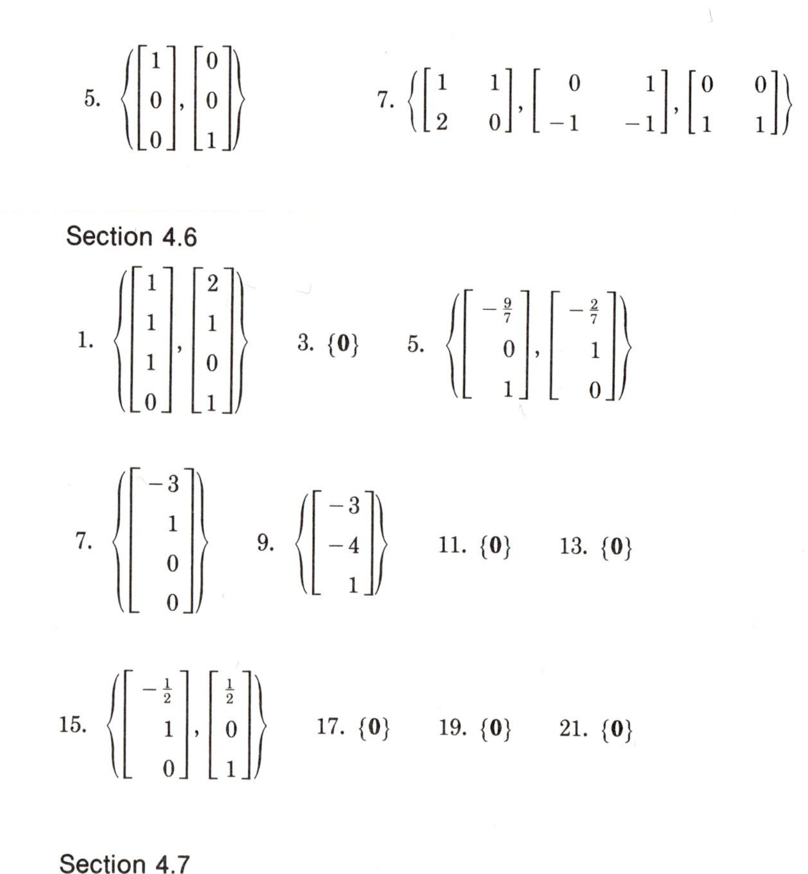

5. $\left\{\begin{bmatrix}1\\0\\0\end{bmatrix}, \begin{bmatrix}0\\0\\1\end{bmatrix}\right\}$ 7. $\left\{\begin{bmatrix}1 & 1\\2 & 0\end{bmatrix}, \begin{bmatrix}0 & 1\\-1 & -1\end{bmatrix}, \begin{bmatrix}0 & 0\\1 & 1\end{bmatrix}\right\}$

Section 4.6

1. $\left\{\begin{bmatrix}1\\1\\1\\0\end{bmatrix}, \begin{bmatrix}2\\1\\0\\1\end{bmatrix}\right\}$ 3. $\{\mathbf{0}\}$ 5. $\left\{\begin{bmatrix}-\frac{9}{7}\\0\\1\end{bmatrix}, \begin{bmatrix}-\frac{2}{7}\\1\\0\end{bmatrix}\right\}$

7. $\left\{\begin{bmatrix}-3\\1\\0\\0\end{bmatrix}\right\}$ 9. $\left\{\begin{bmatrix}-3\\-4\\1\end{bmatrix}\right\}$ 11. $\{\mathbf{0}\}$ 13. $\{\mathbf{0}\}$

15. $\left\{\begin{bmatrix}-\frac{1}{2}\\1\\0\end{bmatrix}, \begin{bmatrix}\frac{1}{2}\\0\\1\end{bmatrix}\right\}$ 17. $\{\mathbf{0}\}$ 19. $\{\mathbf{0}\}$ 21. $\{\mathbf{0}\}$

Section 4.7

1. $\begin{bmatrix}\frac{13}{7} - \frac{3}{7}s + \frac{3}{7}t\\ -\frac{1}{7} + \frac{4}{7}s - \frac{4}{7}t\\ s\\ t\end{bmatrix}$ 3. $\begin{bmatrix}\frac{2}{3} + \frac{2}{3}t\\ \frac{1}{3} - \frac{5}{3}t\\ t\end{bmatrix}$

5. $A\mathbf{x} = \mathbf{b}$ has $x\mathbf{u} + s\mathbf{v}$ as a solution when $s = 1 - x$.

Section 4.8

1. a. 2 b. 6 c. 38 d. 0 e. 3

3. a. $\left\{\begin{bmatrix}\frac{19}{6}\\\frac{2}{3}\\-\frac{7}{6}\end{bmatrix}\right\}$ b. $\left\langle\begin{bmatrix}-1\\0\\1\end{bmatrix}, \begin{bmatrix}1\\1\\0\end{bmatrix}\right\rangle$ c. $\left\{\begin{bmatrix}\frac{1}{2}\\\frac{13}{18}\\-\frac{7}{2}\end{bmatrix}\right\}$

d. $\left\{\begin{bmatrix}-1\\4\\1\end{bmatrix}\right\}$ e. $\left\{\begin{bmatrix}1\\-\frac{7}{3}\end{bmatrix}\right\}$

5. $r = 3$, nullity $= 0$ 7. $r = 3$, nullity $= 0$

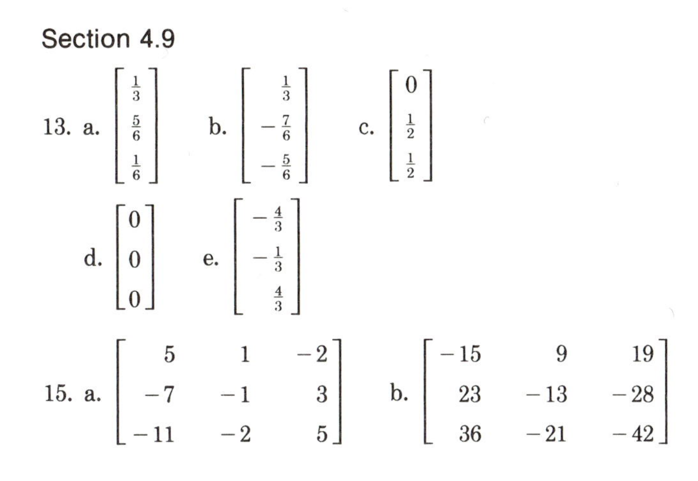

Section 4.9

13. a. $\begin{bmatrix} \frac{1}{3} \\ \frac{5}{6} \\ \frac{1}{6} \end{bmatrix}$ b. $\begin{bmatrix} \frac{1}{3} \\ -\frac{7}{6} \\ -\frac{5}{6} \end{bmatrix}$ c. $\begin{bmatrix} 0 \\ \frac{1}{2} \\ \frac{1}{2} \end{bmatrix}$

d. $\begin{bmatrix} 0 \\ 0 \\ 0 \end{bmatrix}$ e. $\begin{bmatrix} -\frac{4}{3} \\ -\frac{1}{3} \\ \frac{4}{3} \end{bmatrix}$

15. a. $\begin{bmatrix} 5 & 1 & -2 \\ -7 & -1 & 3 \\ -11 & -2 & 5 \end{bmatrix}$ b. $\begin{bmatrix} -15 & 9 & 19 \\ 23 & -13 & -28 \\ 36 & -21 & -42 \end{bmatrix}$

CHAPTER FIVE

Section 5.1

1. -19 3. -9 5. $\frac{1}{6}$ 7. $x = -10.5$

9. $x = 6$, $x = -2$ 11. 0 13. 2

15. 0 17. $16a^2 - 3ab - 16a^2c + 3c^2 + 2ab^2 - 2bc$

27. $\det(B) = c^3\det(A)$, $\det(B) = c^n\det(A)$

29. a. 24 b. $a_{11}a_{22}a_{33}$ c. $a_{11}a_{22} \cdots a_{nn}$

Section 5.2

1. $\det(A) = 36$

3. a. Row 1 = Row 3 b. Row 3 = -2 Row 1

5. The determinant is 9

7. 6 9. 193 11. 0

13. a and b are roots 15. 1 and 3 are roots

23. -37.184

Section 5.3

1. 35 3. -39 5. 0; has no inverse 7. 120; has an inverse

9. $\det(A) = -14$, $\det(B) = -46$

11. $\det(A) = 99$, $\det(B) = -42$

Section 5.4

1. $x_1 = -1$, $x_2 = 1$

3. $\det\begin{bmatrix} 1 & 4 \\ 1 & 4 \end{bmatrix} = 0$, so Cramer's Rule does not apply

5. Cramer's Rule does not apply

7. Cramer's Rule does not apply

9. $x_1 = \frac{10}{33}$, $x_2 = -\frac{1}{99}$, $x_3 = \frac{61}{99}$

11. $x_1 = \frac{27}{10}$, $x_2 = \frac{37}{10}$, $x_3 = \frac{4}{5}$

13. $x_1 = 3$, $x_2 = 2$, $x_3 = -1$, $x_4 = -4$

CHAPTER SIX

Section 6.1

1. $\dfrac{5-\sqrt{33}}{2}, \dfrac{5+\sqrt{33}}{2}$

3. $3, \dfrac{1+\sqrt{13}}{2}, \dfrac{1-\sqrt{13}}{2}$

5. $0, 2, \dfrac{3+\sqrt{5}}{2}, \dfrac{3-\sqrt{5}}{2}$

7. $-1, 2, 2$

9. $-1, 1, 2$

11. $\lambda_1 = 0, \quad \lambda_2 = 1, \quad \lambda_3 = 2,$

$$E(0) = \left\{\begin{bmatrix} -2 \\ 1 \\ 0 \\ 0 \end{bmatrix}\right\} \quad E(1) = \left\{\begin{bmatrix} 1 \\ 0 \\ 0 \\ 0 \end{bmatrix}\right\} \quad E(2) = \left\{\begin{bmatrix} 10 \\ \frac{3}{2} \\ 1 \\ 1 \end{bmatrix}\right\}$$

13. $\lambda = 1, \quad E(1) = \left\{\begin{bmatrix} 1 \\ 0 \\ 0 \end{bmatrix}\right\}$

15. $\lambda_1 = 1, \quad \lambda_2 = 2, \quad E(1) = \left\{\begin{bmatrix} 1 \\ 0 \\ -1 \end{bmatrix}, \begin{bmatrix} 0 \\ 1 \\ 0 \end{bmatrix}\right\}, \quad E(2) = \left\{\begin{bmatrix} 0 \\ 0 \\ 1 \end{bmatrix}\right\}$

17. $\lambda_1 = 2, \quad \lambda_2 = 3, \quad E(2) = \left\{\begin{bmatrix} 1 \\ 0 \\ 0 \\ 0 \end{bmatrix}\right\}, \quad E(3) = \left\{\begin{bmatrix} -3 \\ 1 \\ 0 \\ 0 \end{bmatrix}\right\}$

19. $k = 3$

Section 6.2

1. $D = \begin{bmatrix} -1 & 0 \\ 0 & 3 \end{bmatrix}, \quad P = \begin{bmatrix} -1 & 1 \\ 1 & 1 \end{bmatrix}$

3. $D = \begin{bmatrix} 0 & 0 & 0 \\ 0 & 1 & 0 \\ 0 & 0 & 2 \end{bmatrix}, \quad P = \begin{bmatrix} 1 & 0 & 1 \\ 0 & 1 & 0 \\ -1 & 0 & 1 \end{bmatrix}$

5. $D = \begin{bmatrix} 1 & 0 & 0 \\ 0 & 5 & 0 \\ 0 & 0 & -1 \end{bmatrix}, \quad P = \begin{bmatrix} 1 & 0 & 1 \\ 0 & 1 & -1 \\ -1 & -2 & 0 \end{bmatrix}$

7. $D = \begin{bmatrix} 1 & 0 & 0 & 0 \\ 0 & 1 & 0 & 0 \\ 0 & 0 & -1 & 0 \\ 0 & 0 & 0 & -1 \end{bmatrix}, \quad P = \begin{bmatrix} 1 & 0 & 1 & 0 \\ 0 & 1 & 0 & 1 \\ 0 & 1 & 0 & -1 \\ 1 & 0 & -1 & 0 \end{bmatrix}$

9. Cannot be diagonalized.

Section 6.3

1. $\begin{bmatrix} -\frac{1}{\sqrt{5}} & \frac{2}{\sqrt{5}} \\ \frac{2}{\sqrt{5}} & \frac{1}{\sqrt{5}} \end{bmatrix}$

3. $\begin{bmatrix} \frac{1}{\sqrt{6}} & -\frac{1}{\sqrt{3}} & -\frac{1}{\sqrt{2}} & 0 \\ \frac{1}{\sqrt{6}} & -\frac{1}{\sqrt{3}} & \frac{1}{\sqrt{2}} & 0 \\ 0 & 0 & 0 & 1 \\ \frac{2}{\sqrt{6}} & \frac{1}{\sqrt{3}} & 0 & 0 \end{bmatrix}$

5. $\begin{bmatrix} \frac{5}{\sqrt{30}} & \frac{1}{\sqrt{6}} & 0 \\ \frac{2}{\sqrt{30}} & -\frac{2}{\sqrt{6}} & -\frac{1}{\sqrt{5}} \\ \frac{1}{\sqrt{30}} & -\frac{1}{\sqrt{6}} & \frac{2}{\sqrt{5}} \end{bmatrix}$

7. No

9. No

11. $Q = \begin{bmatrix} \frac{2}{\sqrt{5}} & 0 & -\frac{2}{\sqrt{5}} \\ 0 & 1 & 0 \\ \frac{1}{\sqrt{5}} & 0 & \frac{1}{\sqrt{5}} \end{bmatrix}, \quad D = \begin{bmatrix} 0 & 0 & 0 \\ 0 & 0 & 0 \\ 0 & 0 & 5 \end{bmatrix}$

b. Cannot be diagonalized.

Section 6.4

1. $\begin{bmatrix} \frac{7}{5} \\ 0 \\ \frac{14}{5} \end{bmatrix}$

3. $\mathbf{0}$

5. $\begin{bmatrix} \frac{1}{4} \\ \frac{3}{4\sqrt{3}} \\ 0 \end{bmatrix}$

7. $\begin{bmatrix} \frac{5}{19} \\ \frac{15}{19} \\ \frac{15}{19} \\ 0 \end{bmatrix}$

9. $\begin{bmatrix} -\frac{14}{39} \\ 0 \\ \frac{7}{39} \\ -\frac{21}{39} \\ \frac{35}{39} \end{bmatrix}$ 11. $\begin{bmatrix} 1 \\ 3 \\ 1 \end{bmatrix}$ 13. $\begin{bmatrix} 1 \\ 4 \\ -2 \\ -1 \end{bmatrix}$ 15. $\frac{1}{449}\begin{bmatrix} 318 \\ -311 \\ 372 \\ -629 \end{bmatrix}$

17. $\left\{ \begin{bmatrix} \frac{1}{\sqrt{2}} \\ 0 \\ \frac{1}{\sqrt{2}} \end{bmatrix}, \begin{bmatrix} -\frac{1}{\sqrt{3}} \\ \frac{1}{\sqrt{3}} \\ \frac{1}{\sqrt{3}} \end{bmatrix}, \begin{bmatrix} \frac{1}{\sqrt{6}} \\ \frac{2}{\sqrt{6}} \\ -\frac{1}{\sqrt{6}} \end{bmatrix} \right\}$

19. $\left\{ \begin{bmatrix} 0 \\ \frac{3}{\sqrt{13}} \\ \frac{2}{\sqrt{13}} \end{bmatrix}, \begin{bmatrix} \frac{13}{\sqrt{286}} \\ \frac{6}{\sqrt{286}} \\ -\frac{9}{\sqrt{286}} \end{bmatrix}, \begin{bmatrix} -\frac{195}{11}\sqrt{13}\sqrt{126} \\ -\frac{222}{11}\sqrt{13}\sqrt{126} \\ \frac{333}{11}\sqrt{13}\sqrt{126} \end{bmatrix} \right\}$

21. $\left\{ \begin{bmatrix} \frac{1}{\sqrt{3}} \\ -\frac{1}{\sqrt{3}} \\ \frac{1}{\sqrt{3}} \end{bmatrix}, \begin{bmatrix} 0 \\ \frac{1}{\sqrt{2}} \\ \frac{1}{\sqrt{2}} \end{bmatrix}, \begin{bmatrix} \frac{2}{\sqrt{6}} \\ \frac{1}{\sqrt{6}} \\ -\frac{1}{\sqrt{6}} \end{bmatrix} \right\}$

Section 6.5

1. $A^+ = \begin{bmatrix} \frac{16}{9} & -\frac{10}{9} & \frac{11}{9} \\ -\frac{11}{9} & \frac{8}{9} & -\frac{7}{9} \end{bmatrix}$, $\mathbf{x}_0 = \begin{bmatrix} \frac{14}{9} \\ \frac{59}{9} \end{bmatrix}$

3. Cannot be done since A^TA is singular.

5. $A^+ = \begin{bmatrix} -\frac{10}{26} & \frac{14}{26} & \frac{22}{26} \\ \frac{7}{26} & -\frac{2}{26} & -\frac{5}{26} \end{bmatrix}$, $\mathbf{x}_0 = \begin{bmatrix} \frac{92}{26} \\ \frac{50}{26} \end{bmatrix}$

7. Cannot be done since A^TA is singular.

9. $A^+ = \frac{1}{6}\begin{bmatrix} 7 & 2 & -5 \\ -3 & 0 & 3 \end{bmatrix}$, $\mathbf{x}_0 = \begin{bmatrix} \frac{5}{2} \\ \frac{1}{2} \end{bmatrix}$

11. $A^+ = \frac{1}{172}\begin{bmatrix} 4 & 76 & -20 \\ 3 & -29 & 28 \end{bmatrix}$, $\mathbf{x}_0 = \begin{bmatrix} \frac{122}{86} \\ \frac{27}{86} \end{bmatrix}$

Section 6.6

1. $Q = \begin{bmatrix} \frac{\sqrt{3}}{2} & -\frac{1}{2} \\ \frac{1}{2} & \frac{\sqrt{3}}{2} \end{bmatrix}$, $D = \begin{bmatrix} 1 & 0 \\ 0 & 2 \end{bmatrix}$

3. $Q = \begin{bmatrix} \frac{1}{2} & -\frac{\sqrt{3}}{2} \\ \frac{\sqrt{3}}{2} & 1 \end{bmatrix}$, $D = \begin{bmatrix} 0 & 0 \\ 0 & 4 \end{bmatrix}$

5. $Q = \begin{bmatrix} \frac{2}{\sqrt{5}} & \frac{2}{3\sqrt{5}} & -\frac{2}{\sqrt{6}} \\ \frac{1}{\sqrt{5}} & -\frac{4}{3\sqrt{5}} & \frac{1}{\sqrt{6}} \\ 0 & \frac{5}{3\sqrt{5}} & \frac{1}{\sqrt{6}} \end{bmatrix}$, $D = \begin{bmatrix} 0 & 0 & 0 \\ 0 & 0 & 0 \\ 0 & 0 & 3 \end{bmatrix}$

7. $Q = \begin{bmatrix} -\frac{1}{\sqrt{2}} & 0 & \frac{1}{\sqrt{2}} \\ 0 & 1 & 0 \\ \frac{1}{\sqrt{2}} & 0 & \frac{1}{\sqrt{2}} \end{bmatrix}$, $D = \begin{bmatrix} 0 & 0 & 0 \\ 0 & 0 & 0 \\ 0 & 0 & 2\sqrt{2} \end{bmatrix}$

9. $Q = \begin{bmatrix} \frac{1}{\sqrt{2}} & \frac{1}{\sqrt{3}} & -\frac{1}{\sqrt{6}} \\ \frac{1}{\sqrt{2}} & -\frac{1}{\sqrt{3}} & \frac{1}{\sqrt{6}} \\ 0 & \frac{1}{\sqrt{3}} & \frac{2}{\sqrt{6}} \end{bmatrix}$, $D = \begin{bmatrix} 0 & 0 & 0 \\ 0 & 0 & 0 \\ 0 & 0 & 1 \end{bmatrix}$

11. $Q = \begin{bmatrix} -\frac{5}{\sqrt{26}} & -\frac{1}{\sqrt{195}} & \frac{1}{\sqrt{30}} \\ \frac{1}{\sqrt{26}} & -\frac{5}{\sqrt{195}} & \frac{5}{\sqrt{30}} \\ 0 & \frac{13}{\sqrt{195}} & \frac{2}{\sqrt{30}} \end{bmatrix}$, $D = \begin{bmatrix} -1 & 0 & 0 \\ 0 & -1 & 0 \\ 0 & 0 & 2 \end{bmatrix}$

INDEX